More praise for
STARLIGHT DETECTIVES

"Hirshfeld documents how the practice of astronomy changed between 1840 and 1940 thanks to innovative pioneers whose efforts made it possible to capture and preserve otherwise faint and fleeting images, and to decipher the cryptographic messages found in the light of celestial bodies. His riveting narrative brings to life their challenges, failures, and successes. It will captivate all who have observed the night sky." —**Barbara J. Becker**, author of *Unravelling Starlight: William and Margaret Huggins and the Rise of the New Astronomy*

"Writing this book would ideally require an author with an extensive knowledge of astronomy, including astronomical instruments, a deep understanding of the ways of thought of astronomers, a broad range of historical knowledge, and an exceptional skill at making astronomical ideas clear and engaging. Alan Hirshfeld possesses all of these skills. His *Starlight Detectives* is remarkable."
—**Michael J. Crowe**, author of *The Extraterrestrial Life Debate, 1750–1900*

"A thrilling historical account of the rise of astrophysics, the early years of astronomical photography and spectroscopy, and the innovations that transformed the astronomical telescope in the nineteenth century. Alan Hirshfeld's thoroughly researched narrative is accessible, entertaining, and scholarly, and includes many pioneers who have been overlooked until now. I greatly admire this outstanding contribution to the history of astronomy."
—**Simon Mitton**, co-author of *Heart of Darkness: Unraveling the Mysteries of the Invisible Universe* and author of *Fred Hoyle: A Life in Science*

James Nasmyth's twenty-inch Cassegrain-Newtonian telescope, circa 1845.

STARLIGHT DETECTIVES

How Astronomers, Inventors, and Eccentrics Discovered the Modern Universe

Alan Hirshfeld

BELLEVUE LITERARY PRESS
New York

First Published in the United States in 2014 by
Bellevue Literary Press, New York

For Information, Contact:
Bellevue Literary Press
NYU School of Medicine
550 First Avenue
OBV A612
New York, NY 10016

Library of Congress Cataloging-in-Publication Data
is available from the publisher upon request.

Bellevue Literary Press would like to thank all its generous
donors—individuals and foundations—for their support.

Book design and composition by Mulberry Tree Press, Inc.

Manufactured in the United States of America.
first edition

1 3 5 7 9 8 6 4 2

Paperback ISBN: 978-1-934137-78-9

ebook ISBN: 978-1-934137-79-6

Contents

To Erika, who believed in me. Twice.

The story of scientific discovery has its own epic unity—a unity of purpose and endeavour—the single torch passing from hand to hand through the centuries; and the great moments of science—when, after long labour, the pioneers saw their accumulated facts falling into a significant order, sometimes in the form of a law that revolutionised the whole world of thought—have an intense human interest, and belong essentially to the creative imagination of poetry.

—Arthur Noyes, Prologue to *Watchers of the Sky*, 1922

STARLIGHT DETECTIVES

Introduction

Like seafarers of yesteryear, astronomers explore the vast ocean of space, sailing before the winds of imagination and scientific scrutiny. Through their instruments of observation and analysis, they have transformed the night sky from a dark, depthless field studded with glimmering specks and wisps of indeterminate nature into a multidimensional expanse of stars, galaxies, and electromagnetic waves. Under what circumstances did this transition take place? How did classical astronomy mature into its modern form?

For millennia, astronomers had studied the universe by eye, first without optical aid, then, beginning in the early 1600s, augmented by the telescope. After two centuries of incremental improvements, astronomical instruments were optimized to the needs of celestial cartographers, but were woefully inadequate tools for a meaningful exploration of deep space. The telescope was only part of the problem: the human eye itself was a fundamental roadblock to progress. The eye is an evolutionary artifact, optimized for acuity in the daytime, but ill-suited to the low-light environment of the night sky. It has neither the capacity to accumulate luminous energy over spans of time, nor the ability to delineate a light beam's constituent wavelengths. The radiance of a celestial object forms an ephemeral, nearly monochromatic image on the observer's retina. Once an astronomer's eye drew away from the telescope, no facsimile existed of the cosmic scene, other than an impressionistic précis or pencil-sketch. Until they could generate an objective, permanent record—a photograph—of the object, astronomers remained hostage to the physiological constraints of their eyes and the descriptive limitations of language and art. And until a practical means was developed to distill light into its component colors—a spectrum—the physical processes underlying the glow of a comet, a star, and or a nebula would defy explanation.

Starlight Detectives explores the decades-long bridge of innovation that transformed Victorian-era visual astronomy into the scientific discipline that is observational astrophysics. It is an inspiring tale of practical dreamers—a clockmaker, a chemist, a printer, a physician, a lawyer, a sanitation engineer, a builder—driven by a common desire to explore the night sky in a profoundly different way. Together with a few forward-thinking professionals, these nineteenth-century apostles of technology spurned the traditional study of the positions and movements of heavenly bodies to hunt down clues regarding their chemical makeup and physical conditions. Through inventiveness and unflagging persistence, they turned their backyard observatories into unlikely centers of cutting-edge astronomy—incubators for the fledgling fields of celestial photography and celestial spectroscopy.

Over succeeding decades, the observational techniques advanced by these amateur scientists joined with foundational developments in physics—relativity, quantum mechanics, atomic structure—to create a scientific framework that could scarcely have been imagined a century earlier: the Sun, formerly an impenetrable disk, became a structured, blazing body of chemical elements; the stars, no longer mere gleams of light, became celestial energy factories; and the galaxies, once enigmatic incandescences in the telescope's eyepiece, became titanic stellar vortices populating the void. In blazing the pathway to a more powerful mode of cosmic observation, amateur astronomers and inventors guided their professional counterparts toward the future, and in doing so found themselves unprepared for its heightened technological and mathematical rigor.

The work of two astronomers—Ireland's William Parsons, the Third Earl of Rosse; and American observer Edwin Hubble—effectively serve as "before" and "after" models that make manifest the revolutionary degree to which astronomy changed between the 1840s and the 1920s. Both men surveyed the fringes of the visible universe in their respective times, each employing the largest telescope then in existence: Rosse, his six-foot-wide Leviathan reflector, slung between masonry walls on the grounds of his sprawling estate; and Hubble, the eight-foot-wide Hooker reflector, emplaced high up on Mount Wilson in California.

The Leviathan's yawning aperture was put to immediate use gathering up the feeble light of celestial nebulae. These mysterious, cloud-like luminescences, thousands in number,

William Parsons, the Third Earl of Rosse.

The Leviathan of Parsonstown.

appeared in various forms—some round, some oblong, some ragged-bordered. Upon telescopic magnification, many had revealed themselves to be clusters of stars, whose stellar character strained the limits of visual acuity. Astronomers at the time debated whether *all* nebulae consist of stars, the irresolvable wisps rendered indistinct by virtue of their remoteness. The Leviathan, it was hoped, might prove these distant pockets of efflorescence to be starry as well. While the mammoth telescope did succeed in resolving additional nebulae into stars, this feat was soon overshadowed by a pivotal discovery.

In April 1845, Rosse discerned in the faint glow of the nebula Messier 51 an unmistakable, and wholly unexpected, spiral pattern. The Whirlpool Nebula, as it became known, proved far from unique. By 1850, Rosse's Leviathan had revealed more than a dozen others. Whatever the spirals were, they comprised a populous species within the celestial zoo and demanded further study.

Although equipped with the largest telescope of the day, Rosse was a prisoner of Victorian-era science; like his fellow observers, he had no basis upon which to comprehend the true character of the celestial objects he viewed. He was leafing though a book of the cosmos, indexing its contents, with no understanding of the book's meaning. Rosse's Leviathan was an opto-mechanical dinosaur, successful in its time, but doomed to extinction by its physical bulk, its cloud-swept location, and, most significantly, its allegiance to the human retina. Indeed, every telescope of the era, large or small, was compromised by its dependence on the astronomer's subjective eye and hand. Even as Rosse limned the dim swirls of the Whirlpool Nebula from his darkened aerie, a new technology was sweeping the world: a photochemical process that recorded images on a metal plate.

Seven decades later, Edwin Hubble took up the study of spiral nebulae, and proved them to be galaxies on par with our own Milky Way. The contrast between Rosse's and Hubble's working methods and their overall understanding of nature illustrates the stark differences between classical visual astronomy and modern astrophysical observation. Rosse perused telescopic images by eye and sketched what he saw, whereas Hubble

Rosse's drawing of the Whirlpool Nebula.

The Whirlpool Galaxy (née Nebula), photographed by the Hubble Space Telescope in 2005.

applied the camera and the spectrograph. Rosse, the wealthy gentleman-scientist, hired local laborers to construct instruments of his own design; Hubble, the salaried scientist, employed equipment under the auspices of an institution. Rosse's observatory was utterly Victorian in design and execution: a grand assembly of wood, metal, and masonry, set appropriately on a lawn and operated by ropes, pulleys, and handwheels; Hubble's apparatus was pure industrial chic: a massive steel-girder cylinder, cradled in a steel yoke atop riveted steel piers, hunkered underneath a cavernous, steel-ribbed dome, every movable component electrically driven. Rosse erected his telescope on the grounds of his own home, so it was easily accessible; Hubble's instrument was laboriously hauled piece by piece up mile-high Mount Wilson in California, trading convenience for the chance of clear skies.

Starlight Detectives is a comprehensive history of this remarkable and complex period in the development of humanity's oldest science. Its large cast of characters, besides Rosse and Hubble, features many whose names are unfamiliar even to present-day researchers, but whose contributions proved key to the advancement of astronomy. In the following pages, the foundational observations, the technological tweaks, the serendipitous insights, and the cross-fertilization of ideas that precede every momentous discovery are brought into focus. From our latter-day perch, it is easy to wonder why celestial research was so protracted during the nineteenth century. The retrospective lens of time inevitably shrinks once-lofty barriers to progress and straightens the winding route to discovery. Indeed, it was more than fifty years after its introduction that photography became a regular tool of astronomical research; mere decades before the success of spectroscopic analysis, the determination of the constitution of the Sun and stars had been deemed impossible. The constant in this story of an evolving science is the inventiveness and unswerving devotion of those who strove to illuminate the darkness. Their heroic achievements provided the foundation for our modern-day exploration of the universe.

Part I:
PICTURING THE HEAVENS

By applying a sensitive photographic plate to the telescope instead of the human eye, we have obtained photographs of comets, stars, and nebulae which it was utterly impossible for the eye to see through the telescope . . . [T]he cumulative effects of many hours' exposure reveal depths in our universe undreamed of before.

—William Seton, "The Century's Progress in Science," 1899

Chapter 1

True Eye and Faithful Hand

There is no one "with a true eye and a faithful hand" but can do good work in watching the heavens.

—Agnes Clerke, *History of Astronomy During the Nineteenth Century*, 1902

On the morning of June 16, 1806, the Moon's shadow crept eastward toward the city of Boston like a gathering herald of an apocalypse. The azure sky, cloudless from horizon to horizon, began to dim, at first almost imperceptibly, then swiftly, as though hastening toward night. Every tree became a living camera obscura, its leafy canopy speckling the ground with a multitude of heavenly crescents. An autumn chill infused the air, raising a mist over the harbor. Birds suspended their song, while on Boston's grassy common, a herd of cows, sensing the close of day, ambled out of the gateway toward home. Throughout the city, the regular bustle of human commerce quieted to midnight stillness.

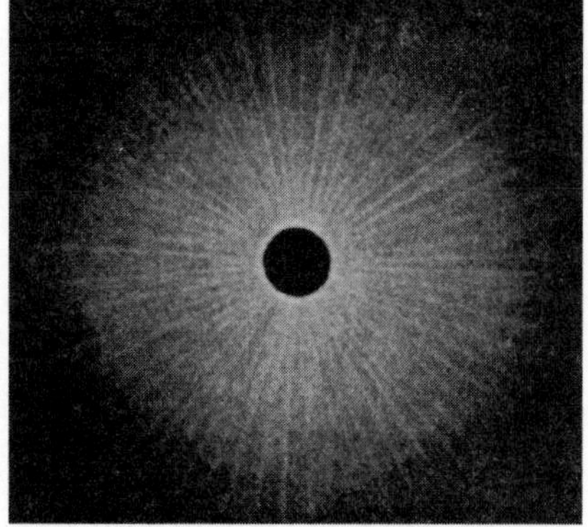

Drawing by Spanish astronomer Joaquin de Ferrer of the solar eclipse of June 16, 1806, as seen from Kinderhook, New York.

Bostonians were well prepared for the "Great Solar Eclipse," as some called it. Already, a month beforehand, they had snapped up three printings of Andrew Newell's fact-filled pamphlet, *Darkness at Noon*. Newell described how the merged celestial bodies would appear as a "dark patch" in the daytime sky, how precaution must be taken to avoid injury to the eye, how an eclipse of such duration—fully four and a half minutes—might not recur over

Boston "for many succeeding ages." Newell was no man of science, but a lesser printer occupying mean quarters on Half Court Square, off Pudding Lane. Nevertheless, his cobbled tract brought out virtually the entire city onto rooftops, street corners, and quays to witness nature's once-in-a-lifetime spectacle.

The few with a spyglass or telescope projected the Sun's gouged image onto a piece of paper. Those without an instrument observed the progress of the eclipse through a smoked glass plate, or lacking that, chanced a direct view of the diminished Sun. At the onset of totality, the Sun's corona extended its crepuscular fingers across the ash-tinted sky. Venus blazed like a diamond in the southwest. Reddish Mars popped into view. The winter stars of Orion and Taurus shone incongruously in June. Nothing in recent memory had presented a more sublime sight. "We seemed to be in the more immediate presence of Deity," remarked an eyewitness.

Four and a half minutes later, at 11:13 a.m., the Sun re-exploded into view over the Moon's receding limb. The umbral shadow swept out to sea, and light returned to the land. It was like a second dawn of creation, someone said. Boston's *Columbian Centinel* would report that if "angels had been in the habit of visiting this nether world, we justly might have expected them on this transporting occasion."

Shouts and applause rose from the city and the surrounding hills. As one, the residents of Boston expressed their gratitude to God, to Nature, to no one in particular for the magnificent interruption in their ordinary existence. They would doubtless recount impressions of this remarkable day to children, to friends. Of course, words and sketches could do only imperfect justice to the celestial tableau now locked away in the private prison of memory. Gifted poets and painters might try to resurrect the all-encompassing wonder of the eclipse, but until accompanied by a true, visual record of the event—a photograph—they would fall short. There was, in 1806, no way to preserve this or any scene for posterity. The only people who could truly comprehend what happened this day were the people who were there.

No one in Boston could have anticipated the solar eclipse more than sixteen-year-old William Cranch Bond, who reveled in its sheer majesty and the irrepressible cosmic engine at its root. Son of a clockmaker—also named William—Bond had reluctantly left school at age ten to work in his father's modest shop, at the corner of Milk and Marlborough (later Washington) Streets, across from the Old South Meeting House, where Sam Adams had spurred the Boston Tea Party nearly seven decades before.

Spindly and shy, William Bond possessed a quick mind, skillful hands, and a horologist's sensitivity to the rhythms of nature. To his friends and his elder brother Thomas, he was the clever craftsman, a reliable producer of animal snares, sports toys, and makeshift "scientific" apparatus. An unlikely clockwork that he fashioned at age ten from wood scraps had kept tolerable time. A handmade astronomical quadrant, of ebony and boxwood, evinced "the neatness, patience, and accuracy of a practiced artist." Despite these precocious glimmers of talent, William Bond felt trapped by his family's near-ruinous finances, having confided to his mother, Hannah Cranch, that he was "in despair of ever being able to accomplish anything."

Today, William Bond stood in the quickening daylight, having witnessed the Great Eclipse from a housetop on Summer Street. The precious minutes of totality had allowed him two simultaneous, yet divergent, views of the event: the panorama of Earth, sea, and sky afforded by the unaided eye; and magnified glimpses of the solar–lunar disk through a family friend's telescope. It's not known which of these prospects left the stronger impression on Bond—the epic sweep of nature's stage, offered equally to all, or his own private telescopic vision of the Moon's mountainous limb silhouetted against the solar corona.

Bond had been warned, of course, not to stare at the Sun, even in its constricted state. His pupil would have widened in the dimness of totality, leaving him defenseless against the Sun's inevitable return. Yet he had been powerless to tear his eyes away from the singular sight. Thankfully, the worrisome dazzle of light and shade that now presented itself everywhere he looked would resolve itself over the coming weeks. On a deeper level, Bond's vision was absolutely clear: no contrivance he might ever generate at his artisan's workbench would be sufficient to satisfy his newfound desire to unmask the mysterious clockwork of the heavens. Science was the only route to this end. Every day forward would be devoted to the pursuit of a goal inaccessible, in the main, to someone without a formal education: "Then and there," Bond's granddaughter Elizabeth writes in her memoir, "he vowed to himself to become an astronomer."

The so-called classical astronomy of William Bond's era was very different from the astrophysical science practiced today. Essential analytic adjuncts to cosmic research—photography and spectroscopy; the physics of atoms, energy, and space; electronic computers—lay far in the future. Telescopes were abundant, but with few exceptions, they were small, crude, and in less-than-capable hands. Examination of lunar and planetary surfaces was largely left to amateur astronomers. Comets, stars, and nebulae were notional rather than physical bodies, both their origin and their action opaque. The measured limits of our Milky Way galaxy were so ragged as to obscure its true extent and form. Nobody perceived that there were other galaxies, much less that these starry islands exist in virtually countless numbers within an expanding universe of finite age.

Lacking the instrumental and theoretical bases to do more, much of early nineteenth-century astronomy was restricted to the determination of positions and motions of heavenly objects. These results, in turn, were applied to tasks such as terrestrial navigation, forecasting eclipses and planetary conjunctions, or predicting the periodical return of comets. That Isaac Newton's mathematical law of gravitation found uniform corroboration within the celestial realm was a marvel of the age. Indeed, Newtonian analysis was a quantitative engine fueled by astronomical data. German astronomer Friedrich Wilhelm Bessel, who in 1838 measured the first distance to a star, asserted that the sole mission of the telescopic astronomer is to obtain the data "by which Earth-bound observers can compute the movements of the heavenly bodies. Everything else that one might learn about these bodies—the appearance and constitution of their surfaces, for example—may be worthy of attention, but it is of no real concern to Astronomy."

In transforming the Royal Greenwich Observatory into a veritable factory of positional astronomy during the mid-nineteenth century, England's Astronomer Royal George Biddell Airy allied with Bessel's narrow view of cosmic studies. The observatory's purpose, Airy asserted, is not for "watching the appearances of spots in the sun or the mountains in the moon, with which the dilettante astronomer is so much charmed. . . . [I]t is to the regular observation of the sun, moon, planets, and stars . . . when they pass the meridian, at whatever time of day or night that may happen, and in no other position."

It was in the 1700s that astronomy and geography were wedded in the name of governmental interests and overseas commerce. Boundary disputes were common between political entities. Many colonial-era land grants in America were based on lines of latitude or longitude, easy

George Biddell Airy, England's seventh Astronomer Royal.

to sketch on a map, notoriously difficult to fix in the field. The decades-long row between William Penn and Lord Baltimore over the extent of their respective colonies was not settled until the 1760s when Charles Mason and Jeremiah Dixon applied astronomical methods to delineate the Pennsylvania–Maryland border. Even a prominent scholar like Friedrich Bessel could be rousted out of his observatory to measure the length of a degree of latitude in Prussia.

Astronomers routinely accepted such earthbound intrusions on their research time, if not for patriotic reasons, then for a simple truth: a surveyor could construct only a relative map of a nation; an astronomer could situate its borders absolutely within the framework of the world. By William Bond's time, fully half of all astronomers were involved in terrestrial position measurement, and more geography-related papers appeared in the astronomical literature than ones on purely celestial topics.

Positional astronomy was likewise applied to transoceanic navigation. "The prosperity of commerce," wrote American astronomer Elias Loomis in 1856, "depends entirely upon . . . the accuracy with which a ship's place can be determined from day to day. Had it not been for the labors of modern astronomers in their observatories, vessels would still, as in ancient times, creep timidly along the coast, afraid to venture out of sight of land; or if they were compelled to venture into the open ocean, they would be exposed to imminent danger in approaching land, not knowing how far distant the port might be."

Sailors' lives and ships' cargoes depended on the accurate delineation of coastlines and shoals. The most effective geo-positioning system for a sailing vessel was astronomical, involving shipboard sightings of the Sun, Moon, or even the configurations of Jupiter's satellites. Given Earth's diurnal rotation, keeping precise track of the passage of time was critical to celestial marine navigation. Monetary awards were offered for improvements in the determination of longitude at sea, as well as for mathematical analyses of lunar motion. Englishman John Harrison's prize-seeking marine chronometer of 1761 deviated a mere five seconds during a transatlantic voyage of 161 days.

One of the hallmarks of classical astronomy was its insistence on exactitude, starting with the precise establishment of the observer's latitude and longitude. The determination of one's latitude is straightforward, from a measurement of the altitude of the celestial pole (approximated in Earth's Northern Hemisphere by the star Polaris). The determination of longitude is more difficult, given Earth's rotation. In William Bond's day, longitude was reckoned astronomically by timing the meridian passages of prominent stars, the celestial analog of surveyors' reference stones.

Transit telescope at the observatory in Besançon, France.

In practice, the astronomer erects a telescope whose axial movement is constrained to the meridian: the north–south arc in the sky that passes through the zenith, directly overhead. Because stars traverse the night sky from east to west (a reflection of Earth's west-to-east rotation), the telescope can be swung around its free axis to intercept each star as it crosses, or transits, the meridian. A chronometer gauges the precise time of transit, whereas a degree-scale on the telescope's axis indicates the star's altitude above the horizon. Mathematical analysis of transit times and altitudes for a set of reference stars yields the observer's geographic coordinates. The local time difference between the occurrence of a celestial event at, say, Greenwich, England, versus Boston reveals the interval in longitude between these two points. That knowledge, in turn, permits coordination of astronomical measurements from observatories around the world. The more precise the observer's transit measurements, the more precise the resultant longitude.

The quest for precision in measurement and analysis had a profound effect on the conduct of astronomy in the 1800s. To its ranks came meticulous, mathematically minded practitioners, eager to embrace an arduous multiplicity of tasks. Their passion extended beyond the study of celestial objects to the identification and quantification of errors in telescopes, chronometers, even the observers themselves. Every telescope, Friedrich Bessel told an audience in 1840, harbors microscopic defects that are revealed only through detailed, systematic observations of the heavens. In Bessel's view, a telescope has to be built twice, "once in the workshop of the artisan, from brass and steel, and again by the astronomer, on paper, through the application of necessary corrections obtained in the course of his investigations."

To ferret out and computationally nullify an instrument's shortcomings, the astronomer turns interrogator: Is the telescope's lens at a precise right angle to the light passing through it? Does the lens sag when the telescope is tipped toward a different direction? Are the mount's rotation axes exactly perpendicular? Does the telescope tube warp under the pull of gravity? Is the instrument level to the ground and aligned north-to-south? Are vibrations of the astronomer's footsteps transmitted to the instrument? Are the markings on the brass coordinate circles equally spaced? Do the circles themselves contract in the cool night air?

Celestial measurement is further muddled by noninstrumental factors that conspire to shift the apparent position of a star in the sky. Earth itself is an imperfect platform from which to observe the heavens. It hurtles around the Sun, spins, and precesses like a top. Its atmosphere swells and agitates the image of a star, whose incoming rays might deflect up to half a degree as they traverse the layers of air. These effects, like instrumental and personal flaws, could be offset by mathematical adjustment of the raw measurements. There were no shortcuts in this line of rigorous observation, nor any promise of fame through discovery, only the chance to make an incremental contribution to the advancement of science.

Fervent as William Bond's cosmic aspirations were, an academic pathway into the profession was nonexistent in early 1800s America. The shuttering of David Rittenhouse's Philadelphia-based observatory upon his death in 1796 left not a single permanent observing facility anywhere on the continent. One astronomical wag defined an American observatory as "a tube with an eye at one end and a star at the other." Attempts by private and public institutions to establish observatories in the United States withered for lack of money. Harvard College prodded wealthy patrons for a research-grade telescope four times before 1825; all of these attempts were unsuccessful. The American Philosophical Society leased space for an observatory in 1817, but failed to raise the added money to buy a telescope. Conversely, Yale purchased a five-inch refractor in 1828, but had no observatory in which to mount it. In 1830, a frustrated president of the University of North Carolina dipped into his own pocket to fund a campus observatory. Its cost: $430.29½.

During their respective terms, Presidents James Monroe and John Quincy Adams petitioned Congress to create a national observatory. Foreshadowing the nationalistic thrust of the Apollo-era race to the Moon, Adams told legislators in 1825:

> It is with no feeling of pride that, on the comparatively small territorial surface of Europe, there are existing upward of one hundred and thirty of these light-houses of the skies; while throughout the whole American hemisphere there is not one. . . . And while scarcely a year passes over our heads without bringing some new astronomical

> discovery to light, which we must fain receive at second-hand from Europe, are we not cutting ourselves off from the means of returning light for light, while we have neither observatory nor observer upon our half of the globe, and the earth revolves in perpetual darkness to our unsearching eyes?

Congress was unmoved, seeing no commercial or political value in governmental sponsorship of basic scientific research; sponsorship of such efforts was the province of states and private institutions. To underscore their opposition, legislators tacked on a proviso to the budget of the U.S. Coast Survey, a mapping project begun in 1807, specifying that "nothing in this act should be construed to authorize the construction or maintenance of a permanent astronomical observatory."

Facilities aside, the United States lacked a vibrant professional astronomical community: although many scientists in the early 1800s credited astronomy's scholarly worth, as well as its practical importance, the nation's full-time astronomers could be counted on the fingers of one hand. The Coast Survey was virtually the only source of employment for the non-academic astronomer. Nor were there any academic training programs in astronomy beyond the general undergraduate curriculum. Absent self-instruction, the aspiring astronomer pursued advanced training through academic apprenticeship, ideally overseas. Europe was the nexus of astronomical studies, primarily Germany, Britain, and France. An 1832 report on global astronomical research by Cambridge astronomer George Biddell Airy does not mention the United States at all.

It would not be until the mid-1830s that American educational institutions embarked on what turned into an observatory building spree. In 1836, Williams College would break ground on a stone building with a thirteen-foot revolving dome to house a pair of small telescopes. Western Reserve College in Hudson, Ohio, would simultaneously embark on its own building program for a four-inch refractor acquired in England. Within two years, Philadelphia Central High School would place a six-inch, German-made refractor atop a domed tower, and by decade's end, West Point would feature three such towers on its grounds. The Federal Depot of Charts and Instruments, created in 1830, would expand over the following two decades into the U.S. Naval Observatory, complete with a dedicated facility, sophisticated equipment, and full-time staff. But for a young, middle-class Bostonian like William Bond, these developments lay in the future. Bond would have to chart his own route into the celestial domain.

Chapter 2

THE INGENIOUS MECHANIC OF DORCHESTER

> No living man . . . has done so much drudgery for science, with so slight a reward, as William C. Bond.
>
> —Astronomer William Mitchell, "The Astronomical Observatory of Harvard University," 1851

IN THE YEARS FOLLOWING THE 1806 ECLIPSE, night became William Bond's refuge from the daytime drudgery of the clockmaker's shop. He evidently complained to no one—save his mother—about the long hours or the relentless pressures of a business to which his father, having failed twice in the lumber trade, seemed no better suited. At his workbench, Bond patiently assembled mechanical implements of time; yet he longed for the end of day, when he could resume his study of the celestial clockwork. In tracking these cycles, he might uncover the mainsprings, escapements, and regulators that make the cosmos run true.

William Cranch Bond.

The stars would have glistened brightly over William Bond's Boston, their radiance unimpeded as they are today by city lights. Bond memorized the constellations and how they cycled through the seasons. He noted the synchronous pas de deux of the Sun and Moon, as well as the stately adagio of planets against the starry backdrop. He gauged separations between stars with a knotted string held up against the sky, as English astronomer William Herschel had done before he became famous. Now and then, a meteor would streak above his head, hell-bent on its rendezvous with oblivion.

Each of these nocturnal communes with nature began the same way, with Bond staring into a well for ten minutes. "[H]is optic nerve became so stimulated," writes his granddaughter Elizabeth, "that he acquired almost telescopic vision and could see stars invisible to others." The spurious claim of telescopic vision aside, once his eyes were adapted to the dark—an essential practice among serious night-sky observers—Bond was well attuned to serendipitous events in the heavens.

On the night of April 21, 1811, Bond noticed a faint whitish blur to the south, a few degrees above the star Sirius in the constellation Canis Major. With the night sky now as familiar to him as his own neighborhood, he knew at once that the object was out of place. It hadn't been there any night before. A longer look brought out the indistinct, yet unmistakable, image of a luminous tail, about a degree in length, projecting from the diffuse core. Bond measured the celestial coordinates of the object. He did the same three nights later and again on several occasions in May before he surrendered to his mounting excitement. The object was moving. He had discovered a comet.

It was only later that Bond learned that what would become known as the Great Comet of 1811 had, in fact, been discovered in Europe a month beforehand. But he was the first observer in America to see it. By autumn, the comet blazed brighter than almost any other in history, extending its tail a full twenty-five degrees. Word of Bond's visual feat reached Professor John Farrar, mathematician and astronomer at Harvard College, and Nathaniel Bowditch, the nation's foremost expert on celestial navigation. Impressed, the two scientists featured Bond's comet observations in their own report to the American Academy of Arts and Sciences in September 1811, introducing him to their colleagues as an "ingenious mechanic of Dorchester, Massachusetts." Congressman Josiah Quincy, who would go on to serve as mayor of Boston and then President of Harvard, encouraged the twenty-three-year-old to pursue his dream. One eclipse and one comet into his calling, William Bond had stepped into the inner circle of American science.

With professional validation in hand, Bond delved into the complexities of celestial position measurement. His first transit instrument, nailed up near the roofline of his family's Dorchester home around 1813, was a homely strip of brass with a sighting hole. Bond would lie supine on the ground, wait for a star to appear in the hole, then record the time of its meridian passage. He could barely contain his excitement upon seeing the moons of Jupiter and the rings of Saturn through his first telescope.

Bond further distinguished himself in 1815 with the completion of America's first sea-going chronometer. Three years in the making, the device was based on a plan by the celebrated eighteenth-century French clock-maker Ferdinand Berthoud. Unable (or unwilling) to obtain the specialized British spring steel during the War of 1812, Bond fabricated a descending-weight mechanism to keep the device running. A voyage in 1818 aboard a U.S. Navy vessel to Sumatra proved Bond's marine chronometer to be as accurate as the world's finest. (The device resides in the Physical Science Collection of the Smithsonian's National Museum of American History in Washington, DC.)

Bond's astronomical bona fides got an unexpected boost in 1815 with the death of his father's brother, a wealthy and childless widower in England. Although the finances were tight, Bond's parents booked him passage overseas to represent his family's interests in the estate. Learning that Bond would be traveling to England—home of the Royal Greenwich Observatory and several noted telescope makers—a group of faculty and administrators at Harvard College revived their moribund plan to erect an observatory on campus. Their stated goal was to purchase a world-class telescope and establish Harvard as a leading astronomical research center. The lofty proposal had been stirring since 1806, but had so far foundered for lack of money. Harvard offered to pay half of Bond's travel expenses if he would make the rounds of British observatories and report back on their design and functionality.

The letter of terms from Professor Farrar specified that Bond visit the Royal Observatory at Greenwich, the Kew Observatory at Richmond, and William Herschel's observatory at Slough. He was to record the size, form, depth, height, and composition of the piers that supported the instruments; width of apertures in the roof, how they opened and closed, how they were protected against the elements; plus every conceivable particular—both optical and mechanical—about the instruments themselves. Bond was also to inquire about the cost of an eight-foot-long transit telescope from England's leading instrument maker, Edward Troughton. Harvard faculty members would furnish letters of introduction they were confident would gain Bond admittance to any scientific facility in Britain. Most significantly, Farrar directed that Bond's report "must be such as to enable you or another person to superintend and direct in the erection of an Observatory." The

Harvard academics declared their full faith in Massachusetts's own "ingenious mechanic."

Upon his arrival in Liverpool, Bond headed to his uncle's house in Kingsbridge, in southwestern England, where his mother had been raised. Entering the garden, he was immediately smitten by his young cousin Selina Cranch standing among the roses. (He returned to marry Selina four years later, in 1819.) After a futile effort to promote his father's claim on the deceased brother's estate, Bond spent his last shilling to reach London, where he was to meet Harvard's local agent and receive his promised travel funds. He was stunned to learn that the agent had gone on holiday, his destination and date of return unknown. Bond knew from his uncle in Kingsbridge that his elder brother Thomas, at that time a sailor, was on a stopover in London. The siblings had been inseparable as children, Thomas marveling at his brother's home-brewed genius. Now penniless, hungry, and alone in a metropolis of a million souls, William Bond was surely eager to find a familiar face. After spending a fitful night on the steps of St. Paul's Cathedral, he managed to locate Thomas and borrow enough money to continue his trip. (Harvard reimbursed him after his return to the United States.)

Bond was warmly welcomed at the Royal Observatory at Greenwich and at a number of private facilities. He duly took notes, made measurements, drew detailed floor plans, and interviewed astronomers and instrument makers about the complexities of building and maintaining an observatory. He spoke colleague-to-colleague to Astronomer Royal John Pond, talked shop with like-minded mechanic Edward Troughton, and was treated to a VIP tour of William Herschel's observatory at Slough by Herschel's sister Caroline. The sight of Herschel's towering reflector telescope—which American writer Oliver Wendell Holmes, Sr., likened to "a piece of ordnance such as the revolted angels battered the walls of Heaven with"—was indelibly impressed upon Bond's memory.

In the end, Bond's overseas trip counted for naught. His 1816 report made clear that the cost of building and running a new observatory far exceeded the resources Harvard had hoped to tap. As Bond would write in his history of the Harvard College Observatory, "The time had not yet arrived when the project could be prudently or conveniently carried forward."

Meanwhile, the Bond family business began to take off. The firm moved to larger quarters on Congress Street, with William now at the helm. Its manufactured and imported chronometers, an essential element of marine navigation, stood at the forefront of the clockmaker's art. In the coming years, Bond's firm would serve the interests of a growing number of New

England ship captains—who were required to purchase their own navigational instruments—as were the U.S. Navy, the U.S. Coast Survey, and the U.S. topographical engineers. It was a lucrative enterprise: chronometers were expensive—up to $300—and maintenance and repair costs were likewise high. Adjustment of a chronometer spring might run five dollars, as much as an entry-level clerk earned in a week. (The company also entered the broader commercial market for precision time; in 1849, the New England Association of Railroad Superintendents would mandate that all station clocks, conductor's watches, timetables, and trains be synchronized with William Bond's timepiece.)

His fortunes secure, Bond married Selina Cranch in 1819. The couple settled into a large clapboard house on Cottage Street in Dorchester, a few blocks from his childhood home. Observations began right away. By late 1820, Bond had acquired two telescopes and was lent a third from Harvard College. Around 1823, an attached parlor was sacrificed to astronomy, with Bond and his brother Thomas sinking a multi-ton, granite-block telescope pier five feet into the earth below the floor and cutting an observing aperture into the ceiling. "His antipathy to an insecure foundation many would have thought extravagant," recalled Bond's son George, "the tremor of an instrument would annoy and fret him as a harsh discord does the cultivated ear of the musician." Smaller telescope-mounting stones dotted the garden and the surrounding fields like a scatter of neolithic monuments. To rest atop these rocky pedestals was a growing array of high

William Bond's house on Cottage Street, Dorchester, Massachusetts.

quality instruments purchased from Europe. By the 1830s, William Bond's private observatory in Dorchester had become America's hub of precision astronomy.

Carving out time for astronomy was a never-ending challenge, with the incessant demands of business, support of his elderly parents, and eventually, six children padding around the house. Bond had assured his wife that he would earn enough from his profession to cover all household expenses. That meant spending full days at the shop plus regular evening hours at home dealing with the steady stream of watch repairs. Only then would he surrender himself to his ruling passion—a pursuit most would have regarded as the acme of tedium.

Bond's son George, who (with initial reluctance) took up cosmic studies, understood perfectly the lure that had drawn generations of precision-obsessed men, from Tycho Brahe to every Astronomer Royal since John Flamsteed. "To watch the motions and record the positions of the heavenly bodies," George writes, "was an occupation perfectly congenial to his tastes. . . . For thirty years this was done, not merely without compensation, but to his manifest pecuniary disadvantage. This consideration, it is probable, never entered his mind."

At the same time, George Bond recognized that his father's enterprise bordered on obsession: "There is something to my mind appalling in the contemplation of my father's labors, from the time when he was first enabled to indulge freely his passion for observation. The accumulated volumes filled with manuscript records give me a shudder at the thought of the weary and straining eye, the exposure [to the elements], and the long, sleepless nights that they suggest."

Through the 1820s and 1830s, William Bond's primary project was to ascertain, through celestial measurements, the position of the granite pier in his parlor relative to the Royal Observatory in Greenwich, England. This he announced in an 1833 report to the American Academy of Arts and Sciences: latitude 42° 19' 20" north; longitude 71° 4' 15" west of Greenwich. So precise was this determination that, in 1838, the Navy Department's worldwide survey of foreign ports referenced their entire set of geographic coordinates, not to Washington, DC, but to William Bond's house in Dorchester. (Although Bond received a nominal sum by the government for his work, he spent around ten times that amount out of his own pocket for the requisite equipment.)

In 1839, fully twenty-four years after enlisting William Bond in its failed attempt to build an observatory, Harvard College again came calling.

Seeking to capitalize on public interest raised by the recent passage of Halley's comet, Josiah Quincy, Bond's longtime champion and now president of Harvard, made an admittedly desperate offer. He invited Bond and his family to take up residence in the Dana House, a recent acquisition at the southeast corner of the Harvard campus. Bond would bring with him all of his astronomical equipment, turning the Dana House into the college's de facto observatory, with Bond as so-named Astronomical Observer. The college offered no salary and only the promise to seek funds for new equipment. Bond declined, replying charitably that "his habits were not adapted to public station [and] that he preferred independence in obscurity to responsibility in an elevated position."

Quincy persisted, despite his own reflections on the absurdity of the job offer: "Mr. Bond was well established in a profitable manufacturing business, happily situated in his domestic and neighborhood surroundings, with an avocation fascinating enough to occupy all his leisure, and a fame extensive enough to satisfy his own modest estimate of his abilities. There was no pecuniary betterment for Mr. Bond in the suggested change. [Harvard] could only offer him what he already had, a family domicile; so that the proposal might warrant an adaptation of Sydney Smith's famous phrase, and be described as an invitation to come to Cambridge and 'cultivate astronomy upon a little oatmeal.'" By way of explanation, Quincy maintained that in Harvard's mean proposal, "there is no disparagement of the college; it was the day of small things, of pennies, not dollars, in the college treasury."

After sustained appeals, Bond relented; in the autumn of 1839, he moved his family and his operations to the Dana House. On December 31, he made his maiden observation as a Harvard astronomer. Bond's children were pressed into service: the mathematically gifted William Cranch Bond, Jr., who died during his senior year at Harvard in 1842; George Phillips Bond, who would eventually succeed his father as observatory director; and Richard Bond, who developed and maintained the timekeeping equipment. The Dana House site was far from optimal for sky viewing, being hard up against neighboring structures, trees, and streets. Among other things, Bond had to pay for permission to bore a sighting hole through an adjacent building to align his telescope to a seven-foot-tall meridian marker twelve miles away on Great Blue Hill in Milton.

Four years into his Harvard appointment, the heavens blessed William Bond with yet a third career-altering spectacle. The Comet of 1843 appeared in the spring, growing and brightening until its two-hundred-million-mile tail could be seen even during the daytime. Not since the solar eclipse of

1806 were Boston's residents so galvanized by a celestial sight. In the comet's wake, contributions totaling $35,000 poured into Harvard's coffers for the establishment of a world-class observatory.

The core specification for Harvard's new telescope was straightforward, if overtly ambitious: a refractor-style instrument with a main (or objective) lens larger and optically finer than any other in the world. The telescope would rest on a sturdy equatorial mount with built-in clockwork to rotate the instrument in synchrony with the nightly movement of celestial objects. After consultations with prominent astronomers and optical craftsmen in Europe, Bond and his Harvard colleagues concluded that the German establishment of Merz and Mahler, celebrated successor of master optician Joseph Fraunhofer, was the only firm capable of meeting their requirements. In fact, the company had just completed a similar refractor, fifteen inches in diameter, for Russia's Imperial Central Observatory in Pulkovo. With lens-making still an imperfect art, Merz and Mahler agreed to fabricate *two* fifteen-inch objective lenses, and Harvard could choose the better—or reject both.

In 1843, Harvard purchased acreage on Summer House Hill, three-quarters of a mile northwest of the main campus, a remote rise surrounded by meadows with few neighbors and only spotty transport to Harvard Square or Boston. The site was an astronomer's dream: pitch dark at night, with an unobstructed view of the horizon in every direction—although the fickle New England climate meant frequent overcast skies.

Construction of the observatory and family residence began immediately. An increasingly harried Bond, now a gray-haired fifty-four, confided his concerns to longtime friend and amateur astronomer William Mitchell: "In regard to the principle features of the observatory, I suppose we must say it is getting on as well as can be expected. Merz is progressing with the second object glass, the carpenters promise to have the new house ready by July [Bond doubted them], but for two years we must be prisoners of hope." He added, with uncharacteristic frankness, that he considered the Dana House period, "in common with the rest of the family, as the most unpleasant in my whole life."

In September 1844, Bond moved his family and his instruments from the Dana House to the grounds of the new Harvard College Observatory. An avid gardener, he covered the yard with flowers, shrubs, and exotic plants given him by noted botanist Asa Gray, his new neighbor. The family residence "was approached by terraces on which grew cherry trees and clumps of peonies. Rose vines and honeysuckle were trained over the house and an arbor vitae hedge screened the lower rooms from view. . . . A barn stood near the kitchen wing, the home of a cow, several horses, pigs and poultry."

Despite Bond's exasperation with the financial and engineering complexities of the observatory project, he was fully devoted to the fledgling institution. When the Naval Observatory offered a substantial sum to recruit him in 1846, Bond turned them down, writing: "An astronomical observer to be useful in his vocation should give up the world, he must have a good eye, a delicate touch and above all, entire devotion to the pursuit." Learning of the Naval Observatory's overtures to the prime mover of its astronomy initiative, Harvard immediately provided Bond with an annual salary of $1,500, plus a $640 stipend for his son George, a recent Harvard graduate, to act as his chief assistant. (In 1851, the salary was made permanent through a $100,000 bequest by George's college classmate Edward Bromfield Phillips, who had committed suicide.)

Through a shipping error, the completed telescope lens arrived by steamer in New York, instead of in Boston, on November 28, 1846. (One day later and the lens would have been subject to an ill-conceived 30 percent tariff enacted by Congress on imported scientific equipment.) The lens was rushed to Cambridge, where it awaited completion of the building, copper-clad dome, support pier, and telescope tube.

Bond made every effort to prepare a solid, vibration-free base for the new telescope, an edifice described by William Mitchell: "An excavation was first made twenty-six feet below the natural summit of the hill; and at the bottom of this was placed a coating of cement intermixed with coarse gravel ten feet in thickness, which, when hardened, formed an entire mass of great firmness. On this bed, the pier, composed of five hundred tons of large granite blocks, well fitted to each other and laid in cement, rises thirty-three feet to the upper surface of the floor of the dome. On the cap-stone of this rests, on three bearers, a solid granite tripod or pedestal, of eleven tons' weight, to the top of which is attached the Great Equatorial."

On Wednesday, September 22, 1847, at 3:30 a.m., Bond turned his new telescope to the Orion Nebula. "[T]he revelation was sublime," he jotted in his observing diary, "the first appearance was like bright clouds—the fifth star in the Trapezium was conspicuous, many stars were seen among the clouds of light, & about the borders." On October 7, he added, "It is delightful to see the stars brought out which have been hid in mysterious light from the human eye, since the creation. There is a grandure [sic], an almost overpowering sublimity in the scene that no language can fully express." (In his enthusiasm, Bond mistakenly believed he had resolved the entire nebula

The Great Refractor of the Harvard College Observatory.

into individual stars. It would be some twenty years before astronomer William Huggins proved spectroscopically that the Orion Nebula is, in fact, a star-studded fluorescent gas.)

Although reluctant at first to join his father's line of work, George Bond became William's steadfast companion in the observatory. The two of them had similar reserved dispositions, as well as a preternatural ability to focus on tedious tasks for long periods. George would eventually pursue projects outside the narrow confines of his father's positional astronomy, but he performed whatever was required of him without complaint. Father and son are curiously intertwined in the observatory's record books and in their joint professional diary, the published title of which is *Diary of the Two Bonds*. Entries are not identified by name, nor was the handwriting always distinguishable. Sometimes the only clues as to who wrote what were subtle stylistic differences.

The wife of Richard Bond, William Bond's youngest son, recalled her childhood impressions of the intimate working relationship between William and George, whom she watched one day drawing sunspots: "One observer, with a sharp pencil, traced the spots as they were reflected on the paper, while the other wrote down any notes or observations, of time, of peculiar appearances, or explanatory of the drawings. But both of them . . . had eye and hand and mind so thoroughly trained, that even to children it was fascinating to watch . . . their enthusiasm and delight in the work, and the quick response and recognition of either to a remark or suggestion of the other."

The public clamored to see the magnificent new refractor, which they were told could magnify celestial objects more than a thousand-fold. Most of the visitors attended the weekly viewing sessions, but a steady stream arrived unannounced at the observatory's doorstep. One couple came all the way from Michigan after reading a newspaper account of the instrument. At first, the good-natured Bonds tolerated these periodic intrusions. The turning point came with the public viewing session on the evening of October 23, 1847. "They came by the Hundreds," William Bond penciled in his diary, "so as to quite overwhelm us." The facility was left strewn with debris, he complained, "something like those portions of the common in Boston, which have been the most crowded, on the morning after a fourth of July." Thereafter, public access to the Observatory was restricted.

The evening's "perfect Babel," as Bond aptly put it, was a fitting coda to a different sort of fiasco that occurred earlier that day. It began with the arrival at the observatory of a lanky, ruggedly handsome man with a fringe beard rounding his jaw. Bond would have recognized John Adams Whipple as the proprietor of a Boston storefront located not far from his own. Although more than twice his age, Bond would surely have felt a kinship with Whipple, who was, like him, a practical-minded mechanic with an eye to nature. Whipple had asked to use the big refractor for an open-ended project that would unwittingly lead to one of the most important advances in astronomy since the introduction of the telescope. Tucked under John Whipple's arm was a wooden box with a leather bellows and glass lens: a camera.

Chapter 3

Writing with Light

> [Photography] seemed to epitomize new means of reaching truth in a form acceptable to everyone. . . . By providing new methods of insights, it seemed to be a manifestation of truth itself—with beauty.
>
> —Richard Rudisill, *Mirror Image: The Influence of the Daguerreotype on American Society,* 1971

The locale was at once familiar, and unlike anything François Arago had ever seen before. There, on the southern flank of the Ile de la Cité, rose the towers and flying buttresses of Notre Dame cathedral. To either side arched the elegant stone bridges Arago had often crossed over the Seine—although today their arrangement was oddly reversed, with Pont de L'Archeveché on the left and Pont Saint-Louis on the right, instead of the other way around. The river itself had a curious unearthly sheen, its normally agitated surface smoothed into a silvery ribbon so still that it mirrored the bridge abutments. Remarkably, the entire scene was rendered in myriad shades of gold-tinted gray, as if nature's pigments had been scrubbed away to reveal only the fine-lined sketch underneath. Yet, even to François Arago's practiced eye, the colors of Paris were hardly missed amid the vibrant interplay between light and dark.

Were he not a man of science—in fact, Director of the Paris Observatory and Permanent Secretary of the French Academy of Sciences—Arago might have sworn that he was witnessing a miracle. But the six-by-eight inch metal slab onto which this diminutive slice of Paris had seemingly been shrunk was the product of human invention. Its secretive creator, Louis Jacques Mandé Daguerre, an acclaimed painter of theatrical sets, must have been pleased with Arago's response to his picture. He had expended years of solitary labor on this endeavor and was ready, at last, to reap the profits. By

Louis Daguerre's photograph of Notre Dame and the Seine, 1838.

the fall of 1838, Daguerre had shown his camera pictures—now numbering around forty—to a privileged few and revealed his photographic technique to no one. He hoped to sell usage rights to the new technology—the daguerreotype, he would call it—to well-heeled visitors at a public exhibition in Paris on January 15, 1839. An endorsement by such a distinguished and well-connected citizen as François Arago—a member of the Chamber of Deputies in the French Parliament—would be a promotional coup.

Arago had requested the meeting with Daguerre, for only by direct inspection could he judge the reality of the rumored process that captured images without paint or brush, but solely by chemical means. If true, it would be a boon to art and to science. With a daguerreotype camera, a small band of workers could record the ancient stoneworks of Egypt, geological features across Europe, flora and fauna the world over, even aspects of luminous bodies in the heavens. Daguerre should be justly compensated for his invention, Arago believed, but the greater good of France—indeed of the world—must take primacy. Once validated, the daguerreotype process must not be patented. It must be freely available to anyone who sought to maximize its potential.

No doubt Daguerre was appraising Arago and his motives, as Arago was surely doing of him. Daguerre offered Arago a magnifying glass and suggested that he take a closer look at the picture. Arago pored over Daguerre's metal plate through the lens. The more he screwed up his eyes, the more he found himself drawn into the encyclopedic still life, whose every recess held a further surprise. Under magnification, every window mullion, every roof tile, every cobblestone was crisply delineated. Notre Dame's varied surfaces became a tumble of tiny Euclidean forms, its spires needlelike slashes against the brilliant sky. Even the shadows underneath the bridges feathered into the gloom. The scene seemed almost too realistic to be real, possessing a fineness of detail that made a mockery of human vision.

Scanning the Lilliputian streets and quays for the telltale forms of people, Arago encountered not a soul. Daguerre had created a barren architectural showcase that drew its peculiar power from the vitality of light playing on forms and materials, but was otherwise drained of any human presence. Arago intuited the reason: every creature or object that had moved during the long minutes of exposure had left no impression on the daguerreotype plate. The pedestrians, barge loaders, cart drivers, street vendors—all had smeared themselves into oblivion by dint of their workaday activity.

Arago put down the magnifier. As to Daguerre's picture, he could draw no other conclusion: regardless of how fine the brush, how keen the eye, how steady the hand, this was no trick of the painter's palette. No artist who ever lived could have created such an excruciatingly accurate scene, so far it stood above human capability. Arago's course was as clear as the wondrous image before him: he had to convince Louis Daguerre to divulge his secret method.

In 1822, Louis Daguerre opened the Diorama, which soon became one of the most popular cultural attractions in Paris. The Diorama featured a central, revolving auditorium, around which were arrayed three skylit stages. Each stage was an illusionist's playground, a wholly artificial reality conferred by outsize murals (up to seventy-two by forty-eight feet), trompe l'oeil paintings, forced perspective, artifacts, sounds, smells, and lighting effects. There were no actors; the visual sleights of hand were the attraction.

Louis Daguerre.

One of the more popular of Diorama presentations, on view for three years, was *Midnight Mass at the Church of St.-Etienne-du-Mont, Paris*, described by one visitor this way:

> As night fell the church gradually darkened, the devout came to take their places, and, to the peals of an organ and the fumes of incense, choir boys lit candles. The Mass was sung, the faithful disappeared, the candles were extinguished and daybreak illuminated the stained-glass windows. The illusion was so perfect that a lad from the country threw a coin onto the stage to find out whether there was space in front of him.

To help establish the proper perspective for his theater sets and Diorama scenes, Daguerre employed a camera obscura, an optical device long used by artists to project images onto a surface for tracing. (Leonardo da Vinci sketched such a device in 1519, featuring a pinhole aperture for entry of light. Giovanni Battista della Porta recommended it as a drafting aid in his 1553 tract, *Natural Magic*. Fifteen years later, Daniello Barbaro, author of a treatise on perspective, replaced the pinhole with a lens.) Daguerre's camera obscura consisted of a wooden box with a converging lens at one end, canted mirror inside, and ground glass screen on top, from which the projected image was traced onto paper. In the mid-1820s, Daguerre considered whether the image might instead be recorded directly onto a chemically treated plate. He found this to be a fertile area of research, with very old roots.

In his *De rebus metallicis* from 1566, Georg Fabricius described a waxy, translucent alchemical product called horn-silver (silver chloride) that was later found to blacken in sunlight. In 1614, Italian chemist Angelo Sala discovered similar behavior in lapis lunearis, or powdered silver nitrate. In what might be considered photography's pioneering experiment, German scientist Johann Heinrich Schulze, in 1727, covered bottles of silver nitrate crystals with paper stencils of handwritten words or sentences. He placed the bottles in a sunny spot, and found that crystals underneath the excised portions of the stencil—those blackened by the Sun—replicated his handwriting. Other silver salts, or halides, including silver iodide, silver chloride, and silver bromide, were likewise observed to be light-sensitive. (Chemical studies reveal that exposure to light liberates the combining element and leaves pure metallic silver. The silver particles form a latent image, which is amplified in the development process.)

During the 1790s, Thomas Wedgewood, son of the famous English porcelain maker, created sunlight-generated silhouettes and contact prints of engravings on paper sensitized with silver nitrate. However, he found that images in a camera obscura were too faint to trigger any photochemical response. A similar abortive attempt was made in 1812 by Samuel Morse, the American painter and inventor of the telegraph, while he was a student at Yale University. Had either succeeded, they would have been confronted by a practical issue: how to "turn off" the photochemical reaction once the exposure is finished. Without such intervention, the entire photosensitive plate inevitably darkens over time, obliterating any picture. As François Arago complained about the inability to fix a photographic image, "What then was there so wonderful, in images of which scarcely a glance could be obtained, and this only by the light of a small lamp, for they disappeared as soon as it was attempted to bring them to day light?"

Incontestable success was achieved by Joseph Nicéphore Niépce, a French lithographer and inventor. In 1826, after more than a decade of experimentation, Niépce took the earliest known permanent photograph from nature. This heliograph, as he called it, was the product of an eight-hour exposure from the third-floor window of his house at Le Gras near Chalon-sur-Saône. The camera plate was an eleven-by-ten inch pewter slab that had been varnished with a suspension of powdered bitumen in lavender oil. Bitumen, an asphalt derivative of petroleum, hardens—it turns insoluble—when exposed to light, a property that drew Niépce's attention. Although bitumen is black in its native form, its derivative varnish turns grayish-white as it dries on the plate.

Upon exposure, the camera image maps itself onto the bitumen-covered plate, creating areas of different hardness, depending on the incident illumination. A latent image results, invisible to the eye. To develop the plate, Niépce applied a solvent of lavender oil and turpentine. The solvent strips away nonhardened and partially hardened areas of bitumen—those that had received lesser illumination—revealing the grayish pewter underneath. By contrast, hardened areas—those that had received the most light—are impervious to the action of the solvent. Thus, the recorded picture consists of brighter tones, rendered in remnant bitumen, against somewhat darker tones of pewter. The fully developed heliograph is a dim, grainy, yet recognizable, reproduction of the actual scene from the window.

While visiting his brother in London in 1827, Niépce tried to enlist the financial support of the Royal Society for his process. He was turned down when he refused to divulge how the pictures were made. (Before returning to France, Niépce left the heliographs with botanist Francis Bauer, who continued to press Niépce's case. Upon Bauer's death, *View from the Window at*

Le Gras passed through a variety of private hands and was considered lost, until discovered in a family trunk in 1952. Photograph collector Helmut Gernsheim donated the picture to the University of Texas at Austin, where it is now on permanent display.)

Niépce was dissatisfied with the narrow tonal range of his heliographs; there was no pure black or pure white, only shades of gray. Having read that silver darkens in the presence of iodine, he replaced his pewter plates with silver-coated copper plates. He took the exposure and developed the plate as before, but then fumed the developed bitumen image in iodine vapor, blackening areas of silver that had been laid bare by the solvent. Finally, he dissolved any remnant of hardened bitumen, revealing the untouched silver below. Now the bright areas of the image were rendered in brilliant silver-white and the dark areas of the image in iodine-blackened silver. The resultant images were a marked improvement over his initial trial. Still, the hours-long exposure times for the bitumen plates were impractically long.

Having learned of Niépce's work through a common acquaintance, Daguerre posted a letter in January 1826 expressing his interest in the subject and suggesting that he and Niépce collaborate. Niépce did not reply, informing his son Isadore that "one of these Parisians wants to pump me for information." A second and third letter followed a year later. (Daguerre's letters were lost when Niépce dropped his pocketbook down a toilet in a London hotel.) After confirming Daguerre's bona fides, Niépce agreed to a series of meetings in Paris. At the time, Niépce had accumulated more than a decade of imaging experience, while Daguerre had spent at most three years working with silver chloride paper and dabbling in color imaging. Nevertheless, Niépce was taken with Daguerre's boisterous sincerity, not to mention his entrepreneurial spirit—"there should be found some way of getting a large profit out of it"—for the two men signed a ten-year business partnership on December 4, 1829.

Niépce's death in 1833 left Daguerre to advance the project on his own. This he did to resounding effect. Daguerre dispensed altogether with Niépce's bitumen substrate, realizing that images could be produced directly through the reaction of silver with iodine vapor. Equally significant, he discovered that exposure to mercury vapor renders the latent silver iodide image visible. (The source of this particular insight is unknown.) And in 1837, Daguerre introduced the all-important means to render his images semipermanent. To deflect any accusations of opportunism, Daguerre later published Niépce's own account of his research, written December 5, 1829. The record is clear that Daguerre's subsequent contributions to the development of the process were significant and that his name is rightly applied to it.

By autumn 1838, having managed some forty successful exposures, Daguerre was ready to market his invention. (The earliest of these photographs, a still life taken in his studio, dates to 1837.) "By this process," Daguerre trumpets in a never-released broadside, "without any notion of drawing, without any knowledge of chemistry or physics, it will be possible to take in a few minutes the most detailed views, and the most picturesque sites, for the technical means are simple, and require no special knowledge to be used. Only care and a little practice is needed to succeed perfectly." Daguerre goes on to suggest that everyone will take a picture of their chateau or country house, and that both science and art will reap the benefits of precise, objective imaging. But he admits that, as yet, human portraiture is not possible because of the sitter's inevitable movements during the half-hour exposures. Daguerre's business proposal was straightforward: one thousand francs bought rights to commercial use of the daguerreotype; two hundred thousand francs purchased the process outright.

Despite all efforts, Daguerre found not a single buyer for the daguerreotype process. Ironically, no one could see the big picture—neither the technology's revolutionary nature nor its monetary prospects. Illustrators and painters were integral to the history and social fabric of France. Why take a proven means of artistic reproduction and fob it off to a chemical process? The notion that the technology itself might spur undreamed-of applications lay beyond the perception of the ordinary businessman.

To the chastened Daguerre, François Arago must have seemed a savior. This man, now poring over samples of Daguerre's unique craft, was by every indication a visionary: astronomer, physicist, writer, lecturer. Who better than Arago to appreciate the value of the daguerreotype? And, with his access to the highest rungs of government, who better to loosen the treasury's purse strings? On Arago's word hinged Daguerre's recompense for all the years of solitary toil, if only he could see the pictures properly in the light of the future.

Arago was indeed impressed by the daguerreotype, which surpassed his every expectation. He grasped its cultural and scientific promise for the nation. Technology of this magnitude, he concluded, was too valuable to be left in the self-interested hands of one man or a cadre of wealthy investors. Let everyone have free access to Daguerre's work; let no restrictive patent hinder its rapid, full flowering. In return, the inventor must be amply paid for his contribution. Arago instructed Daguerre to cancel his January premiere and to suspend all attempts to market his photographic process. Arago would lead the effort to secure fair compensation for Daguerre from the French government. The campaign would begin immediately.

On January 7, 1839, Arago announced and personally endorsed the daguerreotype process before the French Academy of Sciences. He told the body that he himself had taken and developed a picture; with care, he said, anyone can do it. Two noted academy members, Jean Baptiste Biot and Alexander von Humboldt, had independently examined specimens of Daguerre's work, including views of Notre Dame, the Louvre, bridges over the Seine, and other Paris landmarks. They verified that the daguerreotype was a breakthrough and held vast promise for both the arts and the sciences. Symbolic of the latter, Daguerre had even produced a picture of the Moon; although blurred beyond recognition, it was only a matter of time, Arago insisted, before celestial imaging was brought to heel.

Arago revealed nothing of the process itself, only sufficient facts to galvanize the gathered members of the academy and the press. Obligingly, one art critic wrote afterward, "You can now say to the towers of Notre Dame: place yourselves [on this photographic plate], and the towers will obey, brought home in their entirety." In his own way, Arago had outdone Louis Daguerre, promoter extraordinaire.

Arago's surprise announcement made an unexpected ripple in England. Gentleman-scientist William Henry Fox Talbot hastily addressed the Royal Society on January 31, 1839, claiming to be the true inventor of chemical imaging. Talbot's process, the calotype, developed in 1835, produced low-grade, negative images on photosensitized paper. Unlike irreproducible daguerreotypes, a calotype negative could be used to generate any number of positive contact prints. Although Talbot's method of printing from negatives would ultimately become the model for chemical imaging, the calotype did not reach an acceptable level of refinement until 1841. By that time, the daguerreotype had achieved widespread acceptance. It didn't help that Talbot patented the calotype, which further hindered its adoption among professional photographers.

François Arago.

Arago introduced a bill into the French Parliament that would pay Daguerre for rights to his process. He did his utmost to raise expectations of the daguerreotype's aesthetic, scientific, and economic benefits, while coyly avoiding any public exhibitions of the pictures or release of procedural details. At every opportunity, he showed sample images to fellow legislators. Meanwhile, Daguerre found himself the object of international attention. Samuel Morse visited him in March, Sir John Herschel and James Watt in May. Already Herschel had promulgated the catch-all term *photography*—"writing with light" (previously coined by German astronomer Johann von Maedler). All expressed astonishment at the quality of Daguerre's pictures. Herschel supposedly confided to Arago "that compared to these masterpieces of Daguerre, Monsieur Talbot produces nothing but vague, foggy things. There is as much difference between these two products as there is between the moon and the sun." Herschel informed his friend Talbot, adding, "It is hardly saying too much to call them miraculous. . . . I can not counsel you better than to *come* and *see*."

On August 7, 1839, after affirmation by both chambers of Parliament, King Louis Philippe signed the agreement to provide an annual lifelong pension of six-thousand francs for Daguerre (later raised to ten thousand), and four-thousand francs for Niépce's son Isadore in exchange for rights to the daguerreotype. The settlement couldn't have come at a better time for Daguerre; the Diorama, his sole source of income, had burned to the ground the previous March. In voting to grant the pension, legislator Joseph Louis Gay-Lussac, himself a prominent chemist and physicist, acknowledged that Daguerre's process "must have been very difficult to discover, and must have cost, in order to reach the perfection which it has attained from Mr. Daguerre, much time, numberless attempts, and above all a most wonderful perseverance which is strongly excited by failure and which never belongs but to those who are gifted with strong minds."

Arago announced a special joint meeting of the Academy of Sciences and the Academy of Fine Arts for August 19, 1839, at which the secret of the daguerreotype would be disclosed. The auditorium at the Palace of the Institute of France was crammed to capacity two hours before the presentation was to start, with several hundred disappointed spectators relegated to the courtyard. Every once in a while, someone would emerge from the hall and relate to the exiled throng fragments of what was being said inside.

With Louis Daguerre and Isadore Niépce sitting silently at his side, Arago revealed the process step by step:

1. Polish a silver-coated copper plate to a mirrorlike finish, then wash off all residue with diluted nitric acid.

2. Fume the plate in iodine vapor until uniformly golden brown, evidence that the requisite amount of light-sensitive silver iodide has been formed on the plate's surface. (Iodine vapor must be used, as silver does not react with iodine in solution.)

3. Insert the prepared plate into the camera and expose it anywhere from three to thirty minutes, depending on the time of day, season, and weather conditions. Incident light sunders the silver iodide, and leaves a greater or lesser number of elemental silver particles at various locations on the plate, thereby delineating the scene before the camera. The recorded image is as yet latent and not discernible to the eye.

4. Remove the plate and expose it to hot mercury vapor, which renders the latent image visible. The silver and mercury atoms combine into an amalgam, which appears whiter than the plate's original silver coating. The amalgam maps out the highlights of the image, while the iodized silver maps out the shadows.

5. Fix the image by immersing the plate in sodium thiosulfate, or "hypo," which removes any remnant silver iodide. (Both Daguerre and Talbot had fixed their early photographs in hot salt water, but adopted John Herschel's use of hypo when announced in early 1839. In 1840, Arago's twenty-year-old student Hippolyte Fizeau discovered that gilding finished plates with gold chloride is even more effective.)

6. A series of thorough rinses in water completes the process.

Arago admonished would-be photographers to admire their art with the eyes and not the fingers. A completed daguerreotype, he warned, is as fragile as the dust on a butterfly's wing, and must immediately be secured under glass. Varnishes were ineffective and dulled the glorious sheen of the picture.

The conclusion of Arago's speech brought cheers from the normally grave assembly. In the aftermath, one eyewitness reported that "opticians' shops were crowded with amateurs panting for daguerreotype apparatus, and everywhere cameras were trained on buildings. Everyone wanted to record the view from his window, and he was lucky who at first trial got a silhouette of roof tops against the sky. He went into ecstasies over chimneys, counted over and over roof tiles and chimney bricks, was astonished to see the very mortar between the bricks—in a word, the technique was so new that even the poorest plate gave him indescribable joy." However astonishing the results, many tyros were put off by the daguerreotype's complexity, bulkiness, and high cost—around two months' living for the ordinary Frenchman. (Of course, today we recoil at the idea of inhaling toxic mercury fumes.)

Daguerre marketed his own version of the camera, which carried an extravagant seal bearing his signature. At government insistence, he delivered a series of demonstrations of the process and released a seventy-nine page, illustrated instruction manual. By year's end, thirty editions of the manual had appeared in various languages.

Within a week of Arago's lecture, London's *Globe* published a how-to article on the daguerreotype, and within a month, Daguerre's manual arrived in New York. No country saw a more enthusiastic response to the advent of photography than the United States. The editor of the *Knickerbocker* gushed, "We have seen the views taken in Paris by the 'Daguerreotype' and have no hesitation in avowing that they are the most remarkable objects of curiosity and admiration, in the arts, that we ever beheld. Their exquisite perfection almost transcends the bounds of sober belief."

The *Boston Courier* described the daguerreotype as "a mirror which, after having received your image, gives you back your portrait, indelible as a picture, and a much more exact resemblance. . . . They are not paintings, they are drawings; but drawings pushed to a degree of perfection that art never can reach." In Daguerre's pictures of Paris, the report continues, "[W]e distinguish the smallest details, we count the stones of the pavement, we see the moisture produced by the rain, we read the sign of a shop. Every thread of luminous tissue has passed from the object to the [plate]."

Elizabeth Barrett (Browning) confided to her friend Mary Russell Mitford that—damn the painters—she wanted a daguerreotype of everyone dear to her in the world: "It is not merely the likeness which is precious in such cases—but the association, and the sense of nearness involved in the thing . . . the fact of the very shadow of the person lying there fixed for ever!"

Upon receiving his copy of the manual in the first shipment stateside, Samuel Morse daguerreotyped the Unitarian Church near New York University. Studios sprang up in major cities and remote hamlets, photography journals appeared, and hopeful experimenters tried to improve the process. Already by December 1839, a group of enthusiasts in Philadelphia found that silver bromide works more quickly than Daguerre's silver iodide, reducing exposure times and providing a pathway to practical portraiture. (Hippolyte Fizeau, in France, had reached the same conclusion, but delayed publication in deference to Daguerre.) By 1853, there were more than one hundred daguerreotype establishments in New York City alone, and some three million daguerreotypes taken nationwide annually. François Arago's expansive vision of photography's economic payoff was coming true, although in large measure across the Atlantic. Of course, if money can be made, the entrepreneur is not far behind: on June 22, 1846, Mr. A. W. Van Alstin, a photographer in Lowell, Massachusetts, was arrested for taking and distributing "daguerreotype likenesses of naked females."

Although most early daguerreotypists directed their efforts to earthbound scenes, the celestial realm beckoned those seeking a challenge. The obstacles were many. With the notable exceptions of the Sun, which was too dazzling to be favorably captured on a sensitized plate, and the Moon, which was only barely bright enough, celestial objects were too dim to leave any impression at all. To photograph the Sun required a shutter mechanism as rapid as an eye blink. To photograph the Moon meant a long time-exposure, during which Earth's rotation effectively smears out the lunar image. (Just ask Daguerre.) Also, the Sun and the Moon each span a measly half-degree in the sky. The towers of Notre Dame are compelling in a photograph; a tiny circle of light against a featureless background far less so. A camera view through a telescope is required to magnify the Sun or the Moon to reasonable proportions on the plate. Of course, that simultaneously magnifies the smudging effect of our spinning Earth. Despite the hurdles, a few adventurous daguerreotypists raised their cameras to the heavens.

John William Draper.

In March 1840, New York University chemistry professor John W. Draper followed up on Daguerre's failed attempt to daguerreotype the Moon. Draper was an acknowledged expert on the behavior of photosensitive substances. He knew that the daguerreotype plate was only marginally sensitive enough for use in astronomy, where the incident light levels are typically very low. Not only that, the plate is sensitive only to a restricted range of the light spectrum: most sensitive to blue-violet ("chemical") rays, as well as ultraviolet ("actinic") rays, but virtually blind to yellow, orange, or red. How effectively the Moon would register on the daguerreotype plate depended on which colors of sunlight it reflected back to Earth.

Atop a building on Manhattan's Washington Square, Draper secured a camera—actually a cigar box with a lens—at the focus of a five-inch-aperture refractor telescope. A twenty-minute exposure garnered a lunar image about an inch across. Although he tried to track the Moon's gradual movement with his telescope, the coordination was insufficient: only major lunar gradations were recorded; the rest of the features were indistinct. (Draper's pioneering photograph was destroyed in a fire at the New York Lyceum of Natural History in 1866; his subsequent lunar images, in the archives of New York University, have completely faded.)

John Draper is further credited with one of the earliest daguerreotype portraits taken from life, an iconic image of his sister Dorothy, acquired in a sixty-second exposure in 1840. To maximize the concentration of light falling on the plate, he fitted a camera with a wide, short-focus lens and situated the plate at the convergent point for blue-violet light, the camera's so-called chemical focus. For Draper's camera, this was three-tenths of an inch closer to the lens than the "visual focus." What looks distinct to the eye at the visual focus comes out blurred on the daguerreotype.

After several trials, Draper stopped whitening his sister's face with powder and accepted her natural complexion. To keep her from squinting and perspiring, he illuminated her with sunlight filtered through a blue solution of copper sulfate, reducing the solar glare while preserving the primary rays for the daguerreotype. He also secured her head in an iron rest, which became a notorious accessory in portrait studios. Dorothy's exquisite, flower-framed face bears a hint of a Mona Lisa smile—or perhaps it's a sign of resignation. (Draper presented the portrait as a gift to Sir John Herschel. It is now in the Spencer Museum of Art at the University of Kansas.)

Draper's photograph of the Moon was followed by a series of successful astronomical images during the daguerreotype's first—and, as it happens, *only*—decade. The solar eclipse of July 8, 1842 was captured in a two-minute exposure by Giovanni Alessandro Majocchi, professor of physics in Milan. Majocchi's attempt to record the Sun's diaphanous corona failed. (The first successful daguerreotype of the solar corona was obtained in Königsberg,

Germany during the eclipse of July 28, 1851.) John Draper and Edmond Becquerel, a talented Arago protégé, independently photographed the Sun's spectrum in 1843, displaying its famous Fraunhofer lines. On April 2, 1845, two other scientists in Arago's circle, Léon Foucault and Hippolyte Fizeau, obtained in a one-sixtieth-second exposure, the sought-after daguerreotype of the Sun itself, showing a vivid pair of sunspot groups. A lithograph of the picture was featured in Arago's book *Astronomie Populaire.*

The first decade of celestial photography closed with a multiple exposure of the full Moon taken with a portrait camera on September 1, 1849, by Samuel Dwight Humphrey, a daguerreotypist in Canandaigua, New York. More distinct than John Draper's flawed landmark image of 1840, the plate holds nine lunar portraits of various exposures, from a half-second to two minutes, each a mere tenth of an inch wide. The Moon's disk hardly registers in the shortest exposures, and trails into an oblong streak in the longest. But the three-second exposure portrays an unmistakable, alien geography—one that had long been scrutinized by eye, sketched by hand, limned by poets. In Humphrey's daguerreotype, however imperfect, the Moon appears with stark objectivity, its rocky, incontrovertible truth made plain for anyone to see.

Humphrey rushed his Moon portrait to Jared Sparks, president of Harvard University, who pronounced it a "curious and ingenious specimen of art" and foresaw a day when the camera's "exactness" might be applied to lunar mapmaking. Before year's end, Harvard astronomer William Bond and Boston photographer John Whipple affixed a camera to the tail-end of the Great Refractor and pointed the instrument at the Moon.

Samuel Humphrey's multiple-exposure daguerreotype of the moon, 1849.

Chapter 4

Summits of Silver

> It clear'd off afternoon. I came home early to prepare for the reception of the Public. Mr. Whipple has brought out his apparatus for the purpose of attempting to take a Daguerreotype of the Moon and Sun, by aid of the great Telescope.
>
> —From William Cranch Bond's diary entry, October 23, 1847

The door of opportunity swung wide for young John Adams Whipple in 1840 in the form of an overheard conversation at a scientific supply shop in Boston. Recently arrived from rural Grafton, Massachusetts, the ambitious Whipple wondered how he might use his precocious knowledge of chemistry to best advantage. His childhood passion for chemical experimentation had mystified his parents, who feared that every penny their son spent on the noxious hobby was a penny wasted. Of necessity, he had become expert on "doing things with nothing to do with," so when details of Frenchman Louis Daguerre's remarkable photographic process reached tiny Grafton in late 1839, Whipple slapped a lens onto a candle box and recorded an image on the handle of a chemically treated silver spoon. Now, at eighteen, Whipple was a walking compendium of chemical know-how.

John Adams Whipple.

Perusing the wares in the scientific supply shop, Whipple could have regaled the merchant with a rundown of how various compounds change color, smell, froth, explode, even blacken in the light of the Sun. But his ears perked up when a customer requested some "chloride of iodine," a substance recently said to increase the light-sensitivity of daguerreotype plates. A city-wide search had turned up no source of the chemical. The shop owner likewise claimed ignorance. Whipple approached the customer and offered to make chloride of iodine on the spot. A few hours later, production complete, Whipple received his first earnings as a chemist.

By 1842, Whipple was running a successful chemical supply house that catered to the needs of local daguerreotype studios. He took up photography himself, exhibiting his pictures at the city's annual artisans' exhibition as early as 1841. Although he had long tolerated the toxic emissions of his private chemistry experiments, Whipple grew increasingly ill from the fumes of wholesale chemical manufacture. He saw no choice but to recast himself as a full-time portrait photographer. Within a few years, his studio was competing head-to-head with Boston's preeminent firm of Southworth and Hawes. Whipple's chemical and mechanical adeptness proved vital to his business success; yet he also had an artistic eye, whether capturing a thousand-person church gathering at Plymouth Rock or an intimate close-up of a sweet-faced child. Among his eventual portrait subjects were Henry Wadsworth Longfellow, Daniel Webster, Nathaniel Hawthorne, Oliver Wendell Holmes, and Louis Agassiz.

Whipple's studio and display gallery occupied the upper floors of three adjacent buildings on Washington Street, Boston's busiest commercial thoroughfare. Fitting out the premises to suit the opulent sensibilities of his mostly upper-class clients had cost him over $25,000. At the height of the business, he employed nearly forty people.

Boston studio photography during the 1840s was a rough-and-tumble enterprise. Whipple proved to be something of a marketing impresario. He advertised incessantly, featuring his proprietary innovations, such as the steam engine he used to polish photographic plates, run an immense cooling fan for the comfort of clients, heat up cups of mercury, distill water, even revolve the gilded sunburst at the entrance of his establishment. (His catchphrase: "Daguerreotypes by Steam.") Whipple produced a Grand Optical Exhibition in Boston's Tremont Temple, featuring projected views of Rome, Mexico, and Palestine, plus portraits of P. T. Barnum's Swedish Nightingale, Jenny Lind, "thrown upon the scene as large as life."

If possible, Whipple's seemingly boundless marketing energy was exceeded by his practical inventiveness. Through a clever optical enlargement scheme, he offered super-size photographs, among them the largest-ever print of a street scene, nearly five feet by three feet. He developed and patented the crystalotype process to generate, from a single glass negative, unlimited paper prints, which publishers could tip into books. He was first in the world, in 1846, to take a passable daguerreotype through a microscope—a ninety-minute exposure of a spider's mandible. "By this most simple means," Whipple writes, "it is in the power of every Dagerreotypist [sic] to greatly aid the naturalist in his researches, giving him in a few minutes drawings of invisible objects penciled by nature's own hand, which it would be impossible for him to obtain in any other way."

John Whipple's affirmation of the daguerreotype as a scientific tool was not exclusive to the microscopic realm. He certainly contemplated its potential in astronomy, for he tried to photograph the Moon by exposing sensitized plates for three to five seconds at the focus of a small telescope. The results were dismal, yet instructive. Under a magnifier (the images were a mere half-inch wide), he could make out the broadest aspects of the Moon. But because his telescope was stationary, the lunar forms were smeared out by Earth's rotation.

Whipple realized that a more promising instrument for celestial photography stood right across the Charles River in Cambridge: Harvard's Great Refractor. Not only did the vaunted telescope have a far larger aperture—promising shorter exposure times—its clockwork drive was designed to track heavenly bodies, keeping them centered in the eyepiece as they marched across the night sky. With such an instrument, Whipple figured, he could take what was only barely discerned by the human eye and transform it into "a fixed fact for the naturalist to study at his leisure."

On the afternoon of October 23, 1847, William Bond opened up the observatory dome for John Whipple and rotated it until sunlight poured in. Perhaps distracted by conversation or by the evening's public observing session, Bond swung the telescope unthinkingly toward the Sun. The focused beam shot out of the eyepiece and instantly burned a hole in his coat sleeve. It was only when he felt the intense heat on his skin that he snatched his arm away. Clearly, the Sun's light was already sufficiently intense that it didn't need the amplifying power of a fifteen-inch telescope. For safety's sake, the two men decided to conduct their trial with a smaller telescope next door. Several attempts were made to capture a focused image of the Sun, but Bond's diary entry says no more than "Mr. Whipple was satisfied that he should be able to accomplish the object."

According to the observatory's annual report, Whipple and Bond made other attempts to image the Sun's disk in 1848, all of them failures. Still, Bond asserts in his report that "we do not despair of ultimate success, when our time and means are adequate to the requisite expenditure." In private, he was less sanguine. Bond's allusion to adequate time might be an oblique reference to his exasperation with Whipple's pursuit. Testing an untried technology, no matter how promising for the future of astronomy, was not Bond's highest priority. The observatory's stated mission was to conduct and report measurements of star and planet positions, and to use the telescope's unique capabilities to carry out *visual* studies of planets and comets. Every time John Whipple showed up with his camera, the delicate micrometer used to measure celestial positions had to be removed from the rear end of the telescope. Putting it back afterward involved time-consuming readjustment and recalibration. Evidently, George Bond was more tolerant than his father to these periodic interruptions for photography. William Bond would have shunted Whipple to a smaller telescope, but George pressed for use of the Great Refractor. It is the younger Bond who would become the astronomical profession's most vocal evangelist for celestial photography.

On December 18, 1849, while the Bonds were making micrometer measurements of the position of Mars, Whipple and his partner William B. Jones arrived, this time to photograph the Moon, presently in its crescent phase. William Bond's sigh of frustration is practically audible in a note he jotted afterward—the micrometer, he points out, had been secured to the telescope continuously for almost a year. Nevertheless, he removed the fragile measuring device. He makes no mention in his note of the outcome of the evening's picture taking. Presumably, this effort was as unsuccessful as the solar attempt that preceded it.

In an 1853 letter to the *Photographic Art-Journal*, John Whipple writes: "Nothing could be more interesting than [the Moon's] appearance through that *magnificent* instrument: but to transfer it to the silver plate, to make something tangible of it, was quite a different thing." In our age of point-and-shoot digital cameras, it's difficult to imagine the challenges facing a nineteenth-century daguerreotypist. In the realm of early nature photography, the Moon was an obvious target. Big and bright to the eye, it nonetheless defied the best efforts of seasoned photographers, especially those who tried to capture a *magnified* lunar image through a telescope. The inevitable result wasn't worth the trouble: a tiny, oblong, whitish splotch with no discernible features—like a dab of paint on a darkened slab.

In a critical way, the eye is more forgiving than the camera of the hard realities of astronomical observation. The roiling of the atmosphere, the vibrations of footsteps on the observatory floor, the rumble of passing vehicles—all contribute to an incessant shaking of the telescope. Imperceptible to the bystander, these minute jostles are made manifest to the astronomer by the telescope's essential function: to magnify things. Looking at a star in the eyepiece, one sees, not a languid speckle against the firmament, but a crazed firefly, darting about in random fits. Only occasionally does the star image hover at a central spot and is vividly seen. It is these rare moments of supreme clarity that visual astronomers prize: when a double star shows itself as a distinct pair of pinpoint lights instead of a single, elongated glow; when a lunar mountain peak settles into crisp definition, revealing its jagged shadow over the land; when a swirling storm on Jupiter rises out of its turbid cloak. For centuries, observers have coveted these brief, privileged aspects of the heavens. That's one reason why they risk frostbite to study Saturn's rings on a frigid February night or lug their telescope up a mountain to peer through one less layer of our restive atmosphere. And it's one reason why visual astronomers were initially dismissive of the new photography. It conferred, as yet, no advantage over the human eye.

A photographic plate is fundamentally different than the eye: it is endowed with a chemical-based memory for light—once a ray of light triggers the chemical reaction, a visible record of the event is imprinted on the plate. (A chemical fixing agent renders the image permanent.) This means that the nervous jitter of a bright star leaves its trace on the plate, building up a swollen blob of light many times larger than the true pinpoint image of the star. Thus, stars in the earliest photographs registered as bloated variants of themselves—if they registered at all. Likewise, extended objects like the Moon or a planet appeared as fuzzy mirages, all but the grossest details washed away.

Before releasing any astronomical photograph to the world, John Whipple and the Bonds knew that they had to capture an image that rendered detail comparable to that seen by a sharp-eyed observer peering through a telescope. The astronomical community would dismiss anything less. What factors made it so hard, in the 1840s infancy of the photographic art, for an expert like Whipple to take a decent daguerreotype of the Moon through a telescope?

A daguerreotype plate is only marginally sensitive to light, requiring long exposures even in the glare of day. (Early portrait photographers located their studios on the top floor of buildings to gain access to skylights;

nevertheless, an exposure of twenty-seconds duration was required.) Even a relatively bright celestial object like the Moon only grudgingly impresses itself on the plate's chemical substrate. Concentrating the Moon's light by means of a telescope helps, but a significant time exposure is still required to capture the image. Only with the advent of more sensitive photographic emulsions, starting in the 1850s, did "snapshot" lunar imaging begin.

The execution of time-exposure photographs through a telescope is problematic. Every instant the camera shutter is open, the telescope is being whipped around at many hundreds of miles an hour by Earth's rotation. Consequently, a celestial object centered in the telescope's eyepiece creeps toward the edge of the viewing field until it disappears from sight. Most telescopes are equipped with a hand-crank mechanism that nudges them along and restores the straying object to its central place. Of course, the clocklike regularity of Earth's spin lends itself to automated versions of this manual device. Whether an old fashioned, weight-driven gearbox or a modern, computer-controlled electric motor, the turn of the telescope can be set to match the rotation rate of our planet, only in the opposite direction. As a result, the telescope tracks celestial objects as they arc across the night sky. (Strictly speaking, this applies only to objects outside the solar system; the orbital movement of, say, the Moon or a planet causes it to go out of sync with the telescope's clock drive.)

No matter how excellent a telescope's optics, the instrument's photographic potential can be ruined by an inaccurate drive. A drive that is too slow or too fast stretches a pinpoint star image into a line on the photographic plate. Every lurch or stumble likewise leaves its imprint for posterity. Whipple struggled mightily to deal with the Harvard refractor's so-called Munich drive, which was ill-matched to the stringent requirements of time-exposure daguerreotyping. He reported that the friction-based mechanism "had a tendency to move the instrument a little too fast, then to fall slightly behind. By closely noticing its motion, and by exposing my plates those few seconds that it exactly followed between the accelerated and retarded motion, I might obtain one or two perfect proofs in the trial of a dozen plates, other things being right."

Another factor that hindered the advancement of celestial photography was the difficulty in focusing the telescopic image onto the camera's plate. Focusing is accomplished by shifting the eyepiece (or photographic plate) closer or farther from the telescope's objective lens. The adjustment is exquisitely sensitive; a mere tweak of the eyepiece position substantially alters image clarity. Early cameras had a ground glass screen that rendered the image visible before the plate was inserted. Brilliant celestial bodies like the Sun or the Moon were visible on the screen, yet even the brightest stars barely showed up.

Focusing was further complicated by the chemical nature of the daguerreotype plate. The human eye is most sensitive to midrange colors of the optical spectrum, such as yellow and green; by contrast, a daguerreotype plate is most sensitive to blue, violet, and ultraviolet light. This would not matter, except for an inherent property of refractor telescopes of the age, including Harvard's: they were all designed for optimal performance using the human eye, not the camera.

A simple objective lens focuses light of different colors to different distances behind the lens. Thus, the focal point for the red component of a star's light falls at a different position along the telescope's axis than the focal point for the blue component of the star's light. (Light emanating from most celestial objects is a comingling of many colors.) No matter where the viewing eyepiece is placed, the magnified image is marred by colored fringes. This condition, known as chromatic aberration, is to a degree ameliorated by the use of a pair of nested lenses—a doublet—each made of a glass with a slightly different index of refraction. A common *achromatic* configuration is a convex (converging) lens of crown glass coupled with a concave (diverging) lens of denser flint glass; the convergence of various colors by the convex lens is partly offset by a divergence from the concave lens, such that the colors focus more tightly together. However, a doublet can merge the focal points of only a selected pair of colors; although colors adjacent to these remain somewhat dispersed along the axis, image fringes are significantly reduced.

Because the human eye is most sensitive to midrange colors like yellow, nineteenth-century refractors were fitted with achromatic objectives that merged the focal points of colors flanking yellow, that is, red-orange and green. This arrangement reduces the dispersion of all three colors. But such an objective cannot simultaneously accommodate blue and violet rays, which are focused to a somewhat different position. In Harvard's Great Refractor, whose focal length is around twenty-three feet, violet rays converge about an inch behind yellow rays. Placing a violet-sensitive daguerreotype plate at the telescope's *visual* focus yields a compromised photograph. Only after the Harvard photographers positioned their plates an inch or so farther out, at the *photographic* focus, did they produce more vivid renderings of cosmic bodies. (Reflector telescopes, which use a mirror as the objective instead of a lens, do not suffer from chromatic aberration.)

A more general criticism in John Whipple's 1853 letter to the *Photographic Art-Journal* echoes the Bonds' opinion about observing conditions in New England. Whipple writes of "the sea breeze, the hot and cold air commingling, although its effects were not visible to the eye; but when the moon was viewed through the telescope it had the same appearance as objects when seen through the heated air from a chimney, in a constant tremor, precluding the possibility of successful Daguerreotyping. This state of the atmosphere often continued week after week in a greater or less degree, so that an evening of perfect quiet was hailed with the greatest delight." George Bond suggested that future generations of telescopes be placed in locations most advantageous to astronomy. In particular, he cited the dry, high desert regions of western South America. Today, this area is host to some of the largest telescopes in the world.

On July 16, 1850, after Whipple had completed a series of daguerreotype experiments with Harvard's small telescope, George Bond invited him to once more try the Moon with the large instrument. Again the micrometer was removed, this time interrupting nightly measurements of a comet's path in the sky. The camera was secured in its place, without an intervening eyepiece, relying only on the converging power of the objective lens. Whipple's images of the first-quarter Moon were poor. But later that night, Bond pointed the telescope toward brilliant Vega, in the constellation of Lyra, the second brightest star in the sky's Northern Hemisphere. Whipple's ninety-second exposure, aided by his skillful tending of the recalcitrant Munich drive, produced the first-ever daguerreotype of a star other than the Sun. A follow-up exposure of the double-star Castor revealed the pair as an elongated globule of light, the telescope's jittery drive preventing their separation into distinct specks. Attempts to image fainter stars failed, regardless of the length of exposure.

George Bond's announcement of the stellar daguerreotype in the Boston *Advertiser* named Whipple as an assistant, although his expertise was surely essential to the night's success. Bond marveled at how Vega's rays had retained their power to stimulate a chemical reaction despite their long passage through space—a voyage that had begun some twenty years earlier, before Daguerre had developed the technology that now secured the starry portrait. Bond further pointed out what is obvious today, but unproven in his day: the light from the Sun and the stars must be fundamentally similar, for each activates the human retina and the photographic plate. The physical laws governing the production of light are presumably no different in the deep cosmos than in the solar system, a notion already accepted for Newtonian mechanics.

Bond's closing statement was a clear expression of where he thought celestial photography was heading in the long term: "It is our purpose to pursue the subject of daguerreotyping the stars, proceeding step by step from the brightest to those of lesser magnitude. We do not despair of obtaining ultimately, faithful pictures of clusters of stars and even nebulae."

Despite George Bond's enthusiasm, Whipple and his colleague Jones did not return to Harvard for another eight months. On the evening of March 12, 1851, they took six daguerreotypes of the first-quarter Moon through the large telescope, although "much troubled by clouds and with an unsteady atmosphere." To no one's surprise, the pictures were blurry; the camera plate had been held at the telescope's visual focus. Nevertheless, the results were sufficient to indicate an optimal exposure time of ten to fifteen seconds for the next attempt. On March 14, with the Moon swollen to a gibbous phase, a total of thirteen plates were exposed at various distances beyond the visual focus. Each advance toward the telescope's photographic focus produced a more distinct image of the Moon. "The effect was at once apparent," George Bond noted, "in the great improvement of the picture which is now obtained so as to give a better representation of the Lunar surface than any engraving of it, that I have ever seen."

Enlargements of the Harvard lunar daguerreotypes were an instant sensation at the 1851 Crystal Palace exhibition in London, acknowledged to far surpass any previous rendering of the Moon. The *Annual of Scientific Discovery* enthused, "We have rarely seen anything in the range of the daguerreotype art of so great beauty, delicacy, and perfectness, as the pictures referred to. The inequalities and striking peculiarities of the moon's surface are brought out with such distinctness, that the various mountain ranges, highlands, and isolated peaks are at once recognized."

Of a subsequent Whipple–Bond daguerreotype of the quarter Moon, presented at a meeting of the British Association for the Advancement of Science, a reporter lavished fulsome praise: "Fringes of darkness casting themselves off behind the peaks and summits of silver, rounded waves of shadow, filling up cavities in the form of hollow cups as abysses in the midst of this strange surface; triangles of jet, shooting forth like twigs under luminous spots, brilliant as diamonds—this is what the telescope displayed. In the photographic image produced by Dr. Bond, all these details are revealed to the eye. Everything there is so completely and faithfully reproduced, that by the aid of a magnifying glass we perceive new objects, minute details, that had escaped the sight. . . . It is impossible to calculate the services that photography is called to render to astronomy."

Daguerreotype of the moon taken with the Harvard refractor by Whipple and the Bonds in 1852. A similar photograph created a sensation at the 1851 Crystal Palace exhibition in London.

On March 22, 1851, Whipple and George Bond took a daguerreotype of Jupiter that captured two of the planet's broad equatorial belts. Bond was surprised to find that the exposure time required to imprint Jupiter on the plate was nearly identical to that needed for the Moon. He had predicted a much longer exposure. Jupiter is just over five times farther than the Moon, that multiple of distance, in effect, diluting the light emission received from Jupiter's surface. Because both the Moon and Jupiter shine by reflected sunlight, a patch of Jovian surface should appear about twenty-seven times dimmer than the same-size patch on the Moon. (Light intensity diminishes with the square of the distance.) Hence, Bond was mistaken in his expectation that a successful exposure of Jupiter would be of longer duration than a lunar exposure. He drew the conclusion, later confirmed, that Jupiter's surface is much more reflective than the Moon's, offsetting the dimming effect of distance. Bond's observation of this phenomenon is the first scientific result credited to celestial photography. (A full report was published in 1861.)

If William Cranch Bond had a sentimental favorite among the early pictures taken from Cambridge, it might have been the daguerreotype he took with Whipple of the partial solar eclipse of July 28, 1851, some forty-five years after the total eclipse that had sparked his passion for astronomy. George Bond, then twenty-six, was touring the observatories of Europe and observed the July 28 eclipse from a remote village in Sweden, which, unlike Cambridge, lay along the geographic path of totality. George's letter to his father must have awakened the elder Bond's youthful memories of the 1806 eclipse: "The change which takes place in less than a tenth of a second, so entirely alters the scene, that the second which precedes total obscuration gives one no idea of what is to follow. . . . What I then saw, it is utterly beyond my power of language adequately to express. The corona of white light which encircled the dark body of the Moon, resembled the aureola, or glory, by which painters designate the person of the Savior, its radiance extending from the circumference to a distance equal to about half of the Sun's diameter. . . . How shall I attempt to describe . . . these flame-like protuberances projecting from the inner edges of the corona? . . . [The eclipse] was surpassingly beautiful—the most sublime of all that we are permitted to see of the material creation." Lacking a camera, George Bond afterward sketched the eclipse from memory. The daguerreotypist Berkowski, from Königsberg, obtained an acclaimed photograph of the event that clearly depicted Bond's "flame-like protuberances."

Harvard's experiments in celestial photography continued through April 24, 1852, culminating in a superb twenty-second daguerreotype of the crescent Moon. Then the observatory record is silent for five years on the subject of nighttime photography. (There is a mention in 1853 about a failed attempt to take pictures of sunspots.) Evidently, Bond and Whipple concluded that they had accomplished pretty much all they could with the ultraslow daguerreotype process and with the telescope's lamentable Munich friction-drive. It was time to wait until technology caught up to their aspirations.

Whipple's youthful enthusiasm remained high as his initial foray into celestial imaging drew to a close. His photographer's imagination was stirred by the sight of his own Moon pictures, each a crisply defined, two-inch miniature of a familiar, yet alien, world. Just twenty-six, he looked ahead to a time when technological improvements might permit a sequence of lunar photographs "taken as old Sol lights up peak after peak, shining first on this side, then on that, and as these shadows sweep through her immense cavernous valleys, then as the full blaze of the sun penetrates those awfully deep yawning gulfs."

In fact, Whipple's vision would be realized within the decade, the result of a new photochemical process already making the rounds in England.

Chapter 5

The Man with the Oil-Can

> In bringing before the Association the present Report it will be only necessary, after referring briefly to the labours of others, to confine myself to an account of my personal experience; for, although other observers have occasionally made experiments in Celestial Photography, there has not been any systematic pursuit of this branch of Astronomy in England, except in my Observatory.
>
> —Warren De La Rue, Report to the British Association for the Advancement of Science, September 1859

The Great Exhibition at London's Crystal Palace in 1851 was a showcase of humanity's industrial and artistic achievements. Some thirteen thousand displays filled a sprawling cast-iron and glass greenhouse of almost a million square feet. Giant steam engines, Jacquard looms, and power reapers sat imposingly amid samples of silk, furniture, Colt revolvers, even the Koh-i-Noor diamond. More than six million visitors paraded through the vast hall, among live elms, marble statues, and celebratory fountains, all to the reverberant backdrop of the world's largest pipe organ. Among the throng of attendees was Harvard's George Bond, then twenty-six, who enthused in his diary on June 4, 1851, "Anyone who is not satisfied would better find another world to live in. . . . The Arabian Nights are thrown far into the shade by the realities of the Crystal Palace."

Warren De La Rue.

Bond's brother, Richard, was there to

demonstrate his astronomical recorder: a rotating-drum chart-plotter that time-stamped meridian passages of stars at the observer's press of a button. Also in attendance, John Whipple, showcasing the pioneering daguerreotype of the Moon that he and the Bonds had taken through Harvard's Great Refractor. The lunar photograph was one of only three American exhibits to receive the highest award for excellence. In the eyes of wonderstruck viewers, a celestial body had miraculously been brought down to Earth for close inspection.

Down the building's main avenue, positioned between Cruchley's large-scale map of England and Armstrong's specimens of illustrated musical printing, was an especial admirer of Whipple and Bond's daguerreotype. Warren De La Rue, partner in his father's stationery firm, was showcasing his own creation: an envelope-making machine. The device could fold and glue twenty-seven hundred envelopes per hour, as many as an experienced worker could complete in a day. Educated in Paris, the affable De La Rue quickly distinguished himself as a self-taught mechanical wizard, in the mold of William Cranch Bond. "I am the man with the oil-can," he once remarked, characterizing his skill at practical problem solving. That and his keen business sense made him a valuable asset to his father's company, which he joined upon leaving Paris at the outbreak of the Revolution of 1830.

De La Rue's outside interests ranged among the technical disciplines, starting with electrochemistry. A founding member of the Chemical Society of London, he published his first research paper, on the chemistry of batteries, in 1836 when he was just twenty-one. He subsequently studied at the Royal College of Chemistry under the prominent German organic chemist August Wilhelm von Hofmann. De La Rue's chemical studies brought him to the attention of the Royal Society, which elected him Fellow in 1850. But by then, he had virtually abandoned chemistry in favor of a new pursuit: astronomy.

In the late 1830s, De La Rue had been tasked with the design of a production facility for white lead, which was to be used in the manufacture of the company's popular line of playing cards. He sought the advice of industrialist James Nasmyth, inventor of the steam-driven hammer and pile driver. An engineering prodigy, Nasmyth had built his own steam engine in 1825 at age seventeen, and two years later, created a working, eight-passenger "road steam-carriage"—a primitive automobile. That same year, he made a six-inch reflector telescope, the first in a series of progressively larger and more sophisticated instruments that he used to study the topography of the Moon. By the time De La Rue arrived at his doorstep in 1840, Nasmyth was an irrepressible champion of astronomy:

"If I were asked what course of practice was the best to instill the finest taste for refined mechanical work, I should say, set to and make for yourself from first to last a reflecting telescope."

Nasmyth invited De La Rue to watch the casting of his thirteen-inch telescope mirror from molten speculum metal (an alloy of tin, copper, and arsenic). Intrigued, De La Rue returned regularly to observe the grinding and polishing of the mirror, and several times more to watch the fabrication of the telescope's mechanical mount. By the late 1840s, every fiber of De La Rue's mechanical, scientific, and artistic talents was being channeled toward astronomy and its instruments. He commissioned Nasmyth to cast him a thirteen-inch speculum metal disk, which he ground and polished to its proper concave shape with a machine of his own design. The machine produced a more accurate concavity than the best existing devices, and became the subject of De La Rue's first publication in astronomy. In 1849, De La Rue set up the completed telescope in the garden of his home in Canonbury outside London. His skillful drawing of Saturn's rings was exhibited to acclaim at the January 1851 meeting of the Royal Astronomical Society, which elected him to fellowship a few months later.

De La Rue recalled navigating his way through the Crystal Palace in 1851 to the much-heralded lunar daguerreotype of Whipple and the Bonds. Although by then a seasoned observer, with the Moon's terrain securely mapped into his memory, he faced the picture as if seeing that celestial body for the first time. The dusky maria, sharp-rimmed craters, and vertiginous mountain ranges were all there, captured with stunning sharpness. Here

Drawing of the planet Saturn by Warren De La Rue, 1856.

was a picture, De La Rue realized, whose fidelity far surpassed anything he could generate by hand, even with his considerable drafting skills. This was no mere representation of the Moon; it *was* the Moon, as it had appeared on March 14, 1851, chemically frozen in time. In the reserved Victorian vernacular of his day, De La Rue admitted to being "charmed" by the picture. However, his subsequent burst of activity suggests a much more profound impact. The advancement of celestial photography posed precisely the sort of cross-disciplinary challenge that tugged at De La Rue's engineering, astronomical, chemical, and artistic sensibilities. The only question was how best to proceed.

De La Rue was probably unaware at the time, but his future pathway was being laid during the last days of the Great Exhibition. The hastily erected display by London-based sculptor Frederick Scott Archer was too late to appear in the Exhibition's official catalog. The contents of the display were neither carvings nor bronzes, as one might expect from Archer's nominal line of work. They were photographs: remarkably vivid images, not on paper, not on metal, but on sheets of glass.

The first decade of photography, the 1840s, was a contest for dominance between two rival methods: the daguerreotype and Talbot's calotype. Both processes employed light-sensitive silver compounds, required chemical development of the latent image, used hypo to fix the image, and rendered scenes without their natural colors (shades of gold-tinted gray for the daguerreotype, shades of mellow brown for the calotype). Beyond these basic commonalities, each process had its advantages and disadvantages—technical, aesthetic, and legal.

The daguerreotype system produces a direct positive image, although reversed left to right. The clarity of the daguerreotype image surpasses that of the calotype, recording a measure of detail that appears remarkable even today. However, the metal plate has a mirrorlike finish, restricting visibility of the image to a narrow range of viewing angles. Also a single exposure produces a unique, practically irreproducible picture. The only way to make a copy is to take a picture of the daguerreotype itself. This is not an easy proposition, as the plate's reflectivity makes it difficult to illuminate properly.

Like the daguerreotype, the calotype requires a great deal of advance preparation. A sheet of high-quality writing paper is dipped into salt water, brushed on one side with silver nitrate, and dried. The process is repeated several times until the sheet has absorbed a sufficient amount of the chemical. It proved difficult to produce a sheet with uniform sensitivity and to maintain comparable sensitivity levels among different sheets.

The calotype technique generates a matte-finish, paper negative with the correct left-right orientation and no viewing angle restriction. Early calotypes looked murky compared to daguerreotypes, their resolution limited by the fibrous texture of the paper. Given their relatively weak light sensitivity, exposure times far exceeded those of daguerreotypes. On the other hand, any number of positive contact prints can be easily generated from a single calotype negative. Yet even after it was improved, the calotype found only a limited market, in part because of Talbot's vigorous protection of his patent.

Neither the daguerreotype nor the calotype were astronomy-friendly. In fact, they were downright hostile. Even after significant improvements, both technologies required long time exposures at the telescope, during which all the demons of man, machinery, and nature worked their mischief. The longer the camera shutter was open, the greater the risk of failure. That any progress was made in telescopic photography during the 1840s and early 1850s is remarkable in retrospect. The accomplishments, such as they were, speak to the perseverance of astronomers and their photographic allies in the face of fatigue, serial failures, and capricious weather. The frustration Whipple and the Bonds had endured to obtain their celebrated lunar daguerreotype was invisible to spectators at the Great Exhibition, who reflected only on the picture's rugged beauty and perhaps its implications for the future of scientific and artistic representation. Few at the time, except the practitioners themselves, expected that the astronomical daguerreotype had reached its zenith. The vaunted Moon picture represents what could be done *despite* the technology, not because of it. Until exposure times were shortened and telescope drives improved, astronomers would remain hostage to every flutter of the air and lurch of a drive gear. By the early 1850s, the effort to photograph the heavens had reached a technological impasse.

From the start, astronomers affixed cameras to their telescopes in an attempt to generate accurate, objective, permanent pictures of celestial objects—essentially, to unhitch the depiction of the cosmos from the subjective eye and hand. Celestial photography places strict demands on the imaging technology, critical aspects of which must be optimized for the particular challenges involved. Of course, the ability to accurately render detail is paramount. Clarity of the image is compromised by turbulence in Earth's atmosphere, as well as by vibrations of the telescope, effects amplified by the magnifying power of the instrument. By starting with a highest-resolution photographic process—in the 1840s, this was the daguerreotype—the astronomer mitigates, in part, the blurring of images by external agents.

Another major consideration in celestial photography is the faintness of most of the target objects. A telescope operates as an optical funnel: it collects cosmic light over the entirety of its objective lens or reflector, and focuses that energy into a progressively narrow beam that enters the eye or the camera. That's why we are able to perceive heavenly bodies that are invisible to the naked eye. The telescope, in essence, expands the pupil of the eye from its nominal quarter-inch to, say, six inches or six feet—whatever the telescope's aperture. The eye has no means to store the light emerging from the telescope; the retina is a real-time sensor: when stimulated by a pattern of light, it fires off a stream of electrical signals through the optic nerve that the brain synthesizes into an image. Thus, the eye is our visual portal to the world; our brain does the actual seeing.

Viewed through the eyepiece of a telescope, a star appears, not as a composed, fiery speck, but a turbulent bead of light, its edges erupting in spikes, like a cat trying to claw its way out of a luminescent sack. The image goes in and out of focus and flits randomly about the center of the field of view. The eye is adept at tracking these frenetic changes in appearance, ever ready to behold the rare instants when the image snaps into a momentary state of calm and sublime clarity. (Even seasoned observers can't suppress a "Wow!" whenever this occurs.)

A camera is fundamentally different from the eye: it accumulates light as long as the shutter is open, storing the luminous energy in chemical form on the photographic plate. Every succeeding second of exposure reinforces the overall definition of the image, rendering it increasingly vivid as time proceeds. In theory, at least. The camera does not distinguish between intended aspects of an exposure—the way the object is supposed to look—and unintended distortions that afflict the photographic operation. If the telescope vibrates or shifts during the exposure, any newly arrived photons of light stray from the spot on the photographic plate where their predecessors had fallen. The result is an ill-defined disk of starlight, or the planet Saturn with a diffuse, bulging waistline instead of a delicate brace of rings. Everything is recorded, for good or ill. The camera is all-seeing, yet blind to the photographer's intent.

The most an astronomer can do to garner a better celestial picture is to mitigate the factors that tend to ruin a photograph. Of absolute necessity is a mechanical clock drive that turns the telescope to precisely counteract the movement caused by Earth's rotation, movement that would otherwise smear out the image. This was a rare accessory for telescopes of the 1840s when celestial photography was inaugurated. And even if so equipped, the clock drive's accuracy almost always fell short of the stringent demands of a long time-exposure. Witness the Harvard refractor's original Munich drive, which was sufficient for visual astronomy, yet wholly inadequate for photographic astronomy. To achieve their breakthrough images, Whipple and Bond had to treat the drive like an uncooperative child: open the camera shutter while the drive is behaving, close it when it's about to have a fit. They ultimately abandoned their photographic efforts until a better drive—and a better means of photography—could be put to use.

From the very start of the photographic era, a number of practitioners tried to develop a process that would combine the clarity of the daguerreotype with the negative-to-positive print capability of the calotype. The most promising pathway appeared to be replacement of the calotype's paper negative with a glass negative, whose transparency would make for finer-grained, daguerreotype-like prints. The problem was to develop a clear, nonreactive substance that would permit a dissolved photosensitive chemical to adhere to glass. (An early desperate attempt involved the slimy exudate of snails.) In 1839, noted English astronomer John Herschel exposed a glass plate on which multiple applications of a silver chloride solution had been allowed to dry; although successful, he judged the lengthy preparation phase to be impractical.

In the late 1840s, John Adams Whipple and Claude Niépce de Saint-Victor, son of Nicephore Niépce's cousin, independently produced glass negatives using albumen—egg whites—as the photographic substrate. The positive prints derived from these glass negatives were comparable in quality to daguerreotypes. But the long exposure times prevented their application to astronomy.

The breakthrough came from an unlikely quarter: a reserved, pale-faced sculptor named Frederick Scott Archer, who had taken up calotype photography in 1847. Dissatisfied with the quality of prints produced from the calotype's paper negatives, Archer tried albumenized glass. This, too, proved unsatisfactory; the albumen coating was hard to spread uniformly and extremely delicate when dry. Archer read of a recently discovered transparent, viscous substance called collodion, from the Greek "to adhere." Collodion was developed independently in 1847 by Louis Menard of France, and John Parker Maynard, a medical student in Boston, who published the method of production in the *American Journal of Medical Science*. The substance, when dried, was useful as a flexible, adhesive cover for wounds.

Collodion is made by dissolving guncotton—cotton treated with nitric acid—in a mixture of ether and alcohol. (Guncotton, a gunpowder alternative, was discovered accidentally when German–Swiss chemist Christian Friedrich Schönbein, working in his kitchen in 1846, wiped up some spilled nitric acid with his wife's cotton apron; he set the apron to dry near the stove, where it abruptly ignited.) To the collodion, Archer added potassium iodide, then poured the syrupy liquid onto a glass plate and let it dry. When ready to take a picture, he immersed the prepared plate into a silver nitrate bath for a few seconds, the iodine and silver atoms combining to produce light-sensitive silver iodide within the collodion. The plate was exposed *while wet*. Once dry, it lost its photosensitivity and left a silver nitrate residue on the collodion's surface; hence, the hard limit on the duration of time exposures. The plate, still wet, was developed in pyrogallic acid (which precipitates out the silver), then was fixed with hypo.

The wet-collodion photographer had to be near a fully equipped darkroom—or bring one along—as exposure and development took place in rapid sequence. Civil War photographer Mathew Brady toted his equipment in a horse-drawn wagon train. In October 1863, a trio of enthusiasts from Philadelphia ventured into the Poconos for a week, bearing their photographic burden on their backs: cameras, tripods, glass plates, flannel tent, curtains, trays, bottles, rubber-sheet sink and drain hose, stool, plus a carryall box of chemicals. "We will not attempt to state all that the box contained," they reported to the *Philadelphia Photographer*, "it would be easier to enumerate what it did not contain." Or as photography historian Robert Taft puts it, "[T]o be an amateur in wet plate days required fondness for the art verging on fanaticism."

Mobile darkrooms and wet-plate requirements notwithstanding, Archer published a detailed report on his use of collodion in the March 1851 issue of *The Chemist*. His instruction manual appeared in England, France, and the United States in 1852. By year's end, a core of professionals had judged Archer's process to be straightforward and relatively inexpensive, yielding daguerreotype-like detail and calotype-like reproducibility. Collodion plates treated with cadmium bromide (to produce photosensitive silver bromide) were some ten times as fast as daguerreotype plates. The switchover among studio photographers from the daguerreotype to Archer's method accelerated through 1853, as wholesale collodion manufacturers appeared. At the 1856 annual exhibit of the Photographic Society of London, only three of the six hundred displayed images were daguerreotypes.

John Whipple was one of the early adopters of Archer's process, replacing his daguerreotypes and albumenized glass plates with collodion. At the 1853 World's Fair in New York, Whipple received the highest award in photography for his collodion-derived "crystalotype" prints on paper. Samples of his prints were bound that year into issues of the *Photographic Art-Journal.* He became a national authority on collodion photography, teaching it to all comers—for a fifty-dollar fee. With his newfound experience, it was only a matter of time before his attentions were drawn back to the celestial realm.

Archer chose not to patent the collodion process, even though he was first to test and publish its practical aspects. After several years as a landscape and architectural photographer, he died in poverty in 1857 at age forty-four. His wife died a year later. Following Archer's death, the satirical magazine *Punch* commented, "The inventor of Collodion died, leaving his invention, unpatented, to enrich thousands, and his family unapportioned, to the battle of life." In recognition of Archer's contribution to art and the economy, the government awarded his three orphaned children an annual pension of fifty-five pounds.

As a trained chemist, Warren De La Rue would have understood the particulars of the daguerreotype process. And, charmed as he was with Harvard's lunar daguerreotype, he would have grasped the defects of the method when applied to astronomy. His telescope, despite its optical refinement, lacked a clock drive; to track the movement of celestial objects across the sky, it had to be repeatedly nudged by hand. A satisfactory time-exposure photograph would be a challenge with such an instrument, if not a virtual guarantee of failure. At least with the excruciatingly slow daguerreotype technology. De La Rue does not say whether he saw Frederick Scott Archer's wet-collodion photographs at the Crystal Palace in 1851. In any case, he soon learned of the new technique and immersed himself in its astronomical possibilities.

From the start, De La Rue was an ardent advocate for the use of reflector telescopes in celestial photography. Not only is a reflector telescope cheaper to make than a refractor telescope of the same aperture, it focuses all colors of light onto the same plane. That the reflector's visual focus is coincident with its chemical focus obviates the need to guess the proper placement of a photographic plate; wherever the image appears in sharpest focus on a ground glass screen, that's where the photographic plate should go. It was this aspect of the reflector telescope to which De La Rue attributed much of his eventual success in astronomical photography.

De La Rue's original telescope was a thirteen-inch Newtonian style reflector: its solid-metal objective mirror sat at the base of a ten-foot-long wooden tube and converged incident light back up the tube onto an angled secondary mirror, or diagonal, which in turn deflected the light straight out the side of the tube. Here the image was either inspected through an eyepiece or introduced to a camera. The telescope was equatorially mounted—tilted such that one axis is parallel to the axis of Earth's rotation—which allowed it to track stars with a uniform rotation of a single axis. (An alt-azimuth telescope, mounted like a cannon, requires two simultaneous rotations to track celestial objects.)

Warren De La Rue's upgraded thirteen-inch reflector telescope, as depicted in the British Journal of Photography, *1868.*

De La Rue's initial target was the Moon, in part because it was the brightest of nighttime celestial objects, but also because he had a benchmark against which he—and his astronomer colleagues—could gauge his success: the Whipple-Bond lunar daguerreotypes. A major motivation for photographic surveillance of the Moon, beyond humanity's deeply rooted imperative to make maps, was to check for changes in the lunar surface. It was not known at the time whether the Moon had geologically sputtered out or was active, like Earth. Were there intermittent volcanoes, shifting faults, or new impact craters that might show themselves in direct comparison of photographs taken at different times?

Precise hand-guiding of De La Rue's driverless telescope was essential if photographic images of the Moon were to come out sharp. At first, De La Rue used a small guide refractor, complete with crosshairs, like a rifle sight to keep the telescope centered on a target lunar crater. When that proved insufficient, he devised a special plate holder with its own eyepiece and guided by looking directly through the back of the glass photographic plate itself (something that would have been impossible with an opaque daguerreotype plate). However, either of these schemes required the concerted efforts of two people: one to remove the black merino wool cover from the mouth of the telescope—the camera itself had no shutter—and the other to guide the telescope. De La Rue laments, "[I]t was not easy to find a friend always disposed to wait up for hours, night after night, probably without obtaining any result."

In 1852, De La Rue enlisted the steady aid of his collodion supplier, William Henry Thomthwaite, who ran an optical shop in London. By year's end, he had encountered "numberless impediments sufficient to damp the ardor of the most enthusiastic." But he also secured a number of high-quality collodion images of the Moon that he exhibited to the Royal Astronomical Society. Exposure times were typically less than thirty seconds, compared to some twenty minutes or more for lunar daguerreotypes. Even with the improvement in speed, De La Rue had already exhausted the capability of his telescope, as Whipple and the Bonds had done with theirs. There was nothing to do now but join his compatriots across the Atlantic, waiting until their respective telescopes were fitted with precise clock drives.

Chapter 6

THE EVANGELISTS

> The wonderful exactness of the photographic record may perhaps best be characterized by saying that it has revealed the deficiencies of all our other astronomical apparatus—object-glasses and prisms, clocks, even the observer himself.
>
> —Herbert Hall Turner, "Some Reflections Suggested by the Application of Photography to Astronomical Research," 1905

IN MARCH 1857, JOHN WHIPPLE once again marched up Summer House Hill in Cambridge, camera equipment in tow, to Harvard College Observatory. Much had changed during the five years since he taken the final daguerreotype of the crescent Moon through the Great Refractor. Whipple had won international acclaim for his photographic innovations, notably the improvement of paper printing from glass negatives. He was now partnered with James Wallace Black, a former house painter turned photographer who shared Whipple's zeal for invention. Their shop on Washington Street in Boston was a mecca for students eager to learn the latest techniques.

William Cranch Bond, ailing but still the observatory's director, continued his work in precision astronomy and his relentless promotion of Harvard's astronomy facility. Money was always needed for equipment, personnel, and publication of research reports. As his health declined, the elder Bond had increasingly handed off administrative and observing duties to his son George, who had risen to prominence in his own right.

Astronomy at Harvard in the 1850s was a family affair. George Bond had married the daughter of Harvard's librarian in 1853, and now four years later, was the father of two girls. To elder daughter Elizabeth, the observatory grounds were something of a nature park—a meadow-dotted, bird-filled island a million miles from the city. On one side of the director's residence stood the majestic observatory dome, on the other a homely barn with a cow, horses, pigs, and chickens. The family gathered on summer evenings around a large stone behind the observatory to watch the setting Sun. Of her doting father, Elizabeth Bond writes, "Most patiently he taught us the names and the positions of many of the stars and the constellations, and we were always shown anything of special interest in the skies. When a mere baby, not more than three years old, I can remember being held out of an open window in my father's arms . . . to see an eclipse of the moon."

During this period, George Bond worked with his father to complete and publish the first in a series of catalogs listing precise sky coordinates of stars. He himself carried out a long-term visual study of Saturn's rings, which convinced him—erroneously, as it happens—that they were fluid in makeup, not solid as widely believed. (In 1857, Scottish mathematician James Clerk Maxwell proved through a theoretical analysis that the rings must consist of numerous small particles. Astronomer Royal George Biddell Airy described Maxwell's paper as "one of the most remarkable applications of mathematics to physics that I have ever seen.")

There were two circumstances that brought John Whipple back to Harvard in 1857. The first was the long-awaited replacement of the Great Refractor's vexatious Munich clock drive. No more lurches and lags from its out-of-round friction wheel, afflicting the time exposures he and the Bonds had taken during their earlier foray into celestial photography. Now the motion of the telescope was governed by a rigidly precise mechanism, hand-assembled by the Massachusetts firm of Alvan Clark and Sons from William Bond's specifications. The second development that resurrected photography at Harvard arose in Whipple's own realm: the displacement of the daguerreotype by the wet-collodion plate.

The advantage of the new technology was evident from the start. "On a fine night," George Bond told amateur astronomer William Mitchell in 1857, "the amount of work which can be accomplished, with entire exemption from the trouble, vexation and fatigue that seldom fail to attend upon ordinary observations, is astonishing."

Obligatory pictures were taken of the Moon, mostly for public consumption and promotional purposes. Indeed, they were a significant improvement, both in aesthetic and practical terms, over the lunar daguerreotypes of the early 1850s: exposure times were considerably reduced, and prints could be generated from the wet-collodion negatives. Yet to George Bond, the Moon was only a stepping stone to a more ambitious agenda: to identify and explore *quantitative* applications of celestial photography. Bond's belief in the *scientific* potential of the technology required a rapid accumulation of high-quality photographic images that could be subjected to critical scrutiny. Whipple and Black now joined Bond on a regular basis at the observatory, sometimes taking dozens of pictures in the course of several hours. William Bond confided in July 1857 to astronomer Maria Mitchell that "George is, and has been for months, almost hidden from the ken of us mortals in the clouds of Photography."

One of the most promising technical applications of telescopic photography was measurement of the separation and orientation of double stars: pairs of stars that appear close together in the sky. Their adjacency might be spurious—a chance alignment of a star with a faraway counterpart, or it might stem from true proximity to each other—a binary star system bound together by gravity. Compiling a decades-long record of the changing separation and orientation of a double star often reveals whether the double star is an apparent pair—the stars maintain their separation or gradually drift apart—or whether they are a real pair—the stars trace out a mutual orbit.

The first double-star target was the Mizar-Alcor system, a pair of moderately bright stars in the handle of the Big Dipper. Mizar itself has a fainter companion, discovered telescopically in 1617 by Benedetto Castelli, a friend of Galileo's. Bond found that his photographic measures of the separation and position angle between Mizar and Alcor, as well as between Mizar and its companion, agreed with values obtained by direct visual observation. A critical difference, he reminded colleagues, is that he was able to measure the photographs during the day, in the comfort of his office, instead of acquiring the data at the telescope in the often bone-chilling conditions of the observatory. He added that photographs form a permanent record of an otherwise fleeting observation that can be re-analyzed, if needed, at a later date.

Of great concern to professional astronomers was the propensity of their newly minted colleagues to introduce errors into visual measurements of stellar positions. (To prevent contamination of the database, each astronomer was characterized by a unique *personal equation*, a mathematical algorithm expressing the degree to which one's measurements deviate from the communal average.) Bond suggested that extraction of stellar position data from photographs—conceivably by means of a semiautomatic plate-measuring machine—would be less prone to error than subjective eye-estimates of a star's passage across the meridian.

The sensitivity of the new wet-collodion plates was remarkable. A photograph of Vega, which in 1850 had required an exposure of a minute and a half, could now be accomplished in a matter of seconds. Bond and

Whipple's daguerreotypes from the early 1850s had recorded only a handful of the very brightest stars, no matter how long the duration of the exposure. Even Polaris, the North Star, had remained frustratingly out of reach. Now dozens of stars appeared on the long-exposure plates, some of them so dim they are invisible to the naked eye. In Bond's opinion, there was no reason to doubt that further technological refinements of the camera and the chemistry would reveal a multitude of even fainter stars never before seen.

In his Mizar–Alcor paper, Bond anticipated the advent of the astronomical sub-field known as photographic photometry by exploring the means to deduce a star's brightness from the area of its circular image on the camera plate. His comparative study of the bright stars Vega and Arcturus likewise heralded the future of stellar astronomy. Although these two stars appear almost equally bright to the eye—in astronomical parlance, they have the same *magnitude*—Bond noticed that bluish-white Vega registers much more strongly on the blue-sensitive photographic plate than reddish-orange Arcturus. Bond asserted that a photographically determined magnitude gives a more objective and consistent measure of a star's brightness than one derived by eye. In fact, astronomers would come to quantify a star's subjective color by computing a *color index*—the numerical difference between the star's visual and photographic magnitudes.

The Moon, planets, and stars were relatively easy to photograph using the wet-collodion process; more challenging were celestial objects like comets and nebulae, whose light is spread out over a patch of sky. The appearance of Donati's comet in 1858 presented an opportunity to test the camera's potential on a diffuse cosmic body. On September 28, Bond and Whipple succeeded in photographing the head of Donati's comet with a six-minute exposure; however, the comet's faint, diaphanous tail defied the camera. Bond instead sketched the comet by hand. (The night before Harvard's telescopic attempt, William Usherwood, an English commercial photographer, secured a small-scale picture of the comet using a stationary tripod-mounted camera.)

Starting in the late 1850s, Bond published a series of papers that focus less on the photographic particulars of his new observing program, and more on the mathematical elements of extracting—and evaluating—measurement data from wet-collodion plates. He presented direct comparisons of photographically derived double-star measures with those obtained by eye: there was no discernible difference between the two. He highlighted the benefit of collecting the raw data on photographic plates at night, and inspecting them during the day. No more tedious manipulation of the micrometer eyepiece in the half-dark, no more writing down numbers in the shivery cold. The objectivity of the photographic plate, in Bond's view, reduces errors that accompany the challenging environment of visual observing.

In a prescient take on modern institutional astronomy, George Bond envisioned a worldwide network of immense, mountaintop telescopes, each one equipped with a camera. These elevated instruments would be less burdened by the atmospheric turbulence and too-frequent cloudy nights that afflict sea-level observatories. In an 1859 paper, Bond exhorts that the "surface of the globe must be explored for the favored spots where a perfectly tranquil sky will afford the desired field for celestial exploration. If these were occupied, and faithfully improved, the fruits of the enterprise would be beyond all computation rich and interesting." Larger telescopes were essential, Bond added. Telescopes three times the aperture of Harvard's fifteen-inch refractor could be built without delay, if the money could be found. In the near future, wet-collodion plates would be passé, replaced by a soon-to-be-invented chemical process far more sensitive and easier to apply. Bond summed up his passion to William Mitchell in 1857: "There is nothing, then, so extravagant in predicting a future application of photography to stellar Astronomy on a most magnificent scale."

George Bond's evangelical fire about scientific photography found its counterpart in England with Warren De La Rue's almost simultaneous return to celestial imaging. In September 1857, De La Rue renewed his campaign to photograph the Sun, Moon, and planets. His old reflector telescope, now fitted with a high-precision mechanical drive, had been installed in a two-story observatory on his new estate at Cranford, some fifteen miles west of London. The telescope, perched atop a fifteen-feet tall pier, occupied the building's second story; a fully equipped photographic laboratory occupied the space below.

With its integrated darkroom, De La Rue's observatory was the first to explicitly acknowledge the new role of photography in astronomy. Here, the telescope and the human eye serve as mere adjuncts to the camera, which captures the latent celestial image, and to the darkroom below, where the image is chemically revealed. Comfortably nestled on a private estate, the modest structure of brick, wood, and copper was the forerunner of industrial-scale observatories to come.

De La Rue's facility also highlighted a noteworthy difference between visual and photographic observation. If possible, serious visual astronomers

Warren De La Rue's observatory at Cranford, as depicted in the British Journal of Photography, *1868.*

sit while observing, to best steady themselves while peering into the eyepiece. Breathing is periodically suspended, sometimes to the bursting point, to avoid fogging up the eyepiece lens. The importance of enforced stillness is not to be underestimated: William Cranch Bond designed an upholstered, mechanized observing chair for Harvard's refractor that rode on rails around the instrument and cranked up or down to best situate the observer at the eyepiece. (Soften its edges, splash it with color, and Bond's astro-mechano-chair would look right at home in a Dr. Seuss book.)

De La Rue typically stood high on a stage near the focus of his Newtonian telescope, manipulating his photographic equipment while fine-guiding the telescope. Unlike the silent, motionless contemplation of a celestial object by eye, photographic observation requires frequent, active engagement with a host of chemical and mechanical necessities. Absent is the instant gratification of a direct visual sighting; the payoff is delayed until the plate is developed.

De La Rue's upgraded telescope drive proved itself equal to the demands of photographic astronomy, keeping a star on the crosshairs for up to a minute. Of course, De La Rue knew that the Moon doesn't move precisely in synchrony with the stars; it has its own orbital velocity around Earth. Therefore, he added a speed control to the telescope drive, slowing the instrument just enough to follow the Moon. (In fact, the Moon's movement in the sky is more complex and sometimes required a periodic nudge of the telescope.)

By altering the preparation and development of the plates, De La Rue was able to take direct, wet-collodion negatives, which provided finer resolution than his direct positive plates from 1852. (The silver particles that comprise the image are larger on the positive plate than on the negative plate.) With improvements in collodion-plate sensitivity, De La Rue could now produce crisp images of the Moon in as little as three seconds. However, he learned that there was a practical limit to how far he could push the chemical responsiveness of his plates: not only did they tend to fog before being exposed, but spurious "stars" appeared on them as silver particles spontaneously precipitated out of the collodion.

In 1859, De La Rue delivered a lengthy address on celestial photography at the annual meeting of the British Association for the Advancement of Science. Unlike George Bond's brand of advocacy, which was rooted in the research potential of the technology, "the man with the oil-can," as De La Rue called himself, emphasized his working methods. The lecture was a primer on how to apply the camera to the telescope, with sufficient detail that colleagues need not just admire photography from afar, but could actually try it out. But he cautioned them: "[N]o one need hope for even moderate success if he dabbles in celestial photography in a desultory manner, as with an amusement to be taken up and laid aside."

The text of De La Rue's presentation has all the procedural complexity of a Latin Mass: how to clean the glass plate, how to hone the plate's edges, how to mix the silver nitrate bath, how to prepare the collodion, even how far the water faucet must extend over the sink basin. De La Rue treats each plate with a solemnity normally reserved for a holy relic. His photographic laboratory is an 1850s version of a modern industrial clean room. Sanitary agents abound: tripoli powder, spirit of wine, liquid ammonia, alcohol. Hands are washed with a frequency that would delight an obsessive. Dust and lint are banished. Every working liquid is filtered through paper, then filtered again. De La Rue told the assembled scientists with pride, "I have never any failure attributable to a dirty plate."

De La Rue's best lunar images—about an inch across on the original negative—could stand enlargement to an unprecedented eight inches. These paper prints were so fine-grained that inspection with a strong magnifier revealed lunar features in the image that were as small as a thousandth of an inch. Compared to the exquisite photographs, hand drawings began to seem quaint and unscientific, more impressionistic than realistic. In the succeeding years, De La Rue displayed enlargements of the Moon as wide as three feet, as well as a series of striking three-dimensional stereoscopic images of the globular Moon in space—"as if a giant with eyes a thousand miles apart looked at the Moon through a binocular," in the words of John Herschel. De La Rue gained a powerful ally in Astronomer Royal George Airy, who acknowledged that his lunar photographs were much more accurate that any map or hand-drawing. (Airy's interest in photography was more restricted than De La Rue's or Bond's: he viewed photography exclusively as a way to remove human bias from the kind of work done at the Royal Greenwich Observatory—positional astronomy and transit timings—not as a tool of discovery or exploration of the physical properties of celestial bodies.)

Wet-collodion photograph of the full moon by Warren De La Rue, as featured in John Nasmyth's book, The Moon, *1874.*

Among the pressing astronomical issues of the era was the nature of solar surface activity: what it might reveal about stellar physics and whether it influences the atmospheric and magnetic environment of Earth. In particular, there was need to investigate the observation by German amateur astronomer Heinrich Schwabe that sunspots wax and wane in an eleven-year cycle. Even before Schwabe's discovery, John Herschel advocated daily visual monitoring of solar conditions, a proposal he renewed in 1854, but with the application of photography. The British Association for the Advancement of Science entrusted Warren De La Rue with the design and construction of a telescope to be dedicated to the project. Funded by a grant from the Royal Society, De La Rue's "photoheliograph" was installed at Kew Observatory in 1858 and operated there, with few interruptions, until 1872.

The Kew telescope, a refractor, was just over four feet long, with an aperture of 3.4 inches, and featured a specialized lens that focused the visual and photographic rays to nearly the same plane. (Like daguerreotypes, collodion plates were more sensitive to blue and violet colors than to yellow and red.)

A major challenge was the Sun's overwhelming brilliance: only the shortest-duration exposures would render usable images. For the rapid-fire shutter, De La Rue conceived of a lightweight metal plate with a slot cut into it, tensioned by a rubber band and restrained by a thread. The shutter was actuated by burning the thread with a lighted taper. (Later it was simply severed with scissors.) The accumulated photographs helped form the database that led to eventual confirmation of Schwabe's proposed sunspot cycle.

The eclipse of July 18, 1860, offered a timely opportunity to test the worth of wet-collodion photography. There had been a handful of passable daguerreotypes of solar eclipses; however, the technology was too slow to obtain a sequence of images showing the progression of the event. Nor had it captured ephemeral phenomena seen by eye surrounding the Moon's limb during the brief minutes of totality. Visual observers disagreed about the form and nature of these features. English astronomer Francis Baily laid out the issue in an 1846 report about an earlier eclipse: "The accounts [of the corona and prominences] . . . are by no means satisfactory, since they are discordant in many particulars; [especially] the loose description that has been given of them, either by the observers themselves, or by those who drew up the accounts and perhaps did not fully comprehend the intention and meaning of the authors. The difficulty is also very much increased from the want of drawings to represent the exact appearances seen; which are always more readily understood by this method, than by any verbal description."

With the indictment of hand sketches and verbal descriptions, English astronomers mounted an expedition to Spain to photograph the 1860 solar eclipse, and perhaps, to settle various controversies about eclipse-related phenomena. Among the participants were Warren De La Rue and a team of assistants, who had packed up the Kew photoheliograph and brought it to the Spanish village of Rivabellosa. There it was set up in a purpose-built shack with an attached darkroom. In fact, most of the expedition members were visual observers; even De La Rue brought along his sketch pad as a backup.

The photographs of the eclipse were completely successful, capturing a variety of luminous features that extended beyond the Moon's obscuring limb. Among these were the mysterious "red flames," or prominences as they are now called. Astronomers were split about the origin of the prominences: Are they fiery emissions from the Sun or volcanic eruptions from the Moon? (At the time, prominences could be seen only in the diminished glare of a solar eclipse, leading to uncertainty about their origin.)

A single photograph cannot answer this question, for it depicts only the superposition of the solar and lunar disks at a given instant; there is no sense of the crucial third dimension—the distance to the prominences. However, the sequence of De La Rue's photographs showed clearly that the Moon gradually covered and subsequently uncovered the prominences, which remained fixed relative to the Sun: prominences originate in the Sun. The conclusion was affirmed when De La Rue traveled to Italy to inspect collodion plates taken by Angelo Secchi of the Vatican Observatory, who had photographed the eclipse from a site about 250 miles away. If astronomers harbored any doubt about the potential utility of celestial photography, here at least was one counterargument: it had resolved a scientific dispute.

In extracting *quantitative* data from the eclipse pictures—the size of the prominences, relative positions of the Sun and the Moon, and the like—De La Rue first had to determine whether the collodion plates were uniform. The flexible collodion layer might shrink or ripple on the glass, rendering any position or length measurements suspect. De La Rue assessed the stability of the collodion over a year's time and concluded that there was no detectable shrinkage. He also proved, by taking hundreds of photographs of the identical scene, that cardinal points on successive plates differed by less than a thousandth of an inch. Collodion photographs could indeed be trusted for quantitative measurements in astronomy. De La Rue's final report on the eclipse of 1860, including the resolution of the prominence issue and his various detailed measurements, was hailed by the astronomical community and served as a model for subsequent photographic eclipse studies. Two years later, he received the Gold Medal of the Royal Astronomical Society in recognition of the groundbreaking nature of his work.

Photography, once subordinate to the astronomer's eye, was positioning itself to become a catalyst for a new means of discovery. To forward-thinking proponents like George Bond and Warren De La Rue, the design of observational astronomy verged on a dramatic rearrangement: What matters to the photographic astronomer is not what is *seen* through the telescope, restricted to an individual viewer's perception; it's what is *recorded* on the camera plate—"the retina which never forgets," De La Rue called it—which is available for inspection by all. The technology intrigued professional astronomers, but with few exceptions, these seasoned skywatchers were not ready to stake their advancement in the field on a nascent method. Further refinement would have to come from those with sufficient time, skill, and means: self-sustaining amateur scientists outside of academic institutions.

Chapter 7

The Aristocrat and the Artisan

> I take great pleasure in bringing to your notice, the workmanship of Mr. Fitz, as he is an American and a self-taught artist, who places within our reach at home, those instruments which heretofore have been obtained from abroad, at a great cost.
>
> —Lewis M. Rutherfurd, astronomer, 1848

Lewis Morris Rutherfurd was the quintessential nineteenth-century amateur scientist, who crisscrossed the boundaries between practitioner and patron, scholar and craftsman, theorist and experimenter. Born into great wealth, elevated into stratospheric wealth (through marriage to a Stuyvesant), Rutherfurd could have spent his life and his fortune toward any end he might have desired. Yet it would be a mistake to define the man by the obligatory yacht; elite racket club; or slew of senators, justices, governors—even a signer of the Declaration of Independence—dotting his patrician pedigree. Rutherfurd possessed, according to one scientific colleague, an "almost shrinking modesty" and a "singular absence from all ostentation."

Lewis Morris Rutherfurd.

After graduating from Williams College in 1833, at age eighteen, Rutherfurd studied law with William H. Seward, later Lincoln's Secretary of State, and George Wood, whom Daniel Webster regarded as one of the country's finest legal minds. Rutherfurd passed the bar exam in 1837, then practiced law in New York City for the next twelve years, before retiring to Europe in response to his wife's frail health. While in Florence, he renewed his latent interest in science—he had studied physics at Williams, even cobbled together a telescope "from spare parts found in a lumber room"—by studying with the optics expert Giovanni Battista Amici.

Upon his return to the United States in 1856, Rutherfurd erected a small astronomical observatory and workshop behind his mansion in fashionable lower Manhattan, "189 feet N.W. from Second Avenue and 76.3 feet N.E. from Eleventh Street." The tasteful, brick-faced structure was some twenty feet in diameter, with a revolving dome and an attached chamber that housed a transit telescope. (A retail–residential complex occupies the site today.) The urban setting was hostile to astronomy. The lower expanses of the night sky were eclipsed by adjacent buildings and the towering willows of nearby St. Marks Cemetery. The air seemed ever-infused with dust and scattered light. The ground trembled every time a freight wagon rolled by. Rutherfurd brooded over these environmental intrusions. Yet his interests lay more in the engineering aspects of instruments than in astronomical discovery, a line of inquiry more amenable to his city-bound technological oasis.

When it came to fitting his observatory with a telescope, Rutherfurd opted to go local. Henry Fitz, the first American telescope maker of consequence, ran a small shop on Fifth Street, six blocks from Rutherfurd's home. Like Rutherfurd, Fitz had been a precocious tinkerer, concocting his first telescope at age fifteen with a lens scavenged from a pair of eyeglasses. Largely self-taught, he read voraciously about science and mechanics. In 1837, at age twenty-nine, Fitz shunted aside his lucrative lock-making business to help a friend grind and polish a telescope mirror. The project swept him into a new career as an optician. Fitz honed his optical skills in 1839 during a four-month blitz through the workshops of England's master instrument makers. However, he found the quality of English telescopes and raw optical glass not much better than those available back home. The supreme creations were produced on the Continent: telescopes from Germany, optical glass from France.

During the early 1840s, Fitz pursued his off-hours optical hobby while running a successful daguerreotype studio in Baltimore. One of his earliest lenses was fashioned from the sawed-off bottom of a flint-glass tumbler. An observing run in the autumn of 1844 with a friend's German-made refractor—by the peerless optician Joseph Fraunhofer—tipped Fitz's hobby into obsession. The making of an astronomical lens was no trivial venture. One

London wag remarked, "Men have been known to go and throw their heads under waggon wheels, and have them smashed, from being regularly worn out with working an object glass, and not being able to get the convex right."

Fitz's wife, Julia, recalled being awakened one night in January 1845, when her husband excitedly announced that he had crafted a telescope equal to those of his European rivals. "I was soon on the balcony with him," she remembered, "and there sure enough was Jupiter with well defined disc, without a particle of stray light, clear and beautiful. . . . I never saw him in such a glow of enthusiasm, so perfectly radiant with happiness, as that night."

Before the year was out, Fitz shuttered the Baltimore studio and opened his telescope-making shop in New York City. The business was an immediate success, with a variety of individual and institutional clients clamoring for an American-made refractor. Fitz ran a lean operation, employing only two assistants and, eventually, his son Henry G. "Harry" Fitz. He was as intent as a high-end diamond dealer on securing a steady supply of high-quality optical glass, free of streaks, bubbles, and otherwise ruinous flaws. One visitor marveled at the transparency of the raw European crystal: so clear, he noted, that he could see through a sixteen-inch width as plainly as he could see through the air. Between 1845 and 1860, Fitz would produce 40 percent of all telescopes sold in the United States. In 1861, he would complete what was then the largest telescope in the country, a refractor of sixteen inches aperture; the buyer: Buffalo dentist William Vanduzee, who found neither money nor practicality a bar against indulging his daughters' interest in the heavens.

Henry Fitz.

Having ordered several telescopes of increasing aperture, Lewis Rutherfurd became a regular in Fitz's optical shop, absorbing everything he could about the art of lens grinding and figuring. Telescope makers were notoriously secretive about their methods. Yet Fitz evidently had no qualms about revealing his processes to Rutherfurd, whose scientific acumen complemented Fitz's more intuitive approach. Even while Rutherfurd was in Europe, Fitz kept "Friend Rutherfurd" informed of his activities. His letters often overflow onto the envelope with arcane details of the optical business as well as neighborhood chit-chat. Across an ocean and a social divide, the two men bonded over the minutiae of abrasive powders, rivets, and bench tests, both striving toward the same end: technical perfection. In a letter dated November 12, 1849, Fitz tells of the birth of his second son, whom he and his wife named Lewis Rutherfurd Fitz.

From the start, Rutherfurd was an ardent promoter of Fitz's instruments. He closed an 1848 report to the *American Journal of Science* about a recent lunar eclipse with this encomium about his Fitz refractor: "My largest is an achromatic telescope, equatorially mounted, of six inches aperture and eight feet focus. The object-glass is the workmanship of Mr. Henry Fitz of this city, an optician of great skill and rising reputation. . . . [I]t has shewn me at one time last winter, the disk of Jupiter covered with small belts, in addition to the two usually seen, while two of his satellites were plainly seen projected upon the planet's disk, followed by their shadows, which were as distinct as black wafers upon white paper."

Throughout much of 1856, Rutherfurd and Fitz labored side by side to sculpt an eleven and one-quarter-inch objective lens for the newly built observatory on Second Avenue. Just over fourteen feet long, the completed telescope excelled at high-magnification views of lunar and planetary surfaces. Yet it didn't take long for Rutherfurd to chafe at the limits of visual astronomy. He was a Victorian-era "techie," and always alert to the next innovation.

As England's Warren De La Rue had been spurred by Whipple and Bond's 1851 daguerreotype of the Moon, so Rutherfurd found his own muse in the most recent series of wet-plate images from Harvard. In the spring of 1858, he fitted his telescope with a precision drive like the Great Refractor's, and immediately began to photograph the heavens. His initial pictures, while comparable to those obtained by others, were, by his own exacting standards, a failure. His best photographs of the Moon could sustain only a five-times enlargement before lunar features lost definition. Nor did his plates surpass the human eye in "seeing" faint stars. Close-together double stars, which appeared as a pair of luminous points through the eyepiece, were cloaked in an oblong cocoon of light on the plate. Jupiter's moons, visible to Galileo even in a lowly spyglass, eluded Rutherfurd's camera entirely.

Rutherfurd experimented with the preparation of the collodion, at one point bathing the plates in a grape-sugar solution to enhance their sensitivity.

He tested various thicknesses of collodion and tried a host of variations in the development process. "The making of the best negative seems to be a matter of compromises," he concluded. "We cannot have [light sensitivity] and fineness of detail at the same time."

The main culprit, Rutherfurd realized after a time, was not chemical, but instrumental. His telescope, like Harvard's refractor, had been designed for visual observation, not for photography. He and Fitz had shaped the lens components so as to focus the rays of "eye-friendly" colors, such as green and yellow, to the same point. Shades of violet remained uncorrected: even on the clearest of nights, every star in the eyepiece appeared as a scintillating, whitish speck enveloped in a subtle corona of violet. The halo effect was hardly noticeable to the eye and did not diminish the aesthetic pleasure of peering at celestial objects.

This same visual correction was problematic for celestial photography, whose wet-collodion plates—like the daguerreotype before it—were activated by violet light. Whereas the eye sees primarily the point-like core of a star's image, the camera records only the out-of-focus, violet-triggered halo. (Naturally, the halo lacks color on a black-and-white plate.) Through trial and error, Rutherfurd concluded that the best focus for photography—the place inside the telescope where violet rays nominally converge—was seven-tenths of an inch behind the visual focus. Yet, to his chagrin, shifting the plate to this position yielded only modest improvement. In fact, nowhere along the telescope's optical axis did the plate-activating violet rays perfectly unite.

In 1859, Rutherfurd retrofitted his visual refractor with a succession of correcting lenses provided by Fitz. Surely, he believed, some lens or combination of lenses, inserted along the telescope's light path, would impel all shades of violet to a common focus on the photographic plate. (In similar fashion, modern astronomers managed to sharpen blurry images formed by the Hubble Space Telescope's misshapen primary mirror.) Two years of experimentation yielded only partial success: images near the center of the plate were crisp, but those toward the periphery remained indistinct.

By late 1861, Rutherfurd had had enough. He abandoned the refractor design entirely in favor of a mirror-based telescope, which is inherently free of chromatic aberration. But there was a downside. Stars look different in a reflector telescope than in a refractor. English amateur astronomer Andrew Ainslie Common described it this way in 1884: "If we look with a reflector at a bright star, the image is seen as a bright point of light, dazzling to the eye if the telescope is large, and we see rays or coruscations round it of an irregular shape that are never steady. . . . The image of such a bright star in the refractor is quite of another kind: it is seen as a small disk of light of *sensible diameter* surrounded by the well-known system of diffraction rings and outstanding colour. The disk of light, though small, has a different effect on the retina: it can be seen as a shape, pretty steady and free from too much dazzling glare."

In short order, Henry Fitz delivered a lightweight thirteen-inch-wide, ten-foot-long reflector telescope, which he and Rutherfurd strapped papoose-style to the tube of the main instrument. Rutherfurd's frustration only mounted. The incessant tremors of the city kept the reflector's delicately sprung mirror in a constant state of agitation, spoiling the photographic images. And unlike the refractor's maintenance-free lenses, airborne moisture and pollutants attacked the mirror's fragile reflective glaze. Every ten days, Rutherfurd had to remove the tarnished disk and perform a noxious resilvering procedure—"a labor," he confided, "not to be contemplated with equanimity." After just three months, Rutherfurd set the reflector aside.

In his writings from this period, Rutherfurd refers to his visually corrected telescope lens (pejoratively?) as the *uncorrected* objective, revealing his photographic bias: the camera, not the eye, was to be the sole arbiter of image quality. Here emerges an evolving paradigm in astronomical observation. For centuries, the telescope had been considered an optical adjunct of the human eye; now commenced its gradual transformation into a photographic accessory. In essence, the telescope was reimagined by Rutherfurd as but a giant telephoto lens for the camera.

Rutherfurd realized that his photographs would never come into crisp focus until he renounced the human eye's longstanding hegemony over telescope design. The only way to take full advantage of photographic technology was to bypass the eye entirely and design a telescope exclusively for the camera. In such a telescope, the various shades of violet would all focus onto the plane of the photographic plate. That other colors were out-of-focus at this same location was immaterial, as these retinal-stimulating rays do not register on a violet-sensitive plate. Rutherfurd's photographic refractor would be a radical departure from centuries-long tradition: it would be utterly useless for visual observing; nowhere along its optical axis could the eye see images in focus.

Rutherfurd and Fitz turned their working partnership toward the iterative task of creating, testing, and correcting the new type of telescope lens. It was an arduous process: cutting and shaping the best pieces of optical glass that could be had; testing each attempt on an artificial star in the workshop, then on a real star in the night sky; smoothing out of rough patches on the glass surfaces, this last with bare thumbs and a sprinkle of wetted rouge abrasive. By 1863, with the U.S. Civil War in full blaze, Henry Fitz had telescope commissions coming in as fast as his modest workshop could handle. He was training his now-teenage son Harry in the lens-grinding craft and was building a house, with a workshop and top-floor observatory, down the block from Rutherfurd's. In the works was Fitz's most ambitious project to date: construction of an unprecedented twenty-four-inch refractor, for which he planned to sail abroad at year's end to handpick the raw glass disks. The instrument was never realized: Fitz died on October 31, 1863, after a brief illness.

Rutherfurd took on the young Harry Fitz as his new collaborator, simultaneously preparing him to lead his father's firm (which he did for almost twenty years). They immediately resumed work on Rutherfurd's photographic refractor. But what guide can be used to shape a lens whose focusing properties cannot be assessed by eye? That is, how can the crucial violet component of a star's image be isolated so its best focus can be determined? Coincidentally, news had arrived only recently from Germany of a great scientific advance. Gustav Kirchhoff and Robert Bunsen had discovered that the Sun's chemical composition could be deduced by spectroscopic analysis of its light. Rutherfurd realized that a star's spectrum is precisely what he needed to test his photographic lens. A prism, placed at the nominal focus of the lens, will disperse starlight into a narrow spectrum, red at one end, violet at the other. The color at which the spectrum is narrowest is the one that has been brought into most vivid focus. The lens is repeatedly reshaped and retested until the violet segment of the spectrum is a virtual hair's width across.

In December 1864, Rutherfurd replaced the visual objective of his eleven and one-quarter-inch refractor with the photographic lens he and Harry Fitz had completed. The improvement was startling. Now that a star's violet rays were sharply focused onto the plate, images formed some ten times faster than with the visual objective. A three-minute, wet-plate exposure recorded stars six times fainter than any previously impressed on a photographic plate, regardless of exposure time or telescope size. "The power to obtain images of the 9th magnitude stars with so moderate an aperture," Rutherfurd notes in an 1865 article in the *American Journal of Science*, "promises to develop and increase the application of photography to the mapping of the sidereal heavens, and in some measure to realize the hopes which have so long been deferred and disappointed."

Rutherfurd's prime targets were star clusters, whose relatively compact dimensions allowed them to be imaged onto a single plate. What better than a decades- or even centuries-long series of photographs to ascertain the relative positions—and possibly systematic movements—of a cluster's stars? Such movements, if detected, might reveal the distance and the overall mass of the cluster, physical quantities otherwise difficult to obtain.

Between 1865 and 1867, Rutherfurd took forty-five plates of the Pleiades and Praesepe star clusters with his photographic refractor. A decade earlier, astronomers struggled to image even a single star; Rutherfurd's typical Pleiades plate captured around 175. So pinpoint-sharp were these star images that they could not be distinguished from motes of dust in the collodion itself. Rutherfurd was compelled to record *two* photographs on every plate: one long exposure, followed by a thirty-second, closed-shutter pause (during which Earth's rotation slightly shifted the view), then a second, briefer exposure. The result: all of the star images appeared double, while the dust spots remained single.

Despite its astrophysical promise, Rutherfurd's hoped-for photographic revolution stalled. The majority of professional astronomers considered wet-plate photography to be an esoteric practice, more chemical manipulation than science, whose utility remained suspect. The photographic images of star clusters were viewed by researchers as representational art, not dissectible data.

Although Rutherfurd's star cluster photographs passed almost unnoticed, his lunar images fairly burst onto the scene. A two-second exposure of the first-quarter Moon, taken on March 6, 1865, was shown at meetings worldwide, reproduced in books, and sold as large-format prints. The editor of *The Philadelphia Photographer* raved: "[W]e are filled with mingled wonder, and awe, and admiration. . . . The attempts of De La Rue, Bierstadt Bros., and others have all been successful, but in no ways as successful as Mr. Rutherfurd."

Rutherfurd recounted that it was the preternatural stillness and transparency of the air that evening that allowed the crystal-clear image. Photographic pioneer Warren De La Rue agreed that the night must have been superb: "I have made many thousand photographs but never could get one like that and I don't believe if Mr. Rutherfurd makes thousands more that he will ever get such another." From their respective negatives, De La Rue could manage up to an eightfold enlargement before loss of definition, Rutherfurd more than twelvefold.

Professional astronomers likewise marveled at the photograph's aesthetic caliber; on the other hand, they failed to accord it much scientific value. True, a vivid—and completely objective—pictorial record might reveal changes in the lunar landscape caused by, say, lava-spewing volcanoes or meteor impacts. Yet as crisp as Rutherfurd's image was, it lacked the sharpness of a direct view through the best telescopes. The eye makes use of the rare instances when atmospheric turbulence subsides and the lunar surface snaps briefly into crystalline focus. The camera, even with an exposure of just a few seconds, is utterly democratic in recording all the jitters of the image. What appears as a jagged, rock-rimmed crater to the eye becomes, under a magnifier, a velvety, grayish circle on the photographic plate. (Think of a portrait by Vermeer as opposed to one by Renoir.) Rutherfurd had pushed wet-collodion imaging of the Moon to its very limit. It was not until the 1880s, and the emergence of a new photochemical process, that lunar photography resumed its advancement.

In 1868, Rutherfurd replaced his eleven and one-quarter-inch photographic refractor with one of thirteen-inch aperture, whose visual objective could be adapted in minutes for photographic use by attachment of a screw-on supplementary lens. (The lens of the eleven and one-quarter-inch refractor broke in transit to a South American observatory. The thirteen-inch telescope, donated by Rutherfurd to Columbia University, was acquired in 2003 by antique telescope collector John Briggs.)

The thirteen-inch hybrid refractor was a complete success. In 1871, Rutherfurd secured photographs of the Sun's surface that revealed the never-before-seen mottling known as "granulation." (Rutherfurd neglected to announce the result until its independent discovery seven years later by French astronomer Jules Janssen.) Despite the renown of his solar and lunar work, Rutherfurd persisted in his efforts to prove the viability of celestial mapping, or astrometry, by photographic means. Throughout the 1870s, he designed and built a series of increasingly sophisticated table-top micrometers to measure relative star positions from exposed plates. His exhaustive study of the stability of collodion films neutralized critics' claims that archived photographs shrink or warp over time, rendering them useless for measurement.

Over the coming decades, Rutherfurd's most vocal champion, astronomer Benjamin Apthorp Gould, extended the photographic study of star clusters to the Southern Hemisphere. Gould admitted that reduction of the accumulating pile of exposed plates would entail hundreds of hours of laborious measurement. Yet, compared to visual reckoning at the telescope, he declared the task eminently sensible: "[I]nstead of being restricted to favourable nights, at favourable seasons, and painfully made by the astronomer in inconvenient postures and under all the attendant disadvantages, [photographic measures] may easily be performed at any time and place, even in another hemisphere, with all the convenience and comfort which the nature of the case admits, and subject to indefinite repetition. Sufficient material to occupy all the energies of an astronomer for a year or more may thus easily be collected in a single night and reserved for subsequent study."

In 1866, Gould announced to the National Academy of Sciences that photographically derived positions of stars in the Pleiades were in precise accord with those derived visually by the legendary German observer Friedrich Bessel. A follow-up study of the Praesepe cluster was equally positive. Gould postponed formal publication of his results for some two decades in deference to Rutherfurd, who planned to publish his own report of his measurement techniques. He never did, confiding to a fellow astronomer that he did not want to "rush into print." As ill health overtook him later in life, Rutherfurd donated all of his plates and measuring machines to Columbia University, where the task of measurement and analysis was completed by faculty and graduate students. The full record of star cluster positions was finally published in a series of volumes during the 1890s.

Overall, in the two decades following his first lunar photograph in 1858, Rutherfurd recorded 435 plates of the Moon, 349 plates of the Sun and the solar spectrum, and 664 plates of star groups and clusters, including fifty-four images of the Pleiades. *Scientific American* hailed him "by far the most distinguished private scientist in the United States." In fact, Rutherfurd was more an emissary from the world of engineering than a "pure" research scientist. Beauty was to be found in the refinement of an apparatus, in the scientific utility of a method. A picture's worth was to be judged, not by an arm's-length aesthetic assessment, but by a microscopic, pointillist's inspection of the images. The universe, to Rutherfurd, was a test bed for technological innovation.

Perhaps the ultimate sign of Rutherfurd's acclaim was the letter he received in 1886 informing him of his election to the French Academy of Sciences. The envelope was addressed simply:

Lewis M. Rutherfurd, Astronomer
New York, N. Y.

Chapter 8

Passion Is Good, Obsession Is Better

> When Henry and Anna Draper rode home in their carriage from Hastings-on-Hudson to their house at Dobbs Ferry on the night of the first of August, 1872, they had just obtained the final proof, the demonstrably objective evidence, that one star . . . more than a million times further away than the sun, was made of the same atoms that are most abundant in man.
>
> —E. L. Schucking, "Henry Draper: The Unity of the Universe," 1982

In 1856, while Lewis Rutherfurd was sighting the heavens from his new downtown observatory, fellow New Yorker Henry Draper sat hunched over a microscope, taking pictures of frogs' blood cells for his medical school thesis on the spleen. Son of renowned scientist–scholar John William Draper, who first photographed the Moon from a rooftop on Washington Square, it's no surprise that Henry chose to record the diminutive images with a camera instead of a pencil. He had mastered the art of microphotography at thirteen, preparing illustrations for his father's forthcoming book on human physiology.

Henry spent his formative years in a lofty, upper-middle-class atmosphere of culture and scientific inquisition, established by his parents and stoked by five studious siblings. John Draper was a literary dynamo, weighing in on subjects in science, history, sociology, and philosophy. He courted controversy, whether his Darwinian-inspired lecture on the intellectual development of Europe, at the 1860 Oxford evolution debate, or his 1874 book, *History of the Conflict between Religion and Science*, which was banned by the Catholic Church. His "Appeal to the People of the State of

New York, to Legalize the Dissection of the Dead," in 1854, led to legislation that lifted medical school anatomy classes out of the shadows.

Of the six children, Henry was his father's favorite. Henry's niece, Antonia Maury (who would become an astronomer at Harvard), recalled that when John and Henry conversed, "the rest sat silent. For it was well known that everything they said was too important for any word of it to be lost. To his father . . . Henry was the perfect foil, his dark eyes flashing electrically over the latest discoveries in physics or astronomy, or when relating some humorous incident, they brimmed with laughter to the point of tears."

John Draper was omnipresent in Henry's development as a scientist. The three-year-old Henry might well have been present in 1840 when his father photographed the Moon from an upstairs room or when he posed his Aunt Dorothy for her now-famous portrait. *Harper's Weekly* reported that Henry "had for a companion, friend, and teacher, from childhood, one of the most thoroughly cultivated and original scientific men of the present age, who attended carefully to his instruction, and impressed upon him deeply the bent of his own mind in the direction of science. . . . Henry Draper inherited not only his father's genius, but his problems of research." The magazine regarded the Drapers as a latter-day incarnation of the famous father–son team of William and John Herschel. "On one side was the sincerest filial devotion, respect, and admiration," astronomer Charles A. Young noted in 1883, "on the other, paternal pride and confidence; on both sides, the warmest affection, and perfect sympathy of purpose and idea."

Henry Draper.

Two years into his undergraduate studies at the University of the City of New York (now New York University), Henry Draper transferred to the medical school, which his father had cofounded. Having breezed through the entire medical curriculum by 1857, the twenty-year-old Draper found himself fully trained, yet too young to receive his degree.

Instead, he embarked with his older brother John Christopher, a physician and chemist, on a year-long scientific and recreational tour of Europe.

In Dublin that August to attend the meeting of the British Association for the Advancement of Science, Draper accepted Lord Rosse's invitation to view the famous Leviathan of Parsonstown, largest telescope in the world. What the Leviathan lacked in refinement, it more than made up for in immensity and sheer boldness of ambition. Its yawning, six-foot aperture funneled a veritable torrent of cosmic light into the eye, allowing astronomers to see celestial nebulae invisible through smaller instruments. This was a telescope designed to push back the frontiers of deep-space observation. Yet Henry Draper was struck not only by the instrument's immensity and its vivid images of the night sky, but by the intricate machines that Lord Rosse developed to forge, grind, and mount the two-ton, solid-metal mirror. Schooled in the earthbound realm of human disease and experienced only in the cramped world of the microscope, Draper saw a totally different path unfold before him. He would join the ranks of those who directed their scientific attentions upward. Like Rosse, he would build his own metal-mirror reflector when he returned home. The specifications of his telescope were simple: it would be the largest reflector in America; and, unlike the creaky, pulley-driven Leviathan, its optical system and mount would be tailored to celestial photography.

By the time Henry Draper and his brother returned to New York in 1858, the post-daguerreotype era in observational astronomy had already begun. George Bond, at Harvard, and Warren De La Rue, in England, were taking unprecedented wet-collodion pictures of the Sun, Moon, and stars. Even so, the photochemical process remained too inefficient and telescope drives too inaccurate for most astronomers to acknowledge photography's potential for discovery. Its perceived merits during this nascent phase lay in its permanency, its reproducibility, and, most of all, its objectivity.

The human hand, guided subjectively by the eye and the mind, and hindered by darkness and discomfort, had produced drawings of celestial objects for centuries. Astronomers like John Herschel and George Bond were known for their artistic caliber. But many telescopic observers lacked the skill to render on paper more than a stick-figure equivalent of the vision in the eyepiece. Charles Piazzi Smyth, Astronomer Royal of Scotland, acknowledged his colleagues' misguided impulse to heighten or otherwise embellish what they saw: "No astronomical drawing ought, however, to be invaded by any such device for procuring a general effect; no vague expression or semblance of that which exists must be allowed to take the place of painstaking, accurate, and detailed delineation."

The actuality of heavenly bodies was subsumed to various degrees by each artist's imagination and skill. Whatever Lord Rosse spied through his mighty Leviathan telescope, the published sketch of his self-described Crab Nebula resembles more a ragged pineapple than its namesake crustacean. For an 1867 portrayal of the Orion Nebula, Rosse turned over his pencil to a local draftsman. Harvard College Observatory and the U.S. Naval Observatory called upon the gifted pastel artist Etienne Leopold Trouvelot well into the wet-plate era of photography. If not the most accurate depictions of celestial sights, Trouvelot's are certainly the most beautiful. (Trouvelot is less favorably remembered for his introduction of the gypsy moth to North America.)

A thorough hand-rendering of, say, a lunar landscape or a wispy nebula required many nights at the telescope—peering, sketching, adding, refining. A single camera exposure, in principle, could accomplish the same task in minutes. A library of such plates would comprise a permanent, unbiased chemical archive of the astronomical realm. Photographs from different eras might reveal subtle changes unseen by the astronomer's eye: volcanic eruptions on the Moon, the gravitational swirl of a gaseous nebulae, snail-paced rearrangements of the members of a star cluster. Dreams all, in the late 1850s when Henry Draper set to work on his telescope.

During September 1858, Draper crafted a reduced-scale version of Lord Rosse's mirror-grinding machine, whose oscillatory strokes would impart the proper concavity to a speculum-metal disk. The following November he cast a fifteen-inch-wide, two-inch-thick slug of the finest Minnesota copper and Sumatran tin, weighing 110 pounds. Then commenced a seemingly interminable cycle of incremental grinding, polishing, and optical testing. Unlike Lord Rosse, who ran something of a feudal estate, Draper did not relegate work to subordinates (not that he could afford to); he carried out the complex engineering himself, always with minute attention to detail.

Progress on the telescope was glacial: Draper's passion for astronomy was subservient to the need to earn a living. In 1859, he joined the medical staff at Bellevue Hospital, but soon left to become professor of physiology at his alma mater, New York University. By all accounts, he was a popular teacher, committed to the education of his students. The NYU *Quarterly* reported:

> His lectures are so interesting and absorbing to his hearers, that the question of order, which in some recitation-rooms assumes large proportions, is hardly even thought of with him. After class, an eager group surrounds him; and every tap by inquiring students is followed by a rich stream of information from a mind whose varied treasures always lie at instant command.

By 1860, almost a year and a half after it had been cast, the fifteen-inch speculum was nearing completion. On several occasions, Draper had mounted it in a wooden tube and tested it out on the night sky, with mixed results. The correction process was a vitreous Whac-a-Mole: every time he polished out a defect in the mirror's curved surface, another defect popped up somewhere else. On a chill day in February, Draper arrived to find the fragile mirror fractured: water had intruded into its support case during the night, frozen, and heaved. He would have to start over. Draper made no mention of the accident in a midsummer report to the British Association at Oxford; a replacement mirror would be in service soon enough. Presumably, he shared the stoic attitude of many telescope enthusiasts, for whom failure was an inevitable tax levied against progress. When his own telescope lens broke at a remote station in Argentina, astronomer Benjamin Gould remarked, "Lamentations being useless I did the best that I could." Gould secured the glass pieces in a frame and continued his observations.

John Draper might have been three thousand miles across the Atlantic at the time, but he was always in the loop of Henry's activities. He wrote in June 1860 with urgent news. He had told the English astronomer John Herschel of his son's travails with metal mirrors. Herschel pressed Henry to abandon the fraught speculum-telescope design that had endured since the time of Isaac Newton. Instead, he should construct a mirror of silver-coated glass, like the ones only recently developed by Léon Foucault in Paris and Carl August von Steinheil in Munich. (Evidently, Henry had missed Foucault's lecture on glass-mirror telescopes at the 1857 British Association meeting he had attended in Dublin.) Foucault had also published a report describing his so-called knife-edge test, an innovative optical procedure that reveals microscopic defects in a mirror's curvature. For the first time, telescope makers could assess the quality of an astronomical mirror in the workshop.

Casting back some 90 percent of incoming light, silvered-glass mirrors were more reflective than their speculum-metal counterparts, which never exceeded 75 percent. The enhanced reflectivity would no doubt shorten photographic exposure times. Glass was also easier to work than the brittle speculum metal alloys, which—as Draper knew from experience—were apt to split under pressure or cold. The composition of the glass was almost immaterial; a glass mirror's reflective element lay, less than a hair's thickness, on its surface. And, inch for inch, a glass mirror was a mere one-eighth the weight of a metal mirror: Draper's fifteen-inch reflector would shrink from a hefty 110 pounds to less than twenty.

Draper took Herschel's advice. Based on what meager information he could acquire from Europe, he spent the next year experimenting with glass-silvering techniques, achieving success in late 1861 with a process by English chemist John Cimeg. The completed mirror, whose substrate was a piece of glass originally destined for a ship's deadlight, was fifteen and a half inches in diameter. The thickness of its glistening silver coat he estimated at about 1/200,000 of an inch. Within three years, Draper had ground, polished, and silvered more than a hundred glass mirrors, ranging in diameter from a quarter-inch to nineteen inches.

The delicate grinding and polishing process was mechanized, in suitable Victorian-era fashion. Draper and his younger brother Daniel took turns on the treadmill-powered rough-grinding machine, sometimes walking the equivalent of ten miles during a five-hour shift. Hearing that the two men had (unsuccessfully) tried to run dogs on the track, their amused father cautioned his daughter Antonia to refuse if they came for her. With its gnashing gears, articulating rods, and incessant grate of abrasive upon glass, the workshop must have been a raucous environment. Yet Draper seems to have been immune to the mill-like assaults on his senses: "It becomes a pleasant and interesting occupation to produce a mirror."

Draper was acutely sensitive to environmental factors that afflict the final shape of the mirror: "A current of cold air, a gleam of sunlight, the close approach of some person, an unguarded touch, the application of cold water injudiciously will ruin the labor of days. . . . [T]he amateur can only be advised to use too much caution rather than too little." He found that a perfectly formed reflection of an illuminated pinhole flared, even bifurcated, from the warmth of his hand held at the back or the edge of the mirror. The imperceptible warp of the glass lingered after he removed his hand. If the mirror was polished in this distorted state, the warp became permanent, and the highly magnified images of stars were ruined.

Anticipating his telescope's completion, Draper hired a carpenter to erect a small stone-and-wood observatory on the grounds of the family estate at Hastings-on-Hudson, twenty miles north of New York City. (John Draper supplemented his income by leasing out cottages on the estate, which is now a museum and archive.) The building was topped with a lightweight, sheet-metal-clad dome, sixteen-foot across, that could be rotated easily by one hand. Family friend Lewis Rutherfurd could only envy Draper's rural observatory site. Into the distance, the slopes and summits were a rolling sea of trees, broken by the occasional house or clearing. "An uninterrupted horizon is commanded in every direction," Draper enthused, adding that "often when the valleys round are filled with foggy exhalations, there is a clear sky over the observatory, the mist flowing down like a great stream and losing itself in a chasm through which the Hudson here passes." In fact, the idyllic scene would prove less than ideal for celestial photography, once Draper evolved from a dewy-eyed amateur into a seasoned professional. It wasn't long before he started to dream, as astronomers will, of remote mountaintops, swathed nightly in utter blackness and desiccated air, a truly hospitable home for a telescope.

Draper's silvered-glass telescope was, by every optical measure, among the best reflectors in the world. The instrument magnified stars a thousand-fold without significant loss of definition. Draper could clearly distinguish the components of the binary star Gamma Andromedae, although separated at the time by only one ten-thousandth of a degree. The elusive celestial object Debilissima, in the constellation of Lyra, had been resolved into a pair of stars through John Herschel's eighteen-inch reflector and into a

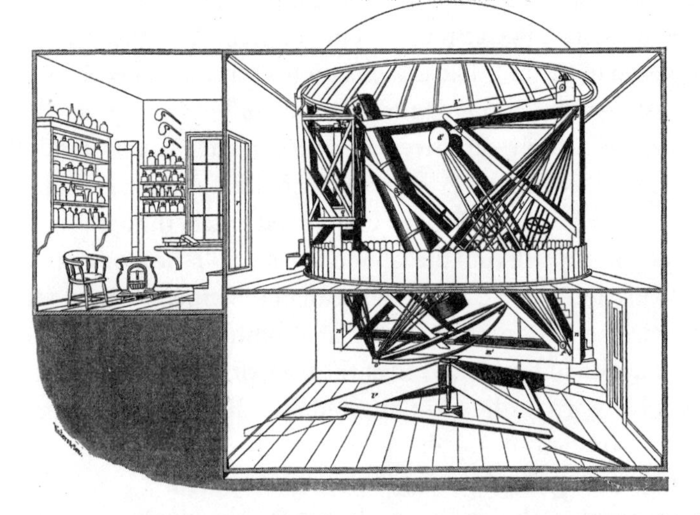

Henry Draper's fifteen-and-a-half-inch reflector telescope, as depicted in his monograph On the Construction of a Silvered Glass Telescope, *1864.*

triplet through Lord Rosse's six-foot-wide Leviathan; Draper reported it to be, in fact, a quintuple system. Jupiter's moons, seen through most telescopes as identical dots of light, showed a range of diameters in the eyepiece of the fifteen-and-a-half-inch telescope. Jupiter's pastel-tinged atmospheric belts shone crisply from equator to pole. (Draper made three fifteen-and-a-half-inch mirrors and adopted the best. Recent tests of the extant mirror at the Hastings Historical Society revealed several optical flaws; it is likely not the reflector he used.)

Although the visual acuity of his telescope was unmistakable, Draper focused on the instrument's original intent. "This is the first observatory that has been erected in America expressly for celestial photography," he had announced to the British Association in 1860, before the instrument and its shelter were completed. He carefully considered his initial photographic subject: the Moon. The Moon's nightly movement across the sky is slightly askew from that of the stars: its own orbital velocity, combined with the inclination of its path through space relative to Earth's equator, alter the Moon's apparent motion in the sky. As Earth turns, a telescope's clock drive (if sufficiently accurate) will keep a star centered in the eyepiece or fixed on the photographic plate; however, the same clock mechanism will slowly lose track of the Moon. Frequent manual adjustments must be made to keep the Moon stationary within the telescope's field of view. Draper envisioned a stable, vibration-free telescope mount that could be moved with no more than a finger's pressure—and without the observer having to step away from the camera. Preferably, the eyepiece or camera would remain fixed while the telescope merely reoriented itself to follow the Moon.

When completed, Draper's telescope mount looked like no other. Conventionally mounted telescopes pivot around a central fulcrum, where their axes of rotation intersect; a heavy counterweight or a long tube must be introduced to equalize the leveraging force of the offset lens or mirror. Draper's design eliminated any such awkward counterbalancing scheme. The telescope's twelve-foot-long tube was suspended inside a cage-like wooden cradle, whose internal counterpoise levers allowed the instrument to point, perfectly balanced, in any direction. The mount was a thing of engineering beauty, its cat's-cradle web of pulleys, levers, and cables a three-dimensional précis of Newtonian mechanics. The mirror itself lay at the bottom of the telescope tube, surrounded by a black velvet curtain and resting like a pasha on an air cushion. Inflation of the rubber sac was controlled by squeezing a bulb near the eyepiece. The cushion counteracted flexure of the mirror under its own weight. (Computer-activated piston-arrays perform the same function in today's large reflector telescopes.)

For lunar photography, Draper brought the telescope to complete rest before exposing, then shifted the camera's plate holder to track the gradual movement of the Moon. "[I]nstead of injuring the photograph by the tremors produced in moving the whole heavy mass of a telescope weighing a ton or more, it only necessitates the driving of an arrangement weighing scarcely an ounce," he reported. During the seconds of exposure, the photographic plate was driven automatically by a clepsydra (water clock), connected to the plate holder by a cord.

By December 1863, Draper had recorded fifteen hundred wet-plate photographs of the Moon. (He ceased operations for five months in 1862 while serving in the Civil War as a Union Army surgeon at Harper's Ferry.) Prints from several of these negatives were regarded as the best lunar pictures ever taken. George Bond proclaimed them magnificent. In reputation, if not surface detail, they would be surpassed only by Lewis Rutherfurd's celebrated shot of the quarter Moon from March 6, 1865. (John Draper reports that both he and Henry were present in Rutherfurd's observatory that night.)

Joseph Henry, Secretary of the Smithsonian Institution, visited the Hastings observatory during the spring of 1863. Impressed with Draper's trove of practical knowledge about telescope construction, he persuaded his host to write a guidebook on the subject for the general audience. Published the following year, *On the Construction of a Silvered Glass Telescope, Fifteen and a Half Inches in Aperture, and Its Use in Celestial Photography* relates Henry Draper's exhaustive, three-year investigation into glass-grinding machines (fully seven alternative models were tried), astronomical mirrors, optical testing, telescope mounts, clock drives, even the workings of a photographic laboratory. The optical

Wet-collodion photograph of the Moon by Henry Draper, 1863.

performance of a professional refractor telescope could be achieved more simply, at lower cost, and to larger scale, with a silvered-glass reflector telescope. Draper hoped that his fifty-five-page treatise might spark a wave of amateur involvement in celestial exploration. From the very first page, he allies himself with the nation's autodidact tinkerers: "[I] can see no reason why silvered glass instruments should not come into general use among amateurs. The future hopes of Astronomy lie in the multitude of observers, and in the concentration of many minds."

On the Construction of a Silvered Glass Telescope was the ultimate how-to guide to an arcane subject, set down with an excruciating specificity that gave the neophyte telescope maker a fighting chance at success. Here the wisdom of experience was neither confided among high-toned attendees at scientific meetings nor abstracted in hard-to-acquire journals; it was available to anyone with the grit and spare cash to take up a most unusual hobby. For the first time, amateur astronomers had been given the means to construct instruments equal to those of their professional counterparts. There was no need for exhortation; Draper's own example provided newcomers with the very model of persistence and high standards in the making of an astronomical mirror. Draper's disciples would come to view the least defect in their silvery surfaces to be as offensive as a blemish on the chin of the Mona Lisa.

After laying out virtually every technological particular he could think of, Draper concludes his manual with a triumphal vision of the future of the glass reflector telescope: "My experience in the matter . . . assures me that not only can the four and six feet telescopes of [William Herschel and Lord Rosse] be equaled, but even excelled. It is merely an affair of expense and patience." Others astronomers, including England's photographic specialist Warren De La Rue, disagreed, claiming that the silver overlay allowed too much photo-activating violet light to pass through. (Draper's subsequent photographs would put this assertion to rest.)

Among those energized by Draper's example was John Brashear, a Pittsburgh laborer who would become one of America's foremost commercial makers of telescopes. Brashear writes in his autobiography:

> I had become acquainted with [Henry Draper] through correspondence . . . and he was never too busy to answer my letters in such a way as to help me solve the problems which were troubling me in the work I loved. His letters found me a toiler in the rolling mill, and together with his work on 'The Construction of a Silvered-Glass Telescope and

> Its Use in Celestial Photography,' they opened a new world, a new heaven to me . . . indeed, his book was of almost inestimable value to hundreds who were enabled to make their own instruments.

With obvious pride in both son and country, John Draper informed Henry from overseas that copies of his monograph were being circulated eagerly among amateur astronomers in England.

While contemplating his next astronomical venture, Draper published a review of American advances in spectroscopy, highlighting his father's contributions and foreshadowing his own entry into the field. In October 1867, he married Anna Palmer, daughter of New York City real-estate magnate Courtlandt Palmer. A gregarious, auburn-haired beauty, Anna insisted from the start that she play an active role in her husband's scientific inquiries. The day after she and Henry were married, they headed downtown on what Anna would later call "our wedding trip": an outing to purchase glass for Henry's next project—a twenty-eight-inch reflector telescope.

The couple moved into the Palmer family mansion on Madison Avenue, between 39th and 40th Streets, near what was then the northern edge of the city. Here they entertained the nation's scientific and political elite. With Anna's money, Draper equipped an astrophysical laboratory, first in a pair of rooms on the third floor of their home, later in a cavernous space over the stables, behind the house. The laboratory featured a roof-mounted heliostat (solar-tracking mirror), gas-powered electrical generators, incandescent lighting, cameras, spectroscopes, induction coils, chemical apparatus, darkroom, machine shop—in short, Draper's home-based facility rivaled those of the era's best scientific institutions.

Having scaled up the various grinding and polishing machines he had used to produce his fifteen-and-a-half-inch mirrors, Draper started work on the twenty-eight-inch reflector. There were no more treadmills to trudge; the new appliances were machine-powered. For the next year and a half, he would grind and polish the big mirror forty-one times before accepting its form. The fifteen-and-a-half-inch reflector, though it displayed stars and planets with remarkable clarity,

Anna Palmer Draper.

was expressly designed to function as a giant lunar-tracking camera. The gaping aperture of the twenty-eight-inch telescope would be better suited to the deep-space realm, gathering up feeble rays from distant stars.

Once again John Draper scouted out the landscape of astronomical research in England and advised his son in 1870, "From what I see here your proper course is to use your telescope first in getting some good lunar photographs. . . . [T]hat done, try your hand at the stellar spectra." With its severe limit on time exposures, the wet-collodion process made it difficult enough to photograph a faint star with tolerable distinctness. John Draper proposed that Henry pass the starlight first through a prism, which diffuses the star's pinpoint gleam into a pale, almost indiscernible spectrum—and to photograph *that*. At the time, no one had succeeded in recording the spectrum of any star other than the Sun. (Joseph Fraunhofer had eyeballed features in stellar spectra through a telescope as early as 1823; England's William Huggins and William Allen Miller had tried—and failed—to photograph the spectrum of Sirius in 1863.) It's no surprise that Henry Draper would have been guided by his father in his choice of research: John Draper was second—after Edmond Becquerel—to photograph the solar spectrum and was a pioneer in diffraction-grating spectroscopy. (A diffraction grating is a finely ruled plate that, like a prism, disperses light into a spectrum.) The two Drapers were, in a sense, sequential collaborators, the younger avidly carrying on the work of the elder.

In fact, the idea for Henry Draper's eventual pursuit of stellar spectroscopy had been percolating for a while. In 1860, word reached the United States from Germany that chemist Robert Bunsen and physicist Gustav Kirchhoff had performed a remote chemical analysis of the Sun's atmosphere based on visual inspection of features in the solar spectrum. If the elemental composition of our nearest star could be deduced from its light, the same spectroscopic process might reveal the makeup of its distant peers. With Draper's photographic expertise and his completion of the nation's largest reflector telescope, the opportunity to obtain and compare a *picture* of a star's spectrum with that of the Sun's must have seemed irresistible.

In August 1869, the twenty-eight-inch telescope was installed in a large dome adjacent to the existing observatory at Hastings. (Draper had offered to place the instrument on Great Hill in New York's Central Park, near 105th Street, but nothing came of the proposal.) To avoid perching atop a ladder to reach the eyepiece, as in the common Newtonian-style reflector, Draper reconfigured the optics to the more convenient Cassegrain layout: light striking the primary mirror converges onto a small secondary mirror,

Henry Draper's observatory at Hastings-on-Hudson, New York.

which in turn reflects it through a central hole bored through the primary. This places the eyepiece—or camera—at the back of the telescope, where it is more easily accessed.

Two more years of testing, alignment, and adjustment followed, during which Draper hand-built—and discarded—six clock drives. In early 1871, having boasted about the still-gestating telescope before a dinner meeting of the Royal Astronomical Society, John Draper pressed his son to finish: "By a little pushing you might have it all in readiness by the time I get back early in April. You will have nothing to do at the University as I will take charge there and so might get to work without interruption or delay. Push the thing a little and you can do it."

On August 1, 1871, Draper began a series of test photographs of the Moon through the twenty-eight-inch. Whereas the fifteen-and-a-half-inch telescope projected an inch-wide lunar image into the camera, the beam of the new instrument was fully five inches across. Yet the photographs were dispiriting: even with all the care lavished on its construction, the twenty-eight-inch telescope proved inferior in definition to its smaller cousin. Draper promptly removed the mirror, took it back to the city, and returned it to the polisher. Finally, in June 1872, now with a *seventh* clock drive installed in its mount, the twenty-eight-inch telescope was complete. Saturn could be now viewed profitably at magnifications up to two thousand. And the latest driving clock, in the opinion of visiting astronomer Charles A. Young, "was as good as any in existence, keeping a star [centered] . . . for an hour at a time."

The lure of the observatory was powerful, and Draper's energy seemed inexhaustible. On clear nights when the university was in session, he sometimes traveled the twenty-mile round-trip to Hastings at the close of the workday. During the summers, when he and Anna were at their country home in Dobbs Ferry, two miles from Hastings, the two of them would head over to the observatory together. "So great was [Anna Draper's] interest," notes Harvard astronomer Annie Cannon, "that he never went to the observatory without her, and in the days of the wet plate, she herself always coated the glass with the collodion. Mrs. Draper told how sometimes after they had been to the observatory and returned to Dobbs Ferry on account of clouds, they would find the sky clearing, and would drive back again two miles to the observatory and recommence work." (In 1878, Anna accompanied her husband—and Thomas Edison—on an expedition to Rawlins, Wyoming, to photograph a solar eclipse. During the minutes of totality, Anna called out the seconds from a clock while sitting in a tent, so as not to be distracted by the heavenly spectacle.)

On August 1, 1872, Draper pointed the twenty-eight-inch telescope, equipped with a camera, toward the bright star Vega. He inserted a small quartz prism into the light path and took an exposure. The recorded spectrum was a hazy slash of light, a mere half inch long and one-thirty-second-inch wide. Microscopic examination revealed the presence of four dark gaps, like those Robert Bunsen and Gustav Kirchhoff had found in the solar spectrum—gaps they had identified with chemical elements found on Earth. On the glass plate was objective, hold-in-your-hand confirmation of what visual spectrum studies had found: atoms in the atmospheres of remote stars are no different than those that constitute the Sun or our own bodies. Frederick Barnard, president of Columbia University, characterized the achievement as "probably the most difficult and costly experiment in celestial chemistry ever made." (The twenty-eight-inch mirror survives in the History of Science Collection at Harvard University.)

Now at the pinnacle of American astronomy, Draper was drafted to coordinate the country's effort to photograph the 1874 transit of Venus across the solar disk. Accurate timing measurements of this twice-a-century event were critical to a geometric determination of the Sun's distance. Photography promised to improve the distance estimate gleaned from visual studies of previous transits in 1761 and 1769. Historian Agnes Clerke summed up astronomers' expectations:

> Observations made by its means would have the advantages of impartiality, multitude, and permanence. Peculiarities of vision and

> bias of judgment would be eliminated; the slow progress of the phenomenon would permit an indefinite number of pictures to be taken, their epochs fixed to a fraction of a second; while subsequent leisurely comparison and measurement could hardly fail, it was thought, to educe approximate truth from the mass of accumulated evidence.

Scientific teams were dispatched from the United States, Germany, Britain, and France, each using different equipment and methods. The Americans adopted the wet-collodion plate, the British and Germans the dry-collodion plate, and the French (unsurprisingly) the daguerreotype. The results were unrelievedly dismal. Given the indistinctness of the Sun's limb, it proved impossible to establish the precise time of contact between the solar and planetary disks. In the end, the Sun's distance was recomputed based on visual observations; the photographs were ignored.

Afterward, an astronomical congress of fourteen nations summarily rejected the use of photography for the Venus transit of 1882. In a statement to the St. Petersburg Academy of Sciences in 1886, Pulkova Observatory director Otto W. Struve summed up the feelings of classical astronomers: "God forbid that astronomy should be carried away by a fascination with novelty." In the realm of the professional observer, the human eye still reigned supreme.

Following the Venus transit debacle, Henry Draper ramped up his spectroscopic examinations of chemical elements in the laboratory and in solar light. He startled the astronomical community by announcing the discovery of oxygen in the Sun's atmosphere, not by its expected dark-line spectral signature, but as an array of bright lines. His explanation posed an alternative theory of solar chemistry, which was met with wide condemnation overseas. (The bright lines proved to be spurious.) In the Hastings observatory, Draper compared the performance of his twenty-eight-inch reflector to a state-of-the-art, twelve-inch visual refractor he had purchased from Alvan Clark and Sons, who had succeeded Henry Fitz as the nation's premier maker of lens-based telescopes. Mounted side-by-side, the two telescopes proved virtually equivalent in their photographic capabilities: the refractor's rigid, unitized construction offset the increased light-grasp of its bigger, but more vibration-prone, rival.

It would be a full four years after Draper's 1872 milestone before anyone else recorded a stellar spectrum. Having succeeded in the photographic challenge of the decade, Draper contemplated the future of cosmic imaging—and it looked bleak. The obstacle was neither optical nor mechanical, but chemical. The wet-collodion process, which had triumphed over the daguerreotype, had by the late 1870s taken nighttime photographic astronomy as far as it could. Its relatively low light sensitivity and hard limit on time exposures drew an effective curtain around Earth, barring detection of objects in deep space. And it was there, in the remote shadows, where opportunities for discovery lay.

For Henry Draper—indeed for the entire astronomical enterprise—the future arrived with surprising swiftness. During the summer of 1879, Draper toured the observatory of his nominal rival, English spectroscopist William Huggins, who alerted him to a new dry-plate photographic process. Not only did the dry plates permit longer-duration exposures than wet plates, they were more light sensitive. And they were being produced commercially in London. When Draper boarded the ship back to America, inside his luggage were several boxes of dry plates, and inside his mind were his next photographic projects. First, he would redouble his efforts to record faint stellar spectra. Then, he would try to capture the face of a celestial object that was, by its very nature, even dimmer: a galactic cloud—a *nebula*. With the approaching dry-plate revolution, the curtain on deep space was about to be lifted.

Chapter 9

From Closet to Cosmos

Collodion—slow old fogey!—your palmy days have been,
You must give place in the future to the plates of Gelatine.

— From the rhyme "Gelatine," *British Journal Photographic Almanac*, 1881

COMPARE THE LUNAR PHOTOGRAPHS of John Draper, Warren De La Rue, Lewis Rutherfurd, and Henry Draper, and the progress over time is clear, like the iconic cartoon of human ascension, from knuckle-dragging brute to quick-witted hunter. The aesthetic and scientific standards for celestial photography rose steadily between 1840 and 1880; the exceptional image of one decade became the norm of the next. To effectively render a lunar crater or resolve an ovate stellar image into a discrete pair of stars, inch-wide negatives had to be examined under a magnifier or enlarged and printed. As often as not, enlargement would reveal the graininess of the photographic emulsion, the celestial target itself camouflaged among a riot of grayish specks.

Much of the advancement in celestial imaging during this period was technological: the introduction of the wet-collodion plate and the simultaneous improvement of telescope drives. But there was also a human element that pressed the effort forward. The cadre of amateur innovators—practical men all—attacked the obstacles as engineers would: coolly analyzing problems; developing solutions on the fly; spending inordinate amounts of time (and money) on the minutest details; and, at least for the most passionate, never giving up.

Visual astronomers praised their photographic colleagues for their celestial curios, even as they judged the images to be proof of the human eye's superiority over the camera. To date, no camera-equipped telescope had discovered anything of consequence in the cosmos. The images merely depicted what had been seen before through the eyepiece. The major advantage of the photograph—its objectivity—was trumped by its failure to surpass the resolving power of the eye, not to mention its inconvenience. Acceptance of the camera as a vital telescopic accessory awaited a dramatic improvement—in fact, a revolution—in image quality, photochemical sensitivity, and ease of use. Until then, it would remain a sideshow to the visual exploration of the heavens.

To the photographic astronomer, the wet-collodion plate was a technological dead end; the limit on exposure time left vast numbers of faint cosmic objects beyond the reach of the camera. Landscape photographers were equally desirous of a better process, unencumbered by the rolling darkroom. The wet-collodion plate had barely arrived before practitioners tried to convert it to a dry process. The experiments were more an alchemical lottery than a scientific investigation, featuring a host of pantry additives: licorice, raspberry syrup, beer, tea, coffee, honey, albumen, grape sugar, gin, and gum—so-called preservatives that promised to eliminate the need to wet the collodion plate prior to exposure. However, the gain in convenience offered by the various dry-collodion recipes was countered by a decrease in photosensitivity, a tradeoff no self-respecting celestial photographer would accept. It was hard enough, even in the 1870s, to get a telescope to accurately track a celestial object; to double or even triple the required exposure time merely replaced a chemical problem with a mechanical one.

The key breakthrough came in the form of a brief report by physician Richard Leach Maddox in the September 8, 1871, issue of the *British Journal of Photography*. Maddox was, by that time, almost two decades and several medals into his noxious hobby of wet-collodion photomicroscopy. This he pursued in an unventilated closet in his home at Woolston, near Southampton, in southeastern England. Long sickened by the fumes of the wet collodion, Maddox had conjured various dry versions, tossing in lichen, linseed oil, quince seed, pulverized rice, and tapioca. The loss in photosensitivity drove him to consider noncollodion alternatives, one of which was gelatin: a flexible, transparent, odorless substance derived from rendered animal tissues. Gelatin had been proposed as a silver-salt adhesive more than twenty years earlier. Maddox softened the gelatin in water, then added a couple of drops of *aqua regia* (nitric and hydrochloric acid). The subsequent addition of cadmium bromide and silver nitrate created a uniform suspension of light-sensitive silver bromide crystals. The milky emulsion was poured onto a glass plate and remained light sensitive, even when dry.

Maddox admitted that his formula was extremely slow, but, deferring to his medical practice, he hoped that his report might spark others to sort out the deficiencies. Experimenters soon found that the presence of the *aqua regia* retarded the action of the emulsion. So did the excess silver nitrate suspended in the gelatin; Maddox's recipe used more than was necessary to maximize the production of light-sensitive silver bromide. The gelatin itself came under chemical scrutiny. Not only was it found that gelatin contains sulfur compounds that enhance the sensitivity of silver bromide, but the substance also protects any unexposed grains of silver bromide from the action of the developer, generating improved contrast over wet-collodion pictures. Gelatin was more than just a passive matrix for the photosensitive compounds; it was a true chemical advance over collodion.

Primitive gelatino-bromide dry plates appeared on the commercial market in small numbers starting in 1873; by decade's end, some twenty brands of precoated plates were available. No longer was a traveling darkroom required; the new dry plates remained sensitive for months, and they could be developed at leisure. Touring the Middle East in 1882, Philadelphia photographer Edward L. Wilson carried his plates for "twenty-two thousand miles on steamer, on donkey-back, on camel-back, and across the Atlantic and the Mediterranean, through the hills of Arabia, in Egypt and other hot countries of the East, and developed eight months afterwards, again in Philadelphia." Echoing the elation of his fellow photographers, Wilson declared, "Blessed be the dry plate!"

A further breakthrough came with the discovery that "ripening" the emulsion for several days at low heat greatly enhances its sensitivity. By the late 1870s, exposure times had plummeted. Photographers could freeze-frame fast-moving objects: the image of water droplets falling from moistened flowers created a sensation at the South London Photographic Society. The hoary thirty-minute daguerreotype had evolved into the dry-plate snapshot. It didn't take long for the rising drumbeat of commercial photographers to fire up their astronomical brethren. Photographic technology had caught up to the demands of the observatory.

During the summer of 1880, no doubt influenced by his friend Lewis Rutherfurd's example, Henry Draper traded in his twelve-inch visual refractor for an eleven-inch photographic refractor from the boutique optical house of Alvan Clark and Sons in Massachusetts. Like Rutherfurd's thirteen-inch refractor, Draper's eleven-inch could be used either as a conventional optical telescope for direct viewing by eye or—with the attachment of a correcting glass—as an outsize telephoto lens for a camera. Like its predecessor, the new instrument was bolted parallel to the existing twenty-eight-inch reflector; an object sighted in one would appear simultaneously in the other. Draper's rustic observatory at Hastings, coupled with its city-based laboratory, had become one of the best-equipped celestial research facilities in the world.

Henry Draper's common-mounted reflector and refractor telescopes, 1880.

Under a moonless sky on the night of September 30, 1880, Draper swung his twin telescopes toward the Orion Nebula, a luminous patch below the trio of stars that define the mythical hunter's belt. Although bright enough to be discerned by the naked eye and a majestic sight in a telescope, the diaphanous billow had so far eluded the photographic plate—a circumstance Draper hoped to remedy that evening.

Ancient skywatchers surely saw the Orion Nebula; yet the oldest extant report of its existence dates to 1611, shortly after the telescope's introduction. The noted Dutch observer Christiaan Huygens published a crude sketch of the nebula in a 1659 book about Saturn. Other depictions followed, from an international roster of astronomical heavyweights: Charles Messier in France, J. L. Schröter in Germany, John Herschel in England, William Bond in the United States, Otto Struve in Russia. The disparity among these hand renderings stoked speculation that the Orion Nebula is coalescing, dispersing, or otherwise evolving.

George Bond at Harvard took up its visual mapping with a particular fervor during the 1850s, after Struve criticized his father's work on Orion. (Although evangelical about photography, George Bond knew that the nebula's diffuse light was insufficient to activate the wet-collodion plates of the time.) Bond's assistant, Asaph Hall, who would later discover the moons of Mars, recalled these marathon sessions with Harvard's Great Refractor: "[H]ow cold my feet were when he was making his winter observations on Orion. I sat in the small alcove of the great dome behind a black curtain, and noted on the chronometer, the transits of stars when Professor Bond called them out. . . . Sometimes I was called to the telescope to examine a very faint star, or some configuration of the nebula. Professor Bond had one of the keenest eyes I have ever met with."

To chemically render the nebula's subtle swells and filaments, a time-exposure of unprecedented length would be required, during which the image had to remain utterly still and in focus. Draper decided to attach his camera, not to his light-efficient twenty-eight-inch reflector, but to the more rigid eleven-inch refractor. The roughly six-fold sacrifice in light-grasp he would offset by lengthening the exposure, confident that his latest hand-built clock drive would faithfully track the nebula as it drifted across the sky.

Draper exposed one of the new gelatino-bromide dry plates from England for fifty-one minutes, yielding a negative that showed clear signs of Orion's ghostly cloud. Convinced that lithographic reproduction of the picture in journals would elide critical details, he sent around photographic prints of the image mounted on six-by-eight-inch cards. On each was the caption: FIRST PHOTOGRAPH OF THE NEBULA IN ORION, TAKEN BY PROFESSOR HENRY DRAPER, M.D. Because stars on the plate were greatly overexposed in capturing the elusive nebula itself, Draper inset a five-minute shot of the cloud's familiar central star quartet, known as the Trapezium. An appended note to his friend Edward S. Holden, then at the U.S. Naval Observatory, conveys Draper's joy in his success: "The exposure of the Orion Nebula required was fifty minutes; what do you think of that as a test of my driving clock?"

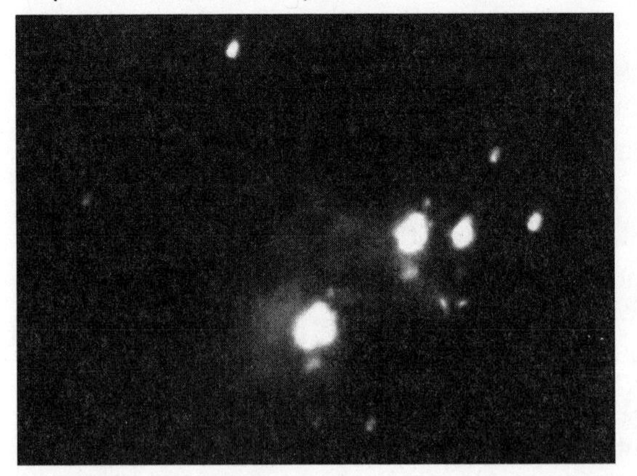

Henry Draper's fifty-one-minute dry-plate exposure of the Orion Nebula, September 30, 1880.

The response to the Orion picture split generally along national lines: American astronomers were delighted, eager to shed their second-rate status in the global scientific community; English astronomers were unmoved, declaring that Draper's photograph showed less detail than contemporary drawings.

The modern eye tends toward the latter judgment: the image smacks more of proof-of-concept than definitive portrait. It conveys only the presence of nebulosity, with no sense of its structure or extent. Evidently, Draper saw it the same way. On March 11, 1881, he made a 104-minute exposure of the Orion Nebula, with noticeable improvement. He tried again just over a year later, on March 14, 1882, when the nebula stood uncommonly vivid against the black drape of the night sky. The thermometer hovered at twenty-seven degrees Fahrenheit; fifteen-mile-an-hour gusts buffeted the dome. Draper upped the magnification of the eleven-inch refractor to 180, attached the camera, engaged the clock drive, then shuddered in the icy gloom while Orion's feeble glow dripped onto the plate for 137 minutes.

Draper's 1882 photograph, although gauzy from a modern perspective, depicts the dark lanes and the winged structure of the nebula so familiar to observers. The picture was hurriedly published as a photolithograph appended to Edward Holden's *Monograph of the Central Parts of the Nebula of Orion*, a commentary on two centuries of visual observations and drawings of the famous celestial cloud. Holden himself had just completed what would become the last great visual study of the Orion Nebula, using the Naval Observatory's twenty-six-inch refractor from 1874 to 1880. Of the

Henry Draper's 137-minute dry-plate exposure of the Orion Nebula, March 14, 1882.

monograph's many images, Holden selects the one compiled over a span of years by Harvard's George Bond as "the best representation of a single celestial object which we have by the old methods." Yet his final aesthetic and scientific verdict is clear: "Dr. Draper's negative was made in 137 minutes, and for nearly every purpose is incomparably better than the other."

In fact, later spectroscopic studies revealed that the eye and the camera had altogether different perspectives on the Orion Nebula. The great cloud is a tenuous assemblage of atoms that, when energized by embedded stars, emit specific colors: primarily a pair of green wavelengths arising from oxygen, plus red, blue, and violet wavelengths from hydrogen. (By contrast, our denser Sun releases a more or less continuous rainbow of light, in the manner of an incandescent bulb.) The human eye is most sensitive to oxygen's green emissions, whereas cameras of the 1880s responded more to the blue and violet of hydrogen. Any difference in the distribution of these two elements would present contrasting views to the visual and the photographic astronomer. No wonder skilled artists rendered details in the Orion Nebula that were not seen in photographs.

If Henry Draper's 1882 photograph did not settle the longstanding matter of whether the nebula had changed over time, it did herald the imminent end of reliance on subjective drawings of astronomical objects. In an 1886 report on celestial photography, Harvard researcher Edward Pickering alerts astronomers to the presence on Draper's plate of a particular faint star, only barely visible to the eye through a comparable telescope. Pickering concludes, "The photographic plate . . . had now become as efficient an instrument of research as the eye itself."

During the summer of 1882, Henry Draper resigned his medical professorship to devote his full attention to astronomy. Although the Orion Nebula continued to beckon, Draper planned to widen his photography of the spectra of stars. Already he had recorded the spectral-line patterns of the brightest luminaries: Vega, Arcturus, Altair, and Capella. And in his laboratory, he had photographed the analogous features for common chemical elements and the Sun. Identification and cross-comparison of the various line patterns, he knew, were key to the establishment of a star's elemental makeup. In principle, astronomers could analyze starlight to the same effect as a chemist analyzing a sample of the star's atoms in a laboratory. The practical hurdles were daunting; except for the Sun, the hair's-breadth gaps in a star's spectrum were barely visible through the eyepiece. However, a time-exposure photograph would enhance these elusive patterns, render them countable, identifiable, measurable. With the new dry-plate technology,

Draper considered the marriage of the camera and the spectroscope to be absolutely vital to the advancement of science.

Only by photographing the spectra of hundreds of stars could Draper assess the variation in chemical composition of the stellar species. To carry out such a project would require years, if not decades, along with significant improvements to the observatory at Hastings. His first priority was to design an even better clock drive: during a two-hour exposure, a star's image could deviate no more than $^{1}/_{300}$ of an inch at the telescope's focus, lest it miss the narrow entrance slit of the spectroscope. In April 1882, Draper reported to the National Academy of Sciences that he had succeeded in photographing the spectrum of a star of tenth magnitude. "It is only a short time since it was considered a feat to get the image of a ninth magnitude star, and now the light of a star [some two-and-a-half times dimmer] may be photographed even when dispersed into a spectrum." At the same time, he dreamed of the heightened images he would capture of the Orion Nebula in the coming months. "I think we are by no means at the end of what can be done," he confided to Edward Holden. "If I can stand 6 hours exposure in midwinter another step forward will result."

In September 1882, Draper accompanied a pair of army acquaintances, Generals Randolph Marcy and William Whipple, on a two-month riding expedition through Wyoming and Montana. A snowstorm overswept the trio during their return leg, and they were forced to camp overnight without shelter above the tree line. The ordeal sapped the strength of the normally vigorous Draper, who arrived back in New York wan and exhausted.

On November 15, Henry and Anna Draper hosted a fifty-guest reception for the National Academy of Sciences at their home. As a novelty, Henry had installed in the dining room a set of Edison incandescent lights—several immersed in decorative, water-filled bowls—powered by the gas-driven dynamo in his laboratory. After the dinner, he grew feverish and short of breath, and was carried to his bed. Diagnosed with pericarditis, he died at four in the morning on November 20, 1882, at age forty-five. As his colleague Charles A. Young wrote in *The Critic* shortly afterward, "It is hard to avoid the appearance of exaggeration in writing of one like Dr. Henry Draper."

Chapter 10

Leaves of Glass

> The camera is an encroaching instrument. So surely as it gains a foothold in any field of research, so surely it advances to occupy the whole, either as adjunct or principal.
>
> —Agnes Clerke, "Sidereal Photography," 1888

The push into deep-space photography, exemplified by Henry Draper's final exposure of the Orion Nebula, accelerated through the 1880s. The dry plate's sensitivity emboldened amateur astronomers, who sought to surpass the eye's limited capacity to register the ghostly wisps of the night. With Draper's death, the locus of innovation shifted from the United States to England, where successors of Warren De La Rue bypassed the snapshot simplicity of lunar photography for the greater challenge of nebular imaging. Even with the recent chemical advancements, an hour's exposure barely fleshed out the Orion Nebula, much less its dimmer counterparts. Significant engineering hurdles remained. Large-aperture astronomical mirrors could now be ordered from a catalog (at substantial cost), but prefabricated telescope mounts were rife with deficiencies. Celestial photographers required a telescope with a vibration-free pedestal, superb balance, silky-smooth bearings, and a dead-on clock drive—all of it exposed to the adverse conditions of the open-air observatory. It was here that the engineering prowess of English amateur astronomers came to the fore.

Andrew Ainslie Common was drawn to astronomy in 1851 at age ten, inspired by the technological daring of Lord Rosse and his Leviathan reflector. Common's elder brother John remarked that, as a youth in the Northumberland town of Morpeth, Andrew "was always at the telescope," a small refractor their mother had borrowed from a local doctor. The flirtation with astronomy was brief: Andrew's father, a surgeon, died in 1852 and the once

Andrew Ainslie Common.

star-struck Andrew turned his attention to his education and his future in a trade. For several years, Common labored at a mill in Gayton, then moved to London in 1864 to join his uncle, Matthew Hall, as a sanitation engineer. Common proved to be a skilled designer and manager. By the mid-1870s, he was running the company's day-to-day operations and became its chief executive upon his uncle's death. (The firm evolved into the global design-construction giant SPIE Matthew Hall.)

Although far from a blowhard, Common was an imposing figure, with a broad, bearded face, a wrestler's build, and a hunger for challenges. A friend described him as "full of enterprise . . . ready to tackle anything": he once tried to hold up a bicycle at arm's length, only to wind up in a sling. The same predilection to test one's limits inevitably found itself applied to a long-deferred interest. Common returned to astronomy in 1874 with the purchase of a five-and-a-half-inch refractor telescope. Marginal attempts at celestial photography convinced him that he needed a bigger instrument. Much bigger.

In 1877, after an abortive attempt to grind his own seventeen-inch mirror, Common bought an eighteen-inch silvered-glass reflector from George Calver, an East Anglian cobbler-turned-optician, whose instruments had garnered acclaim among English amateur astronomers. Like his American contemporary, Henry Draper, Calver issued a guidebook on the making of an astronomical mirror. Draper had hoped to inspire neophytes to construct their own instruments; Calver was more business-minded: having inundated readers with the complexities of the process, he included a catalog of his own telescopes—with prices.

Common built a tube for the eighteen-inch mirror and mounted it in a brick-and-glass garden shed in his backyard at Ealing. (The site was ill-suited to astronomy; one contemporary writer described it as "half submerged by the fogs of London.") Neighbors might have wondered at the mortar-like muzzle that jutted heavenward each night after the shed's sloped roof was rolled aside. But they would have habituated to the sight of a portly figure atop a ladder, peering intently into the side of the broad cylinder.

Had they wandered over, they would have encountered a genial guide to nature, a man to whom "the world was naturally a delightful place." In the autumn of 1877, Common tracked the moons of Mars and Saturn by eye, then gradually turned to planetary photography. The results, although better than before, underscored the root problem: even an eighteen-inch aperture was too small for Common.

This time, George Calver embarked on a thirty-seven-inch reflector, four-and-a-half inches thick and weighing more than four-hundred pounds. Four grinding machines scraped away before one was found that could handle the load and not fragment the glass. The raw disk was secured to a tilt-table, laid flat for grinding and polishing, tipped vertically for optical testing. "The work of correcting was tedious and trying," Calver told the Royal Astronomical Society, "especially in the latter stages, when for every few minutes' polishing, the whole preparations for testing had to be repeated, and the settling of the mass into its normal state had to be patiently waited for, and often days passed before further advance could be made." Calver seems to have been blessed with the patience of Job; yet he did confess that his flagging spirits were buoyed by a sit-down with Henry Draper, who commiserated with his fellow mirror-maker. (It was during this 1879 visit to England that Draper learned of the dry-plate photographic process.) Having completed his so-called three-foot reflector, Calver correctly divined the progression of astronomical telescopes: "I see no obstacles to the construction of glass mirrors of very large sizes."

Andrew Common prepared for the arrival of his outsize mirror by building a house-like enclosure that rolled aside on iron rails. He mounted the telescope as an equatorial—like a swiveling cannon support, but heeled over so its azimuthal motion paralleled Earth's rotation. To ensure ease of movement, the ponderous equatorial axle "floated" in a concentric casing filled with mercury. From this base structure rose an eighteen-foot-long steel-strut tube, whose upper end held the accessory optics, including a plate holder. No fan of heights, Common built a broad, enclosed scaffold that raised him safely to the telescope's eyepiece.

A daytime snapshot, presumably taken from an upper story of the Ealing house, shows Andrew Common in the middle distance, back to the camera, a mere appendage to the giant optical machine into which he is gazing. The backyard vista captures the Victorian era's energetic amalgam of the agrarian and the industrial. We see grass, shrubs, stone pathways, gardening tools casually laid aside for their next use; over the fence, a pastoral backdrop of tilled fields and rolling hillocks. And, peering out of their respective shelters

View of Andrew Common's eighteen-inch and thirty-six-inch telescopes, as seen from his house in Ealing, near London.

like a pair of mechanical beasts, there were the instruments of scientific exploration—one man's idea of a proper English garden.

The three-foot reflector was completed in late 1879. Like his Hudson Valley colleague Henry Draper, Andrew Common was drawn almost immediately to the Orion Nebula. The distant cloud represented the photographic frontier, both in terms of literal reach into space and the state of earthbound technology. Common tried to photograph the nebula on the night of January 20, 1880, nine months before Henry Draper's initial success. The result was dismaying: "The stars were seen as lines," astronomer Edmund Stone told the Royal Astronomical Society, "and the nebula proper presented merely a faint stain upon the plate."

The three-foot telescope was nearly undone by Common's self-designed mechanics. By placing the entire instrument above and forward of its base, he had created a giant, tottery lever requiring a massive back-end counterpoise. (Typically, a telescope's mirror sits aft of the base, helping balance the long tube.) Every time the plate holder was attached or removed, lead slabs had to be added to or withdrawn from a pair of counterpoise boxes. Common's innovative mercury flotation system was a "delusion," according to astronomer James Keeler, who used the telescope after it was donated to Lick Observatory in California. The fluid did little to buoy the dead weight of the mount, which bore down relentlessly on a lone steel pivot. The telescope's clock drive was deficient as well, unable to move the heavy instrument in synchrony with the stars. Tracking of celestial objects was handled by a 1,440-tooth gunmetal gear, forty inches in diameter, which was turned by the slow descent of a weight down an eight-foot shaft. A good photographic telescope in the 1880s could track a star accurately for at least two minutes; Common's gear-driven behemoth barely managed two seconds.

Common's engineering intuition told him that no machine would ever drive the full mass of his telescope with the requisite precision to capture Orion. Instead, he designed a mechanism akin to one used by Henry Draper, which shifted the camera plate itself at the proper rate. To the plate holder, Common affixed a high-power optical sight, allowing him to apply a corrective nudge whenever a chosen guide star strayed from the crosshairs. The faulty clock drive lumbered away, keeping the telescope roughly on target, while Common manually did the rest. After two years of development and testing, Andrew Common was ready to try again.

On March 18, 1882, four days after Henry Draper took his third and final photograph of the Orion Nebula, Common succeeded in imaging the object himself. But it was his follow-up exposures of thirty-seven minutes on January 30, 1883, and one hour on February 28, that captured Orion's ethereal splendor far more vividly than any previous rendering by hand or by camera. The direct telescopic view, aquiver from mechanical vibrations and atmospheric disturbances, was here stilled. In this photograph,

Andrew Common's dry-plate photograph of the Orion Nebula, taken in 1883.

the observer could linger on the whole or on the details, absent the chill winds, errant clouds, or dew-flecked lenses of the observatory. William Abney, president of the Royal Astronomical Society, called Common's picture "epoch-making."

Nor could one fail to appreciate the picture's aesthetic dimensions. This was a technical image that could justly be contemplated by the artist. In its luminous gradations and multitude of forms was visual poetry, if not heavenly associations. The foggy pleats posed a nest of mysteries, each opaque cloud-front begging speculation as to what lies beyond. (Modern astronomers see through such barriers by capturing radio waves or infrared light that penetrate the gas and dust.)

Despite its nonuniformity—in bringing out faint details, stars and bright regions were grossly overexposed—Common's photograph evinced the authority of a scientific document. Not only was it arguably more objective, but more acute, than the most detailed sketches. Although subtle distinctions in appearance were expected—the human retina and the dry plate have different color sensitivities—disputes over which of history's drawings best matched Orion's "true" face were effectively mooted by a single photographic frame. In 1887, popular science documentarian Agnes Clerke featured Common's one-hour exposure as the frontispiece of her *History of Astronomy During the Nineteenth Century*. Photography, she declared, had "assumed the office of historiographer to the nebulae . . . [T]his one impression embodies a mass of facts hardly to be compassed by months of labour with the pencil."

In 1885, again itching for more aperture, Common sold his three-foot reflector to British amateur astronomer Edward Crossley for the present-day sum of two million dollars. Common next embarked on a tortuous quest to construct his own five-foot reflector. A near-plunge from the elevated scaffold persuaded him to reconfigure the partly completed instrument with the eyepiece closer to the ground. Five years in the making, the finished product was mediocre, both mechanically and optically. After taking a few photographs of Orion and other nebulae, Common abandoned his creation to develop telescopic gunsights for the Royal Navy. (The three-foot telescope was donated by Crossley to California's Lick Observatory in 1895; the five-foot went to Harvard in 1904 after Common's death.)

Andrew Common's immediate successor in the British "grand amateur" tradition was Isaac Roberts, a Welsh farmer's son who made his fortune in the Liverpool building trade. Inspired by Common's celebrated image of Orion, Roberts purchased a twenty-inch photographic reflector, and in

March 1885, affixed it to a superior equatorial mount. The telescope itself was the camera: within the tube, a photographic plate holder replaced the standard secondary mirror that deflects light through an eyepiece. Although Common's instruments were larger, the twenty-inch tracked celestial objects far more precisely. Immediately Roberts was taking time exposures up to an hour's duration; before long, three- and four-hour exposures were routine.

Isaac Roberts.

Among Roberts's initial targets was the Pleiades, a loose cluster of stars famously poeticized in Tennyson's "Locksley Hall" as a "swarm of fireflies tangled in a silver braid." In 1859, the keen-eyed comet hunter Wilhelm Tempel perceived the metaphorical braid in the form of a tenuous veil overlying the stars. The full extent of the Pleiades nebulosity appeared in a series of striking photographs by Roberts in 1886, described memorably by Agnes Clerke: "'Streamers and fleecy masses' extend from star to star. Nebulae in wings and trains, nebulae in patches, wisps, and streaks, seem to fill the system, as clouds choke a mountain valley."

The twenty-inch photographic reflector telescope of Isaac Roberts.

That same year, Roberts produced an hour-long exposure of the Orion Nebula that revealed subtle swirls of nebulosity in regions long assumed to be voids. And on the original negative, the nebula's central region, bleached out in Common's now-famous picture, resolved itself into "cloud-like, curdling masses." Three years later, in 1889, Roberts presented to the Royal Astronomical Society seven more plates of Orion, ranging in exposure time from five seconds to almost three-and-a-half hours. The short exposures resembled Herschel's, Rosse's, and Bond's drawings. But with each succeeding jump in duration, Orion's boundaries swelled and its substance thickened. Once-ghostly streamers jelled into luminous tentacles that ensnared neighboring clouds of matter. What cosmic explorers had taken for an islet within the ocean of space was, in these photographs, a celestial continent.

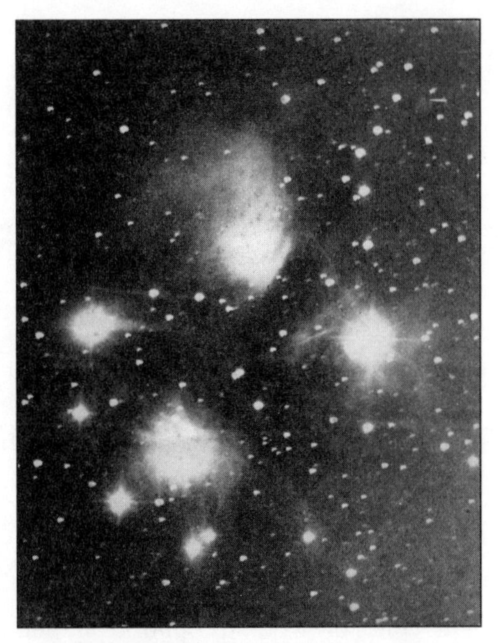

The Pleiades, in a photograph taken by Isaac Roberts in October 1886.

Roberts closed his presentation with a nod to his visual predecessors, while simultaneously elbowing them aside: "[W]e ought, with all gratitude, to admire the patient, long-suffering endurance of those martyrs to science, who during the freezing nights of many successive winters plotted, with pencil in benumbed fingers, the crude outlines which have been handed down to us as correct drawings of this wonderful nebula, which we can now depict during four hours of clear sky with far greater accuracy than is possible by the best hand-work in a lifetime."

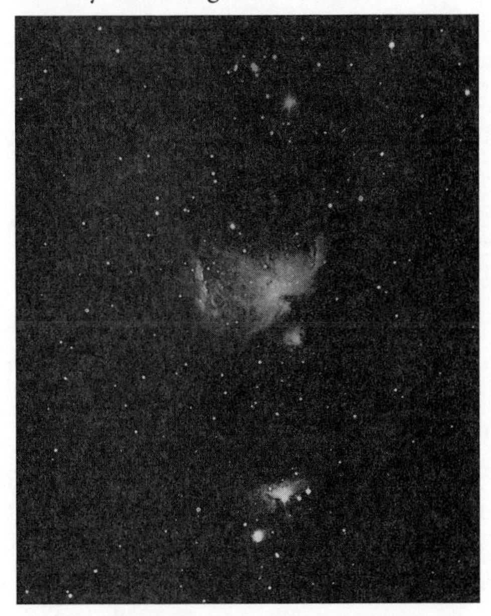

The Orion Nebula, in an eighty-one-minute exposure by Isaac Roberts on December 24, 1888.

Among the night-sky wonders, the Orion Nebula shares top billing with its more northerly counterpart in the constellation of Andromeda. More than five times the span of the full Moon, the Andromeda Nebula is seen as an elongated luminance, bright near the hub, feathering to invisibility in the outskirts. (Having aspired to astronomy as a youngster in light-bound New Jersey, I didn't see the Andromeda Nebula—or the Milky Way, for that matter—until a family vacation took me to the inky skies of Colorado.) Andromeda's ghostly presence was noted by tenth-century Persian astronomer Abd-al-Rahman al-Sufi, and it appears on a Dutch sky chart from the year 1500. Galileo's contemporary Simon Marius first observed the object through a telescope in 1612, aptly describing its pale glow as that of "a candle shining through horn." In the 1700s, French comet hunter Charles Messier included the Andromeda Nebula as the thirty-first entry in his catalog of diffuse celestial objects; hence, its oft-used alias, M31.

As with the Orion Nebula, historical depictions of Andromeda range widely. Written accounts convey more a sense of befuddlement than agreement about its appearance and nature. Messier saw it as "two luminous cones or pyramids opposite at their base . . . without any appearance of stars." Observing with Harvard's Great Refractor in 1847, George Bond drew attention to a pair of nearly parallel dark lanes that stretch lengthwise along the nebulosity. Deep-sky expert Reverend T. W. Webb considered all the available data as of 1882 and concluded that the Andromeda Nebula was "a mystery never in all probability to be penetrated by man." The reverend's lack of faith in the scientific enterprise proved to be misplaced.

There were audible gasps from the audience when Isaac Roberts projected his three-hour exposure of the Andromeda Nebula before the Royal Astronomical Society on December 14, 1888. Herbert Hall Turner, Savilian Professor of Astronomy at Oxford, called the sight a "revelation," while Norman Lockyer, longtime editor of the scientific journal *Nature*, judged the image "one of the most valuable photographs which I suppose has ever been taken." What the attendees saw in this expansive depiction of the Andromeda Nebula was not its accustomed amorphous glow, but a wholly new structure with profound connotations.

A celestial photograph is, by nature, a two-dimensional record of three-dimensional reality: it depicts an object's form as though flattened onto the illusory plane of the sky. Informed by visual cues in the image, the human brain reconstructs the sensation of depth and physical bulk—at least according to one's preconceptions and real-world experience. (How easily the brain can be fooled by optical illusions.) The Andromeda in Isaac Roberts's photograph virtually exploded into the third-dimension to reveal a series of concentric,

The Andromeda Nebula, in a four-hour exposure by Isaac Roberts on December 29, 1888.

highly foreshortened annuli of diffuse matter, all tilted at an extreme angle to the line of sight. George Bond's dark lanes, here vividly seen, were gaps between the adjacent rings of matter. No doubt every astronomer in the room sailed their mind's eye into space to picture Andromeda from the *face-on* perspective: a majestic, multiarmed spiral, like Lord Rosse's Whirpool Nebula and some fourteen others seen through his Leviathan telescope.

An even more stunning four-hour exposure was featured in the mass-circulation magazine *Knowledge* in February 1889. Its editor, A. Cowper Ranyard, pointed out one detail astronomers had evidently missed: "Lines of small stars shown in Mr. Roberts's photographs lie along the edges of the dark rifts following all their sinuosities. . . . [O]ne hardly knows whether to describe them as minute stellar points or as regions of greater nebulous brightness." These stars, Ranyard implies, are not foreground bodies seen in projection, but clearly belong to the spiral nebula itself. At least some fraction of Andromeda's light springs from stellar sources. Could it be that its broader glow is the "united luster of millions of stars," as the famous eighteenth-century observer William Herschel suggested?

The controversy over the nature of nebulae had simmered since their earliest discovery. One theory posed that nebulae are gaseous aggregations scattered among the stars of our Milky Way. Spiral nebulae, with their distinctive whirlpool form, elicited visions of matter cascading inward and condensing into new planetary systems, consistent with the so-called nebular hypothesis proposed by Pierre Simon de Laplace in 1796. The photographs of Andromeda, Isaac Roberts announced, were a clear demonstration of Laplace's idea: "Here one might see a new solar system in process of condensation from the nebula." Roberts bolstered his Laplacian vision in April 1889 with a crisp, four-hour exposure of the Whirlpool Nebula, the definitive face-on spiral.

The alternative theory of the nebulae alleged that at least some of these supposed clouds of gas are "island universes," distant Milky Ways whose remoteness renders their stars irresolvable. The observations of pioneering English spectroscopist William Huggins only fueled the dispute. Huggins confirmed in 1864 that spectra of diffuse nebulae generally mimic features seen in a spectrum of a tenuous, incandescent gas in a laboratory. However, he subsequently found that spectra of *spiral* nebulae, notably Andromeda, indicate a stellar origin—that is, the collective light of individual, unresolved stars. And if Andromeda could hide its spirality through a simple accident of orientation, how many other covert spirals might there be in the night sky? The competing beliefs about the spiral nebulae posed very different concepts of the universe: one, a lone assemblage of stars—the Milky Way—surrounded by a void; the other, a vast population of star systems like our own, strewn throughout what might be an infinite space.

In 1893, Roberts published a photographic album of deep-space species, a first step in realizing Andrew Common's dream of a new kind of "library,

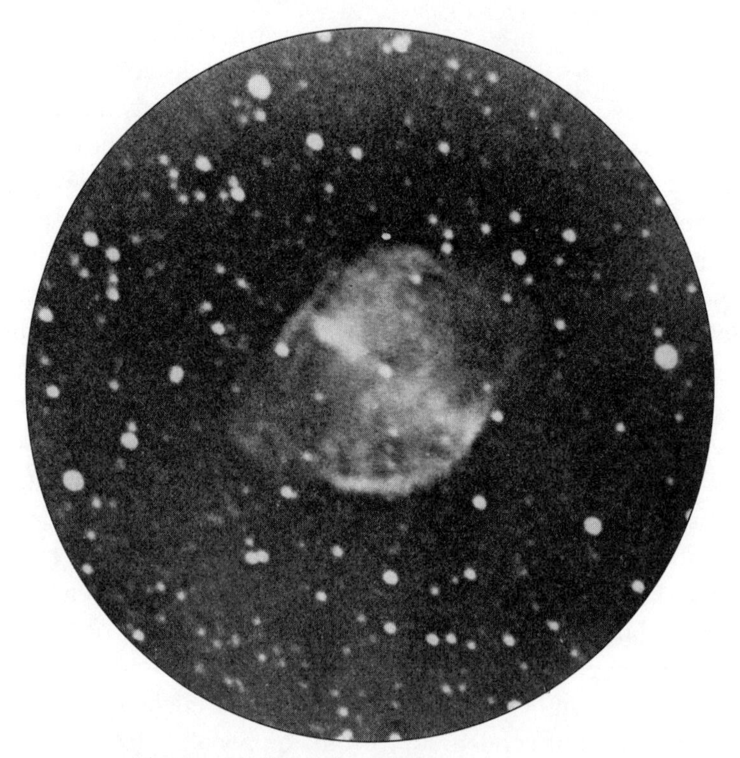

The Dumbbell Nebula, in a three-hour exposure by Isaac Roberts on October 3, 1888.

not of books full of descriptions and figures . . . but of pictures written on leaves of glass by the stars themselves." Even the most ardent visual astronomers had to acknowledge the practical import of these accumulating chemical masterpieces: in probing nebulae, the eye had become the poor stepchild to the photographic plate. In his 1888 memoir of Henry Draper, University of Pennsylvania physicist George Barker writes, "[T]he facility of reproduction by photographic means so far surpasses that by drawing or sketching, and is, moreover, so much more accurate a method of delineation, that the evidence given by an untouched photograph is everywhere accepted as *prima facie* proof."

The camera—and *only* the camera—had the ability to disclose Andromeda's true form. And, going forward, the camera would be the essential agent of nebular imaging. Yet before the human eye was shoved entirely out of the picture, the institutional elite had to be convinced that photography was relevant to their work. The majority of these professionals were based at universities and government-sponsored timekeeping facilities. Like generations of forebears in Britain, France, and Germany, they were practitioners of so-called exact astronomy: the mathematical analysis of positions and movements of celestial bodies. To these number crunchers, nebular exposures were "soft" data, the antithesis of the quantitative ledgers they patiently compiled and scrutinized. To add a decimal place of precision to a star's position was a badge of honor within this fraternity. Their commitment to mathematical rigor and clockwork accuracy was born of tradition dating back more than a century. Positional astronomers were not about to suspend ongoing visual projects and tramp over to the observatory with a camera in hand. At least not until 1882. In that year, a single photograph, taken on a whim by one of their own, made them see the light.

Chapter 11

THE GRANDEST FAILURE

> If we could first know where we are, and whither we are tending, we could better judge what to do, and how to do it.
>
> —Abraham Lincoln, Illinois Republican Convention, June 16, 1858

DAVID GILL WAS ASTONISHED by the brilliance of the comet as it hovered above the mountaintops of False Bay, opposite the Cape of Good Hope. Nothing in his decade-long career, first as director of the Dun Echt Observatory in Scotland and now as Her Majesty's Astronomer at the Royal Observatory in South Africa, had approached the lustrous majesty of this interplanetary visitor. Discovered in the predawn hours of September 8, 1882, by his chief assistant, William H. Finlay, the comet had brightened over the past week and a half until it was visible in broad daylight, even near the Sun's edge. (Andrew Common independently sighted the comet on the morning of September 17 from the Northern Hemisphere, just hours before it rounded the Sun's fiery surface.)

David Gill.

A watchmaker's son, David Gill was an astronomer of the "rigorously orthodox kind," charged with meticulous measurement of positions of celestial bodies in the southern sky. As to why delineation of the heavens is important, Gill once explained that if, "of two points marking a frontage boundary on Cornhill, one were correct, the other 10 feet in error, what a nice fuss there would be! what food for lawyers! what a bad time for the Ordnance Survey Office! Well, it is just the same in astronomy." Like many positional astronomers, Gill obsessed over the detection and correction of telescopic deficiencies. He advised a colleague, "[H]owever perfect an instrument may be (and it is the astronomer's business to see that it is perfect), it is the astronomer's further business to look upon it with complete and utter mistrust." Gill was no stone-faced automaton. He admitted that astrometry—celestial position measurement—offered "no dreamy contemplation, no watching for new stars, no unexpected or startling phenomena." Yet he reveled in the grinding routine of the observatory, frequently cutting the monotony with a pipe or cigar. He counted himself among the intrepid observers who "betake themselves to bed, tired, but (if they are of the right stuff) happy and contented men." Many nights, Isobel Gill recalled, her husband arrived home singing.

David Gill's telescope of choice was an unconventional, split-objective refractor called a heliometer, whose lens-halves could be offset by means of a screw to reveal the separations between objects in the eyepiece. The instrument's complexity confounded all but a handful of astronomers, and Gill was acknowledged to be the world's expert in its use. His greatest claim to fame was his 1877 determination of the Sun's distance, which became the worldwide standard into the twentieth century.

When not immersed in the statistical minutiae of observing and data reduction—or in the tennis and hockey contests he loved to play with the observatory's staff—Gill would plead with the Admiralty to increase the meager funding for its Southern Hemisphere station. Yet in the morning twilight of September 18, 1882, one day after perihelion, the beck of research, money, and sport faded as the billowing comet awakened Gill's poetic muse. "There was not a cloud in the sky," he reported, "only a merging into a rich yellow that fringed the blackish blue of the distant mountains, and over the mountains and amongst the yellow an ill-defined mass of golden glory rose with a beauty I cannot describe. The Sun rose a few minutes afterward, but to my intense surprise the comet seemed in no way dimmed in brightness. . . . I left Simon's Bay and hurried back to the observatory, pointing out the comet in broad daylight to the friends I met along the way."

An observer in India likened the comet's graceful arc to that of an elephant's tusk. By October, the tail lengthened to fifteen degrees, and took an hour to fully rise above the horizon. A few enthusiasts managed to photograph the comet using tripod-mounted cameras, although the nucleus was inevitably smeared out by Earth's rotation. (The spurious elongation happened to align with the comet's tail, masking the flaw; nevertheless, at least one amateur retouched the diaphanous tail with a paint brush.) Hearing of these images stirred another dormant muse in David Gill. In 1869, back in his native Scotland during the wet-collodion era, he had dabbled in lunar photography with a second-hand reflector telescope. It was, in part, the acclaim following one of these pictures that had led to his appointment as director of Lord Lindsay's Dun Echt Observatory. Now, with the spectacular comet gracing his southern sky, Gill suspended his heliometer studies for what would become a brief but momentous return to celestial photography.

Without a camera or any practice in the new dry-plate process, Gill had no choice but to follow the example of his Harvard predecessor William Bond during the daguerreotype era: he sought the help of a professional photographer. E. H. Allis, from the nearby village of Mowbray, proved to be as able and enthusiastic a collaborator as John Whipple had been to Bond thirty-five years before. On October 19, 1882, Allis arrived at the observatory with his portrait camera, which Gill mounted on a stout board and clamped to the counterweight of a six-inch refractor: the telescope became a high-power sight for the camera. Allis operated the camera shutter, while Gill adjusted the slow-motion controls to keep the comet's nucleus centered on a pair of crossed spider webs at the telescope's focus. Six photographs were taken during October and November, with exposures ranging from thirty minutes to two hours and twenty minutes. The best of the pictures showed the comet in sharp focus, with structural details evident in both its tail and its nucleus. In a confluence of human invention and cosmic clockwork, a technology had matured in time to record one of the brightest comets of the modern era.

Yet even as Gill rushed his comet portraits to colleagues in England and France, he was drawn to an incidental—and wholly unanticipated—feature of the photographs: the profusion of background stars. The faintest of these he estimated at tenth magnitude, some forty times dimmer than the minimum light detectable by the unaided eye. What amazed Gill was that the stars speckling his comet pictures had been recorded with a studio camera whose aperture was a mere two-and-a-half inches. (In fact, photographic pioneer Warren De La Rue had imaged the Pleiades star cluster with a portrait camera in 1861, during the wet-plate era.) Tenth magnitude also happened to coincide with the brightness cutoff of the famed *Bonner Durchmusterung* (*BD*), a visual census of more than 324,000 Northern Hemisphere stars compiled by midcentury German astronomers. An extension to just below the celestial equator raised that number to 458,000.

In 1850, a relatively small number of stars had classical names or other

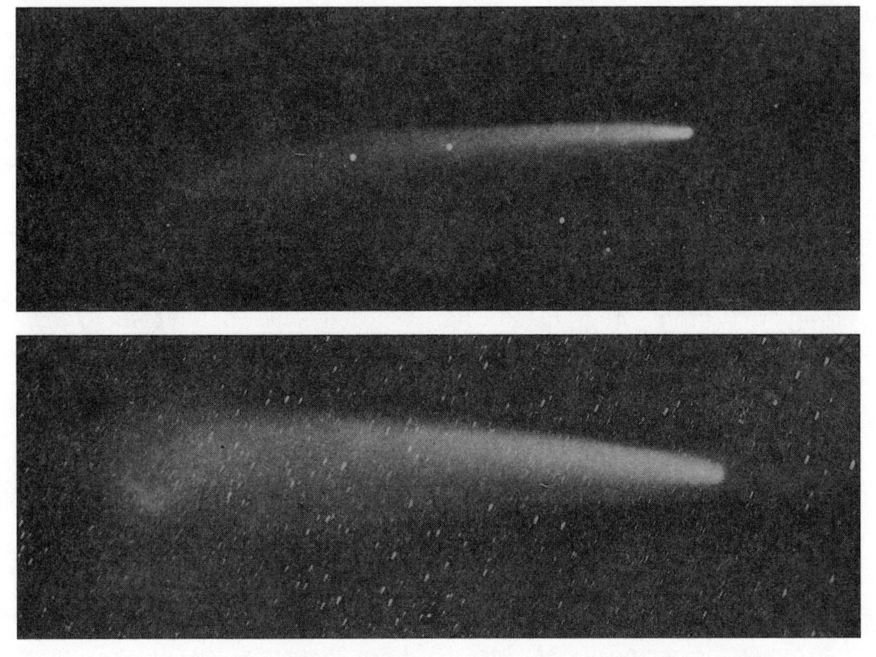

The Great Comet of September 1882, photographed by David Gill at the Royal Observatory, Cape of Good Hope.

designations; the multitude of fainter stars were anonymous. The need for unique stellar identifiers and accurate sky charts had become critical. Already a dozen asteroids had been discovered, whose orbits could be derived only by tracking their movement against a well-defined grid of stars. Variable stars had to be sorted out from their nonvariable neighbors, so they could be monitored for changes in brightness. "Stars not entered in [the *BD*] have no official existence," Agnes Clerke wrote in 1888. "Should they fade and vanish, the fact cannot be attested; should they brighten into conspicuousness, we are obliged to regard them as 'new' for lack of previous acquaintanceship. Whatever is known of the distribution of stars in space is founded on this grand enumeration."

With his own years of telescopic experience, David Gill was no stranger to the squint-eyed fatigue of visual star-position measurement. (The *BD* observers were relieved every one-and-a-quarter hours.) He realized that what the German astronomers had painstakingly assembled star by star might be captured in rapid fashion through a series of wide-angle photographs. Stitched together, the mosaic of chemical images would form a map of any region of the night sky. He envisioned a photographic extension of the *BD*, spreading its coverage all the way to the south celestial pole.

In late November 1882, with the great comet still prominent in the sky, Gill wrote to J. H. Dallmeyer of the noted optical firm in London to propose development of a low-distortion, wide-angle objective lens: "If we can get over the distortion in a *reasonable degree*, so as to get sharp . . . pictures of stars over a field of 10° square, here is a very easy way of making star maps for working purposes." Dallmeyer delivered a four-inch photographic objective within four months and a six-inch version the following year. In 1885, the Royal Society provided Gill with a three-hundred-pound subsidy to begin photography for his proposed *Cape Photographic Durchmusterung* (*CPD*). The exposed plates were shipped to the Netherlands, where the star images were machine measured by prison inmates under the supervision of the astronomer J. C. Kapteyn at the University of Groningen. When the Royal Society withdrew its funding in 1887, Gill paid for completion of the project out of his own pocket (after consulting his wife). The *CPD*, the first comprehensive *photographic* star catalog, listing positions and magnitudes of nearly 455,000 southern stars, was published in installments between 1896 and 1900.

Among the recipients of Gill's 1882 comet prints was Rear Admiral Ernest Mouchez, director of the Paris Observatory. Mouchez, too, contemplated the implications of the pictures' starry backdrops. His conclusions about the future of celestial photography would quickly be underscored by the activities of two of his staff astronomers. Paul and Prosper Henry were inseparable in their professional lives. A coworker recalled that the brothers were "so united that often at the Observatory we saw in them but one person. . . . [I]t is really impossible to discern what may belong to each one in their common work." Fourteen asteroid discoveries are credited to the Henry name: seven to each brother, in alternate succession, between 1872 and 1882. During this period, the Henrys were immersed in the creation of a visual map of stars along the ecliptic (roughly, the plane of the solar system), where asteroids—and possibly new planets—would most likely be found. Having arrived in their survey at the point where the ecliptic crosses the Milky Way, the brothers were stymied by the sheer number of stars to be hand-plotted: up to eighteen thousand on a thirteen-inch-square chart. The new dry-plate photography, they reasoned, might provide a route through these star-choked regions.

During their off-hours in 1884, the Henrys designed and built a six-inch refractor, mounted a camera, and trained it on the picturesque double star cluster in Perseus. The result was astonishing: star images reportedly so pinpoint that skeptical astronomers came from around Europe to inspect the plates. Seeing these finely rendered, twin celestial hives, Admiral Mouchez ordered the immediate development of a larger instrument. Within a year, the Henry brothers completed a boxlike thirteen-inch refractor capable of photographing sizable regions of the Milky Way with minimal distortion. A one-hour test exposure produced a mini-skyscape strewn with dots as small as a thousandth of an inch across: stars of magnitude fifteen. The same field, mapped sequentially by eye, would have involved months of labor at the telescope. (To the amusement of visitors, the Henrys optical "laboratory" was a rude shed on the observatory grounds.)

Although both David Gill and Ernest Mouchez had been steeped in traditional exact astronomy, they shared a progressive—and monumental—vision of the next step in celestial imaging: a comprehensive, pole-to-pole photographic chart of the heavens. This freeze-frame of the night sky at the end of the nineteenth century was meant to be their generation's legacy to astronomers in centuries to come. If stars vary over time, if they appear or disappear, if a faint planet lurks among its starry lookalikes—all would become evident when comparing observations from some future epoch to the archived photographs.

Gill and Mouchez knew that no single observatory or proximate group of observatories could assemble such an all-sky map. Earth's spherical form dictated that participants had to be recruited from around the world. Gill wrote his French counterpart in March 1886 to suggest that they convene an international gathering of astronomers to launch the vast project. For his part, Mouchez circulated among the astronomical community a "Proposal for Photographing the Heavens." The Astrographic Congress—the first-ever international conference of astronomers—took place in Paris in April 1887. Fifty-six delegates from nineteen nations gathered for eleven days to sketch out the creation of the great star map. Procedural standards were developed, committees were formed to supervise the work and adjudicate any technical or scientific disputes, progress reports and future meetings were scheduled. Each facility would bear its own costs for equipment, personnel, and eventual publication of results.

As to the map itself—the *Carte du Ciel*—there would be twenty-two-thousand long-exposure plates, each six inches on a side, covering a two-by-two-degree square on the sky and rendering of some twenty million stars down to fourteenth magnitude. Measuring the position and brightness of twenty million stars was not practicable, even if every astronomer on Earth pitched in. Therefore, a duplicate set of short-exposure plates would be obtained, providing measurement data for about two million stars to eleventh magnitude: the *Astrographic Catalogue*. After much wrangling over the merits of refractors versus reflectors, the delegates agreed to adopt the Henry brothers' thirteen-inch astrographic telescope as the project's standard.

Despite decades of experiments by George Bond, Warren De La Rue, Lewis Rutherfurd, Benjamin Gould, and David Gill, among others, affirming the absolute stability of photographic emulsions, it was decided that each plate would bear a photoprinted coordinate grid, which would shrink or expand in synchrony with the star images. Such measures were still insufficient to mollify critics who questioned whether stellar positions determined by multiple instruments and observers could be reliably knitted together. Other issues as fundamental as the chemical development of plates, exposure of the coordinate grid, and measurement of stellar brightness were not settled at the Congress (in the last instance, until decades afterward).

Swept up in a shared scientific–technological euphoria, eighteen observatories in Argentina, Australia, Brazil, Chile, England, France, Germany, Italy, Mexico, Russia, South Africa, and Spain agreed to participate in the massive mapmaking effort. Asked how long the project might take, Mouchez declared that six to ten years would bring the *Carte du Ciel* to completion. Gill estimated twenty-five years for the *Astrographic Catalogue* alone, at an annual cost of £10,000. Having neither the inclination nor the resources to pursue such a labor- and money-intensive collaboration, not a single American observatory signed on. Their research was already trending toward the astrophysical: solar studies, photometry, spectroscopy. As Edward S. Holden, director of the Lick Observatory in California, put it, "We should hardly be willing to bind ourselves to a programme which exacted so much routine work for so long."

In a further transatlantic affront, Harvard's Edward C. Pickering secured a $50,000 bequest in 1889 to build a twenty-four-inch photographic telescope in what the English journal *The Observatory* called a "rival scheme" to make his own celestial map. Its editors, Andrew Common and Greenwich astronomer Herbert Hall Turner, were particularly incensed that Pickering had made no mention of the Astrographic Congress in his 1888 campaign for support:

> . . . thus neither treated the resolutions with the respect to which they are entitled, nor put the case before the public in a fair manner. . . . Should any evil-disposed person or persons charge Prof. Pickering, in thought, word, or deed, with deliberately excluding such mention

> of the Conference from his appeal as likely to be prejudicial to his chances of success, we cannot acquit him of having materially contributed to this result. . . . He has tacitly condemned the work of the International Conference before it is commenced, and claimed success for himself.

Pickering warned his benefactor, Catherine Wolfe Bruce, of the brewing storm, adding, "It is the first time I have ever felt obliged to apologize on account of the expected *excellence* of a piece of work." Bruce's response was reassuring, if eccentric: "[W]hatever can be done to promote the work of that Pickering telescope will be clear gain. You know you have entire and undisputed control over that chunky Photo-Sterescope-Telescope."

The strong-willed Pickering never shied from controversy: he had only recently concluded a squabble with William Huggins over the wavelengths of spectral lines. (Modern reassessment of the data shows that both men were wrong.) Pickering's reply to the *Observatory* editorial is quietly defiant. He defends his fund-raising, as well as his science, point by point, citing his published intention to map the sky in the *Observatory*'s own pages in 1883. Nor does he see any scientific rationale against duplication of research with instruments of different optical designs. "It is difficult to believe," he concludes generously, "that the editors of the 'Observatory' intend their implication with regard to the work of the Bruce telescope in anything but a friendly spirit, or that they wish to discourage an experiment whose success can be determined only by trial."

David Gill wrote sympathetically to Pickering on November 4, 1889: "I am disgusted with the miserable, carping, envious attack made upon you in the 'Observatory' article . . . and still more so with its outrageous discourtesy and unscientific spirit. Your quite dignified article in reply I much admire." A year earlier, Gill and Mouchez had had their own heated exchange with the editors of *The Observatory*, who claimed that the Astrographic Congress had officially sanctioned only the *idea* of a star catalog, not its actual production. "[S]uch protests," Gill scolded, "may be

Edward C. Pickering.

couched in language less contemptuous and more conducive to good feeling amongst astronomers than that which you have employed. . . . I dispute the facts of the explanation, believing them to rest upon a misunderstanding, on your part, of what really took place at the Congress."

Each side stuck fast to its own interpretation of what had been agreed to at the Congress of 1887. That David Gill was an evangelist for celestial photography in general and for the *Astrographic Catalogue* in particular is certain. That he sped the development of the *Catalogue* faster than more conservative astronomers wished is probably true as well. A colleague remarked that Gill "had the consciousness of a power in him to accomplish great things. He felt that this gave him the right to demand all possible assistance to that end. And he was full of the indomitable energy which compelled support to his projects."

Against this contentious backdrop, and with a host of technical matters unresolved, the epic mapping project began. As the plates accumulated, the conversion of these chemical images into hard data proved troublesome. The stellar positions published by some institutions were afflicted with the sort of systematic errors the skeptics had foreseen. The Congress had tabled the issue of photometry—stellar brightness measurement—anticipating research that would define the relationships between length of exposure, size or opacity of a star's image, and the actual magnitude number astronomers use to quantify a star's brightness. These studies took far longer than anticipated. (In an ironic turnabout, the supervisory committee of the *Carte du Ciel* would ultimately adopt Edward Pickering's list of "standard stars" as its photographic brightness gauge. Pickering's "rival scheme" proved beneficial after all.)

As years, then decades passed, the financial and scientific burdens of the *Carte du Ciel* became increasingly oppressive. Participating observatories had saddled themselves with an expensive, laborious, and mundane task that effectively

Nuns measuring star positions on Carte du Ciel *plates at the Vatican Observatory around 1920.*

barred them from new avenues of research. Four observatories withdrew, forcing others to take up their assigned zones. Most of the participants were still working on their part of the *Astrographic Catalogue* well into the twentieth century, with the last remnant completed in 1964. The final product, scattered among 150 printed observatory reports, contains data for 4.6 million stars. By the original agreement, stellar positions are listed as a pair of x and y coordinates, relative to each plate's center. That is, an astronomer cannot simply look up the celestial latitude and longitude of a star, but has to compute them from a mathematical formula: a galling task during the precalculator, precomputer age. Until recently, little use was made of the data.

The *Carte du Ciel* fared worse than its catalog counterpart: only half of the plates for the great star map were ever exposed, and of these, just a small fraction actually engraved and printed. Meanwhile, twentieth-century technology overran the restrictive instruments and methods adopted for the *Carte du Ciel*. By 1916, astronomer Frank Schlesinger, at the Yerkes Observatory outside Chicago, was photo-mapping with a telescope whose field of view was six times the area of a *Carte du Ciel* plate. The larger plates captured a greater number of well-observed reference stars, against which the positions of others are gauged.

As the *Carte du Ciel* lumbered into the twentieth century, observatories in the United States pursued privately funded programs of astrophysical research and telescope building that arguably brought them to world dominance in these areas. In 1895, Edward Pickering articulated the American viewpoint as the explicit "advancement of the physical side of astronomy. While precise measures of position have not been neglected, the policy has been rather to undertake such studies of the physical properties of stars as would not be likely to be made at other observatories."

The *Carte du Ciel* proved that old astronomy plus new technology does not necessarily equate to new astronomy. Its hard-won photographic charts yielded no major discoveries and no insight into the physics of the stars themselves. Nor did the plethora of positional data leverage early twentieth-century science into more productive directions. (Tycho Brahe's sixteenth-century planetary observations—superficially, a pointless heap of numbers—allowed Johannes Kepler to derive the laws of orbital motion.)

Whereas traditional exact astronomers were, at base, mathematicians with a telescope (and now maybe a camera), the upcoming generation of scientists had scant interest in adding the nth decimal place to a star's position. Celestial cartographers viewed stars as luminous markers on a coordinate grid, while proponents of the so-called New Astronomy envisioned stars as distant suns, whose mass, chemical composition, energy sources, even life histories, might be divined through nascent technology. These astrophysicists were convinced that cosmic light held clues to the physical state of stars and nebulae. Their objective was the development of instruments, methods, and theories to best capture and interpret a celestial body's emissions.

Already by 1887, Princeton astronomer Charles A. Young ventured that "inquiries of this [physical] sort receive to-day fully as great an expenditure of labor, time, and thought as the older work of position-observation, and they are pursued with even a heartier zest." As headline-making astrophysics gradually displaced the more sedate classical astronomy, the *Carte du Ciel* receded in significance. What began as a grand boulevard to the glorification of positional astronomy narrowed over the decades into a patchwork path to the scientific hinterlands.

Despite its manifest failure as a wellspring of scientific progress, the *Carte du Ciel* did elevate celestial photography in the minds of professional astronomers. The institutional attention legitimized its use and made its perfection through experimentation an acceptable end in itself. Meetings of the *Carte du Ciel* committees swelled into educational forums in emerging methods and gave rise, in 1919, to today's International Astronomical Union. Even for positional astronomers, with their long tradition of visual observation, there was no turning back. The 1890s became an interregnum between the reign of the eye and that of the camera: twentieth-century astronomical research would be photographic.

By 1996, the entire *Astrographic Catalogue* had been keyboarded—in an earlier phase, keypunched—into the computer. This made possible a series of statistical analyses and wholesale corrections of the data, including many typographical errors. (The new *Astrographic Catalogue* is accessible at the website of the U.S. Naval Observatory.) The restructured database has allowed astronomers to measure with great precision the movements of stars since the turn of the previous century. Thus, to a modest degree, David Gill's vision for his great celestial map has been achieved. Nevertheless, astronomy historian David Evans wonders whether the *Carte du Ciel* was Gill's "greatest triumph or his grandest failure."

The *Carte du Ciel* sprang from the unexpected appearance of stars in a photograph of a comet. From this, astronomers were inspired to use the camera as an accessory for celestial measurement. But the close of the nineteenth century also highlighted the camera's potential as an agent of discovery. In

March 1887, during a long-term project to photograph and classify stellar spectra, Edward Pickering found that the most prominent spectral line of the star Mizar was, in fact, a double-line. The source of the spectrum, Pickering surmised, is not a single star, but a pair of stars—a binary system—so close together that they appear as a unitary disk in the telescope. Only through a photographic time exposure could this subtle spectral anomaly have been detected. A second spectroscopic binary was found two years later by Pickering's assistant Antonia Maury, niece of Henry Draper and one of Harvard's first woman astronomers.

On December 22, 1891, Maximilian Wolf, at the University of Heidelberg, noticed a small streak on an otherwise blemish-free picture of a star field. The streak, he realized, was an unknown asteroid, its orbital movement having traced out a stubby hairline on the time-exposed plate. Within two years, Wolf had photographed eighteen more asteroids, and during his lifetime 228. The original he named Brucia, after his American benefactor, Catherine Wolfe Bruce, who also funded Pickering's work. Other astronomers joined the photographic production line, boosting the languid rate of asteroid discovery beyond the capacity of the mathematicians charged with computing their orbits. (So distressed was one data analyst by the torrent that he dubbed these minor planets the "vermin of the skies.") And if asteroids were insufficient to motivate astronomers to take up the camera, might not the additional incentive of discovering a trans-Neptunian planet?

American astronomer Edward Emerson Barnard is credited with the first photographic discovery of a comet in a picture taken October 12, 1892, at Lick Observatory. The camera also revealed deep-space nebulae that had gone undetected by eye. The Orion Nebula and the cirrus-like veil of the Pleiades star cluster were fully delineated only with the advent of dry-plate photography. The famous Horsehead Nebula in Orion was first seen by Williamina Fleming, another of Edward Pickering's assistants (previously, his housemaid), on a photographic plate from 1888. Two years later, Maximilian Wolf discovered and named the continent-shaped North American Nebula.

But it was pictures of the controversial spiral nebulae that were to cement the acceptance of photography by professional astronomers. During the 1890s, Isaac Roberts published his glorious album of celestial images, including several spiral nebulae. Simultaneously, Andrew Common's cast-off three-foot telescope began its new life as a workhorse photographic instrument on a California mountaintop. Here it would harness the exquisite power of the chemical plate to open a new window onto the nature of the universe.

Chapter 12

An Uncivil War

> It is a remarkable and highly significant fact that most of the nebulae photographed with the Crossley reflector seem to be spirals.
>
> —James E. Keeler, Lick Observatory, 1899

Telescopes can have a long life. They might serve their original owner for decades, then fall into the hands of another starry-eyed enthusiast, a school, or a local astronomy club. But rarely has any backyard telescope traveled an ocean to occupy a mountaintop perch at one of the world's premier research institutions. Andrew Common's pioneering three-foot reflector, which thrust the photographic Orion Nebula into the public consciousness, made such a journey during the summer of 1895. From its second home, on Edward Crossley's Yorkshire estate, the telescope and its dome were dismantled—sixty-five-thousand pounds of iron, steel, and wood—and shipped to Lick Observatory on Mount Hamilton in California.

Lick Observatory on Mount Hamilton, California, in the 1890s.

The Crossley telescope was to be a working monument to its donor's generosity (which awoke only after Lick refused to meet Crossley's asking price). Although reflectors were still considered inferior to refractors, the largest-in-the-nation Crossley reflector was portrayed as a worthy complement to Lick's largest-in-the-world refractor, also three feet across. In appearance, there was no charitable comparison: the stubby Crossley was a bulldog, its long-tubed counterpart a greyhound. Yet the age of the large refractor was coming to an end; graceless as it was, the Crossley telescope represented the future of astronomical observing.

When James E. Keeler took up the directorship of the Lick Observatory in 1898, he walked into a civil war. On one side stood Edward S. Holden, Keeler's controversial predecessor, an energetic administrator and fund raiser, but at best, a mediocre research scientist. In his decade-long tenure, Holden had managed to alienate most of the observatory's staff, who variously referred to him as "the Dictator," "Prince Holden," or "that contemptible brute." On the other side of the conflict stood the ambitious William J. Hussey, lured to Lick from Stanford University with expectations, if not promises, of a fast-track to professional prominence. Hussey's unspoken goal—and the source of a desperate urgency—was to return to Stanford with funding for a world-class observatory.

And in the middle of this high-flown battlefield, an objective fought over but unwanted was Crossley's telescope. With proper restoration, Holden believed, the Crossley might become one of the premier photographic instruments in the world. Unlike the lens of a refractor, a mirror does not absorb the energetic violet rays that stimulated nineteenth-century photographic emulsions. Secondly, although their apertures were identical, the Crossley's compact form (technically, its fast focal ratio) was more favorable to photography than the stringbean shape of Lick's big refractor. Nor does a reflector suffer from chromatic aberration, which causes some refractors to cast rainbow fringes around celestial images. Nevertheless, Holden had no intention of tackling the project himself. Instead, he ordered his new recruit Hussey to bring the Crossley into operation.

Edward S. Holden.

Hussey was shocked: he had expected to continue Edward Barnard's photographic survey of the Milky Way

(Barnard resigned in 1895 over tensions with Holden), or perhaps even carry out observations with Lick's vaunted refractor. Hussey took one look at the homely reflector from England and pronounced it "a pile of junk." In its idiosyncratic mount—"as antiquated . . . as Noah's ark"—he saw only a fifteen-thousand-pound impediment to his career plans. In fact, the narratives of innovators like Henry Draper, Andrew Common, and Isaac Roberts did portray reflectors as a mechanic's telescope, in frequent need of manual corrections, optical alignment, and resilvering. Hussey's antipathy toward the Crossley was founded in the experiences of others, as well as in the mount's documented inadequacies. C. Donald Shane, Lick's director from 1945 to 1958, admitted that the original mount of the Crossley reflector was ". . . a monstrosity. Despite the elaborate system of platforms and ladders, it was almost impossible at times to reach the eyepiece. One observer suggested that the dome be filled with water so the astronomers could observe from a boat."

Hussey informed Holden that he would use the Crossley telescope—after its mount was replaced. Holden refused. The standoff escalated to the point that Hussey stopped speaking to Holden, communicating instead through terse, handwritten notes (as had Barnard before he left). The staff broke into factions, the majority allied with Hussey. Eventually the board of directors stepped in to mediate. The *San Francisco Chronicle* got wind of the impasse and ran a series of devastating articles, telling readers that a "peace commission is already on its way to the scene of the hostilities." When Holden had the Crossley's dome roped to the telescope mount to prevent it from blowing away in a storm, the *Chronicle* opined that the board of directors "rather wish the accident had occurred," taking the telescope with it.

News of the controversy raced across the Atlantic, where the editors of *The Observatory* magazine declared, "[T]here is civil war on top of Mount Hamilton." They highlighted the affront to Andrew Common's achievement and to Edward Crossley's generosity, then added in a metaphorical stretch, "Why should a Derby-winner end his days between the shafts of a four-wheeler? "

The board of directors eventually sided with Holden, but to no real effect. Hussey made a desultory attempt to observe with the Crossley telescope, but only visually, ignoring its photographic potential. On January 1, 1898, after staff relations worsened and with the threat of a lawsuit against him, Holden resigned his post. Hussey stayed on at Lick until 1905, when he left to oversee the expansion of the observatory at the University of Michigan. His first purchase: a three-foot reflector telescope.

Holden's successor, James Keeler, was no stranger to Mount Hamilton when he took up the Lick directorship in June 1898. He had been a staff astronomer there from 1886 until 1891, setting up Lick's regional time service, witnessing construction of the thirty-six-inch refractor, and fabricating a spectroscope for its use. The lowlands-bred Keeler would have reveled in his lofty vantage point on the Milky Way, which rises vertically from the horizon every August. All around the close-packed hive of high technology was a "sparsely wooded mountain land where are met only deer, skunks, squirrels, roaming astronomers, coyotes, an occasional mountain lion, a few horses, and, in summer time, rattle snakes, while the big, black buzzards sail high in the air on the outlook for a little food."

With a dual degree in physics and German from Johns Hopkins, graduate education in Heidelberg and Berlin, and a string of noteworthy publications from his tenure as director of Allegheny Observatory, Keeler was hailed as one of America's premier astrophysicists, equally adept in the theoretical and observational realms. He was the rare scientific administrator who engendered respect through his own professional accomplishments, and spurred his charges through a combination of modesty and charming inquisitiveness. As one colleague put it, "Keeler doesn't claim to know everything."

Keeler's top priority was to settle the Crossley debacle, which had escalated beyond all reasonable proportion. The Lick Observatory was portrayed in the popular press as an unproductive aerie of bickering hacks, so self-absorbed, claimed one newspaper, that they would miss a comet if it flew right over their heads. And the reputation of Edward Crossley, who sat on the Lick's board of directors, continued to be tarnished by the all-too-public squabbles over his eponymous telescope. Within a month of his arrival, Keeler defused the situation: he reassigned William Hussey to more desirable duties and took charge of the problematic reflector himself. An immediate test of the mirror's figure revealed it to be nearly flawless. Under the limpid skies of Mount Hamilton, the telescope might yet fulfill its promise, if only unfettered by its scandalous mount. The Crossley, Keeler suspected, was a thoroughbred trapped in the body of a nag.

In 1900, Keeler published a twenty-five-page article in the *Astrophysical Journal* detailing a host of incremental refurbishments, parts swaps, and procedural tweaks applied to the Crossley telescope. The long

James E. Keeler.

tally of labor leaves no doubt that Keeler had stripped the ungainly beast down to its bones and observed every component in action—or inaction. A new, more powerful clock drive advanced the massive tube without its predecessor's frustrating palpitations. Whatever irregularities remained were overcome by the more or less constant intervention of an assistant. Keeler's article left the impression of an almost comic duo: the immobile astronomer, peering into the camera's guiding eyepiece; and his frenetic assistant, clambering over the superstructure to apply the requisite adjustments.

Keeler's appraisal of the telescope's checkered history is diplomatic in the extreme. In a footnote, he neatly absolves everyone of responsibility in the Crossley affair, and pays a tangential compliment to Andrew Common for having accomplished what he did: "The difficulties here referred to, about which a good deal has been written, seem to have had their origin in the fact that it was impossible, at the time of the preliminary trials, to provide the observer with an assistant, while the Crossley reflector is practically unmanageable by a single person."

Once its mechanical demons were exorcised (or at least subdued), Keeler confirmed his suspicion that, as a photographic tool, the Crossley telescope was superior to Lick's great refractor. Not only did it have a wider field of view—one degree across—its compact optical path more effectively concentrated light onto its 3¼" × 4¼" plates: a three-hour exposure of the Orion Nebula with the refractor showed less detail than a ten-minute exposure through the Crossley. Keeler also noticed a photographic boost from the mirror's silvered surface, which, unlike a glass lens, absorbs few of the violet rays that activate the plate's chemical emulsion. He writes, "On one of the fine nights . . . when the Milky Way shines with astonishing splendor and the whole heavens look phosphorescent, the photographic activity of the reflector is remarkably increased."

In its mountaintop setting, Andrew Common's twenty-year-old telescope was reborn with a vigor that would have delighted its creator. The central star of the Ring Nebula in Lyra, a challenge to photograph at the time, rendered a distinct image in just one minute. Stars of magnitude seventeen, beyond the light grasp of most 1890s-era telescopes, appeared on the plate in only ten minutes. A thirty-second exposure—practically a snapshot in the astronomical sense—was sufficient to capture the subtle nebulosity enveloping the Pleiades. Nearly fifty-four-hundred stars speckled a two-hour plate of the globular star cluster Messier 13 in Hercules.

Keeler coaxed the Crossley to successively longer exposures, culminating in a series of celestial portraits, each four hours in duration. The nominal

Lick Observatory's thirty-six-inch Crossley reflector on its original mounting, 1898.

targets were a gallery of well-known objects, yet Keeler's eye was drawn to the large number of lesser wisps surrounding them. Like the spray of stars that had unexpectedly dotted David Gill's photographs of the comet of 1882, each of the Crossley plates displayed a host of never-before-seen nebulae: typically eight to ten, with one exposure showing thirty-one. Keeler estimated that the nebular population within range of the Crossley over the entire sky surpassed 120,000, some ten times the number previously cataloged. And, if the new images were any indication, the vast majority of these presumed deep-space denizens were spiral in form.

The release of Keeler's photographs galvanized the astronomical community. The spiral nebulae, a celestial species once considered rare, turned out to be among the most populous breeds in the astronomical zoo. Their explication took on an urgency, spurred by their sheer weight of numbers. The dark and seemingly empty fields between the stars now sparkled like untapped veins of scientific opportunity. In these mysterious whorls of light lay, in Thoreau's phrase, "winged seeds of truth." But of what truth? If a reflector of such humble origin and scale uncovered hundreds of thousands of spiral nebulae, how many more might be revealed by a generation of larger, more sophisticated instruments? Might the universe be characterized as much by the ubiquity of its spiral nebulae as by that of its stars? Fundamentally, does our cosmos consist of a single island of luminous matter amid the void or is it a vast ocean of space containing multitudes of galactic systems like our own?

Having proven the Crossley's worth as a surveyor of spiral nebulae, Keeler next made plans to equip it with a spectrograph. The telescope's fast optics and mountaintop site would confer a distinct advantage over its sea-level rivals in recording the feeble trace of a nebular spectrum. Telltale features in the spectrum, Keeler hoped, might disclose something of the spirals' physical nature, specifically, whether they consist of stars or diffuse gas. The spectrograph was completed on July 30, 1900, but Keeler never got to use it. He descended the mountain that day to seek medical treatment for a longstanding heart ailment. Eleven days later he was dead. James Keeler's final journal publication bore the title "The Crossley Reflector of the Lick Observatory"—a fitting tribute to the indomitable bond between an astronomer and his telescope.

From its crude beginnings in the 1840s, celestial photography evolved over a span of six decades into an essential research tool in astronomy. Its earliest practitioners proceeded on the dubious, yet irresistible, prospect that their technological prowess could surmount the manifold hurdles of low-light imaging. Their grainy frames were remarkable for the time, each one prized as a chemical reliquary of a star's luminous essence. The initial decades were the province of amateur enthusiasts, whose progress stemmed from equal measures of ingenuity, perseverance, and cooperation. The heavenly scenes grew richer—first approaching, then challenging, and finally surpassing the capability of the human eye at the telescope.

With the application of the dry plate in the 1880s, what had been a noisome, exasperating art became a predictable, mainstream technology that would eventually recast the telescope as an adjunct of the camera. One-time Lick Observatory director Edward Holden noted that the camera "does not tire, as the eye does, and refuse to pay attention for more than a small fraction of a second, but it will faithfully record every ray of light that falls upon it even for hours and finally it will produce its automatic register . . . [that] can be measured, if necessary, again and again. The permanence of the records is of the greatest importance, and so far as we know it is complete . . . We can hand down to our successors a picture of the sky, locked in a box."

Mindful of the accomplishments of their venturesome colleagues, professional astronomers increasingly embraced celestial photography as an essential observational astronomical tool. Together, Isaac Roberts's deep-sky portraits, Edward Emerson Barnard's panoramic views of the Milky Way, and James Keeler's multitude of spiral nebulae brought a new clarity to the architecture and demography of the universe. As the twentieth century unfolded, photography's expanding reach spurred development of a new field of cosmic exploration: extragalactic astronomy.

Yet pictures alone were insufficient to answer the sorts of questions posed by practitioners of the "New Astronomy." These academically trained scientists sought to apply the laws of physics to the inner workings of stars and nebulae. They stood ready to pivot on new methods or cross disciplinary boundaries, if it enabled them to decode the cosmic light swept up by their telescopes. As crisp as it might appear, the image of a heavenly body betrays the fundamental limitation of a picture: No matter how much it magnifies a distant world, it reveals little to nothing about the chemical composition or physical state of a planet's soil, a comet's tail, or a star's atmosphere. The indiscriminate cluttering of light upon the photographic plate—the visual equivalent of simultaneously tuning in every radio station on the dial—renders only the *form* of a celestial object.

While celestial photographers labored to improve chemical emulsions and telescope drives, their astrophysical counterparts explored a promising, often confounding, realm of analysis: dispersing starlight into its constituent colors before it strikes the plate. Astronomers suspected that this faint array of colored bands—the star's spectrum—might be used as a laboratory-like probe into the star's elemental makeup and physical properties. Decades of research proved that to be the case. Modern astronomy owes its refinement not only to the introduction of the camera, but to the concurrent development of an equally marvelous instrument: the spectrograph.

Part II
SEEING THE LIGHT

The physicist and the chemist have brought before us a means of analysis that . . . if we were to go to the sun, and to bring away some portions of it and analyze them in our laboratories, we could not examine them more accurately.

—Warren De La Rue, "Proceedings of the Chemical Society,"
June 20, 1861

Chapter 13

The Odd Couple

> The most important discovery made by Bunsen during the short duration of his residence in Breslau was the discovery of Kirchhoff.
>
> —Henry Enfield Roscoe, Bunsen Memorial Lecture, Chemical Society, London, 1900

Robert Wilhelm Bunsen was fearless in the laboratory, even after the chemical explosion in 1836 that shattered his face mask and cost him the sight in his right eye. Bunsen routinely investigated toxic substances like the arsenic compound whose smell, he reports, "produces instantaneous tingling of the hands and feet, and even giddiness and insensibility . . . [while] the tongue becomes covered with a black coating." The arsenic study nearly killed him. On at least one occasion, he had to be roused from unconsciousness brought on by fumes from his noxious brews. So calloused were his hands from chemical exposure that he declined protective aids. Bunsen's protégé, English chemist Henry Roscoe, remarked that his mentor "had a very salamanderlike power of handling hot glass tubes, and often at the blowpipe have I smelt burnt Bunsen, and seen his fingers smoke!"

Having established his bona fides at Göttingen, Cassel, Marburg, and Breslau, Robert Bunsen was practically a legend when he was brought in to lead the chemistry institute at the University of Heidelberg in 1852. At twenty-three, Bunsen had developed an antidote to arsenic poisoning by chemically rendering the toxin insoluble in bodily fluids. His exhaustive study of organic compounds known as cacodyls had placed him in the top rank of experimentalists by the mid-1840s. He had reshaped the economics of the iron-making industry in both his native Germany and in England by improving the efficiency of blast furnaces; explored the geochemistry of Icelandic volcanoes and geysers, descending into the vents to take temperatures; and created a low-cost carbon-zinc battery, as well as a forerunner of the carbon arc lamp. It was at Heidelberg that Bunsen would fashion the hot-yet-colorless laboratory burner that bears his name—and where he would help transform the astronomical telescope from a mere eye on the heavens into a cosmo-chemical probe.

The depth and breadth of Bunsen's published works evinced a relentlessness of purpose perhaps unique in the chemical world. In 1844, Swedish chemist Jöns Jacob Berzelius, the foremost experimentalist of his time, urged a colleague to pursue an investigation "with the true perseverance of a Bunsen." By midcareer, the Heidelberg chemist had risen to cultural icon. A character in Ivan Turgenev's 1862 novel *Fathers and Sons* confesses how she longs to travel, not only to Paris, but to Heidelberg. "Why Heidelberg?" asks her puzzled companion, to which she replies, "Good heavens, Bunsen's there!"

During construction of Heidelberg's new state-funded chemistry building—which he negotiated into his hiring package, along with the university's second highest salary—Bunsen set up in an abandoned monastery. The age-old house of God was repurposed with the secular implements, rituals, and agents of modern science. The former refectory became the main laboratory, while the chapel served as both lecture hall and storeroom. To accommodate the rush of students eager to work with Germany's star chemist, the cloisters were subdivided into experimental alcoves. All of the specialized glassware was manufactured on site, much of it hand-blown by Bunsen himself. Until the installation of gas and running water, the researchers heated their concoctions with spirit lamps and charcoal fires and worked the pump handle in the courtyard. "Beneath the stone floor at our feet slept the dead monks," was Henry Roscoe's dispassionate take on the abbey-turned-laboratory, "and on their tombstones we threw our waste precipitates."

Bunsen, a lifelong bachelor, was married to his occupation. He rose before dawn to pen scientific papers, delegated little work to assistants, and dedicated considerable vacation time to industrial or geochemical field studies. Even in repose, Bunsen was on the job: having drifted off during a colleague's lecture, he jerked awake and whispered to Henry Roscoe with relief, "I thought I had dropped a test-tube full of rubidium onto the floor!"

A broad-shouldered six-footer, "Papa" Bunsen, as he was affectionately known to his students, was the amiable lord of the laboratory, ambling from table to table, admonishing trainees for their sloppy technique or alerting them to potential dangers. On occasion, experimenters arrived in the morning to find their project several steps advanced from where they had left it the night before. Bunsen radiated enthusiasm about even seemingly mundane aspects of scientific exploration. "It was quite in keeping with his nature," recalled a former student, "that others should partake of the infinite pleasure he had experienced."

When a second laboratory explosion in 1869 threatened Bunsen's remaining eyesight, a crowd gathered on the plaza outside his home until his physician announced from the balcony that the patient was all right. That evening Bunsen's students returned in a torchlight procession and serenaded their beloved professor with a rendition of *Gaudeamus Igitur*.

At its peak, the Heidelberg laboratory held upwards of seventy students and visiting chemists from around the world. Among them were such chemical giants as Dmitri Mendeleev and Julius Lothar Meyer, cocreators of the periodic table of the elements. Of the estimated hundred Americans who passed through the laboratory, thirty-three obtained their doctoral degree under Bunsen, and forty-six became chemistry professors themselves.

Bunsen had neither the time nor the inclination for a family life. Asked by one faculty wife why he hadn't married, he replied, "Heaven forbid, when I should return at night, I should find an unwashed child on each step." (The staircase in Bunsen's home rose twenty-five steps to the second floor.) Housekeeping was a task reserved for the laboratory. At home, Bunsen tossed unopened mail into a vacant room, to be sifted through every few weeks by a subordinate. Arrayed around the baseboard of another room was his accumulation of well-worn boots. So rarely did he host social gatherings that a group of faculty wives commandeered his house and invited him to his own party.

Modesty was Bunsen's hallmark. Only on mandatory occasions did he don his waistcoat full of medals; these he hid from public view by buttoning up his overcoat, even during summers, and avoiding well-traveled streets. "The only value such things had for me," he offered late in life, "was that they pleased my mother, and she is now dead." (Celebratory gatherings were not completely wasted on Bunsen: he gathered up used champagne corks to bring back to the laboratory.) In his lectures, he never took credit for his own discoveries, declaring instead, *It has been found . . .* The prominent English physicist John Tyndall summed up Bunsen as "the man who came nearest my ideal of a University teacher. He was every inch a gentleman, and without a trace of affectation or pedantry."

At the same time, Bunsen had a knack for the bon mot. When a student tried to explain his repeated absences by insisting that he was hidden by a pillar, Bunsen replied, "Ah, so many sit behind the pillar." On a trip to England in 1862, he was mistaken by a woman for the deceased Baron von Bunsen, a noted Prussian scholar and diplomat. Asked whether he had completed his great work, *God in History*, the chemist Bunsen lamented, "Alas, no, my untimely death prevented me from accomplishing my design."

Once at Heidelberg, Bunsen used his carbon-zinc battery to electrically decompose chemical compounds into their constituent elements, producing ultrapure samples of chromium, manganese, magnesium, aluminum, sodium, barium, calcium, and lithium. Despite appearances, Bunsen was no production-line automaton, content to churn out reference data and chemical specimens; his thinking was interdisciplinary, and he reached freely into physics, geology, and mathematics. (His math teacher was the eminent Carl Friedrich Gauss.) "A chemist who is not a physicist is nothing," Bunsen asserted. Pursuing his physics muse, Bunsen joined Henry Roscoe in the darkened attic of the chemistry building in what would become a decade-long study of the photochemical action of sunlight. The interaction of light and matter intrigued Bunsen, who wondered whether elements might be uniquely identified by the color of light they emit when heated to incandescence. It was for these explorations that he developed his eponymous burner, with its hot, yet virtually colorless, gas flame.

From the start, Bunsen leveraged his renown to promote the growth of the physical sciences at Heidelberg. Through the 1850s, fully 97 percent of the university's state funding went to chemistry. In addition to being director of the chemistry laboratory, Bunsen was dean of the philosophy faculty (precursor to the modern-day college of arts and sciences), a position through which he exerted considerable power to shape Heidelberg's scholarly future. Mid-nineteenth-century Germany was fragmented into a confederation of kingdoms, grand duchies, and electorates—Baden, Prussia, Bavaria, Hesse-Darmstadt, plus a host of others—each vying for cultural supremacy, if not outright power. In this rivalry, the reputation of universities served as a proxy for more typical economic or military measures of success. Prestige was conferred upon a state by the eminence of its resident scholars—especially scientists. Substantial salaries, cultural amenities, and research support were proffered to convince elite scientists to join the "team." Personnel shuffled from Heidelberg to Munich, Munich to Berlin, Berlin to Potsdam, as the respective state treasuries sweetened their enticements. No academic was more politically adept than Bunsen at securing government funds toward this end. At times, he led hiring forays into new areas of research; when instead he sensed the Baden government trending toward established fields, he promoted candidates for these positions.

In 1854, when a faculty slot opened up, Bunsen pressed the Baden government for the appointment of a novel sort of researcher: equal parts physicist and mathematician, a scientist unconstrained by the experimental laboratory or by disciplinary boundaries. In this effort, he was opposed by his colleagues, who advocated for an industrial or a biological physicist. When the haggling was over, Bunsen's candidate prevailed: Gustav Kirchhoff, a thirty-year-old physicist at the University of Breslau in Prussia, whose mathematical skill and cognitive potency had impressed him when their paths had crossed two years earlier. Together, the pair would form one of history's great scientific partnerships—and would alter the course of cosmic exploration as profoundly as the advent of the telescope.

Gustav Kirchhoff was destined for an extraordinary life—at least so he had thought when not wracked with self-doubt over his height and his mathematical aptitude. He had written dutifully to his parents and to his brother Otto, revealing cracks in a facade otherwise burnished by academic excellence. "I am currently quite annoyed at my slight stature," Kirchhoff once confided to Otto when he was eighteen, "and would enjoy the university more if my size and age were in better agreement." This adolescent lack of confidence persisted throughout Kirchhoff's adulthood, according to physicist Emil Warburg, who knew him for many years.

Burdensome as they were, Kirchhoff's neurotic demons never managed to seize up his creativity. While an undergraduate at the University of Königsberg in 1845, Kirchhoff had developed the now-famous set of algebraic rules that enable the calculation of voltages and currents in multiloop electric circuits. So impressive was this achievement that the university awarded him a doctoral degree a year later at the age of twenty-two.

Following a two-year stint as an unpaid lecturer and scientist-gadabout in Berlin, Kirchhoff found himself, in 1850, teaching introductory physics at the University of Breslau (today's Wrocław, in Poland). Compared to the scientific bustle and urban energy of Berlin with its half-million inhabitants, his new locale was a letdown. Kirchhoff labored in anonymity on his electrical research, finding few like-minded colleagues among Breslau's faculty. But a job is a job, he had reluctantly come to admit, and bread has to be put on one's table.

Kirchhoff had been hired as an experimental physicist, a far remove from the abstract, mathematical physics that fired his curiosity and filled his brief résumé. At the time, Germany's academic centers valued laboratories, equipment, and measurement; the exploratory potential of paper-and-pencil theoretical physics had yet to be established. To Kirchhoff, mathematics was the language of nature, its symbols and operators the means by which the physicist inquired of nature's design. Through mathematical modeling and analysis, the theorist develops testable predictions of the behavior of, say, matter or light. If verified by experiment, the proposed theory is broadened or extended to more complex situations. Thus, science is a cyclical process of experiment driven by theory and theory driven by experiment, leading to discovery of the laws that govern the physical universe.

But experimental, not theoretical, research was what Breslau was paying Kirchhoff to undertake. Disheartened, he wrote hopefully to his father in Königsberg, "It will do me some good to move into experimental work, just as a plant grows stronger when it's moved into new soil." Despite his limited prospects, Kirchhoff's tepid optimism would prove prophetic. The new soil in which he would flourish was soon to be tilled, not in Breslau, but in Heidelberg by Robert Bunsen.

Kirchhoff would undoubtedly have noticed the middle-aged, bearish man who materialized in his introductory physics class in 1851. To count the renowned Herr Bunsen among his listeners—not only that afternoon, but as it happened, for the rest of the semester—was a distinct honor for a young professor barely a year into his post. It was also one of few bright spots in what had otherwise been a depressing term at Breslau, a circumstance compounded by a conflict with his senior colleague in physics. "My stay in Breslau has recently become more pleasant," Kirchhoff briefed his mother in May 1851. "At the beginning of the semester arrived the new chemistry professor, Bunsen, previously in Marburg. I find his manner very appealing and we have had wonderful times together. He is a man of extraordinary kindness."

Kirchhoff and Bunsen stood at opposite poles, both physically and temperamentally. The sight of the pair strolling the streets of Breslau—and later Heidelberg—drew attention, as though one were witnessing the encounter of two different species of man. The mountainous Bunsen stood nearly a head taller than Kirchhoff, even without his trademark stovepipe hat. Where Bunsen was voluble and imposing, Kirchhoff was soft-spoken and wiry as a gymnast (which he was in his youth). Bunsen radiated confidence, while Kirchhoff held close his conflicted self-assurance. But in the arena of scientific debate, Kirchhoff could dispatch any of his peers with authority, yet without offense, a talent that delighted his gregarious companion.

Each day, after Kirchhoff's lecture, the two men gathered other professors and headed over to the beer hall for an evening meal and quaff. Kirchhoff and Bunsen were inseparable: they took daily constitutionals to discuss developments in science, sat next to each other at the theater, and traveled together during school holidays. It was during his walks with Kirchhoff, Bunsen reported, that his best ideas came to him.

Gustav Kirchhoff, Robert Bunsen, and Henry Roscoe, photographed in Manchester, England, in 1862.

Kirchhoff was distressed when Bunsen left Breslau in August 1852 to head the chemistry institute at Heidelberg, some five hundred miles away. The two corresponded frequently until 1854, when Bunsen led with stunning news: he had persuaded the Baden government to appoint Kirchhoff to Heidelberg's recently vacated senior faculty position in physics. Kirchhoff would become director of his own physical institute, subordinate to no one. (Aware of Kirchhoff's insecurities, Bunsen advised him to stifle his habitual self-effacement when dealing with the ministry.)

Before extolling nominee Kirchhoff's virtues, the ever-shrewd Bunsen informed the Baden government, with evident regret, that the "first notables of science" in Berlin, Göttingen, Königsberg, and Vienna would never decamp to Heidelberg, given its deficient experimental facilities. However, a junior faculty member—namely, Kirchhoff—might be induced to accept a job offer despite the shortcomings. Kirchhoff's résumé was sold on his dual strength in physical theory and experiment, a combination that opened the door to fruitful, cross-disciplinary research collaborations with Bunsen. In his request to the ministry, Bunsen quoted a remark by Wilhelm Weber, a prominent physicist, that "two scientists who by working together multiply their achievements . . . lends a special radiance to the university where it is achieved." In fact, Weber's hyperbole underestimated the synergistic power of the collaboration.

In every way, the Heidelberg position was a step up for Gustav Kirchhoff: in title, from assistant professor to full professor; in salary, from 1,050 florins to 1,600 florins, plus 400 florins for housing; in prestige, from a regional university to an internationally acclaimed academic center; in focus, from experimental physics to theoretical physics—or any combination of the two he desired. In October 1854, with anxiety tempering his joy, Kirchhoff set off for Heidelberg and his much-anticipated reunion with Bunsen. (Three years later, Bunsen further elevated Heidelberg's scientific stature by hiring Hermann von Helmholtz, who performed seminal studies on the conservation of energy, fluid mechanics, electromagnetism, and thermodynamics.)

Kirchhoff and Bunsen took up their social routine where they had left off, their daily outings now relocated to the so-called Philosopher's Walk on the southern slopes of the Heiligenberg, across the Neckar River from the university. In 1857, when Kirchhoff married Clara Richelot, daughter of his former math professor, Bunsen found himself a welcome adjunct to the new couple. The three of them joined a dramatic reading group in which they played roles from Goethe, Schiller, and Shakespeare. To the Kirchhoffs' children, Bunsen was "Onkel Hofrat"—from his honorific as a government councilor—who arrived every Christmas Eve bearing an armload of gifts. Clara tried to wrest Bunsen from his near-total immersion in his work, with marginal success. If anything, she was undermined by her own husband, whose research partnership with Bunsen evolved into an all-consuming passion for both men.

During the late 1850s, driven in part by his photochemical investigations with Henry Roscoe, Bunsen sought to identify chemical elements by their distinctive color when ignited in his gas burner or an electric arc. The results proved equivocal. Whether viewed directly or through filters, the various colors of combustion were typically too similar to differentiate between substances. Kirchhoff observed the experiment and at once recognized its shortcomings: The luminous emission of any incandescent element is an admixture of many colors, or wavelengths, all of which flood the eye simultaneously. What Bunsen needed was a spectroscope to sort out the constituent colors by their wavelength *before* they enter the eye. Instead of perceiving a riotous blend of colors, observers would see an ordered spectrum: various hues situated in different positions along a viewing plane, akin to reassembling a group of people alongside each other according to height.

Bunsen's goal—now shared by Kirchhoff—was to demonstrate that the spectrum of a given element is unique and can therefore be used to identify an element remotely by its light alone. Yet for Kirchhoff, there was an additional challenge: to explain the physical basis of the emission and absorption of light from matter. Here, in 1859, at the crossroads of chemistry and physics and of theory and experiment, the duo embarked on their first scientific collaboration.

Chapter 14

WHAT'S MY LINE?

> The world is moved along, not only by the mighty shoves of its heroes, but also by the aggregate of the tiny pushes of each honest worker.
>
> —Helen Keller, *Optimism: An Essay*, 1903

RAINBOWS ARE SEEMINGLY MAGICAL APPARITIONS: fleeting, insubstantial, their twin footings at once anchored to the horizon yet responsive to one's movements. Poised between stormy gloom and sunbeam daylight, rainbows infuse the sky with ghostly colors, arranged with a geometric perfection that charms the eye and stirs the soul. No wonder people have puzzled over their origin and infused them with religious and mythological significance. Through the ages, proto-scientific observers have postulated a host of physical causes, some far afield, others quite close to the mark. A rainbow is nature's most flamboyant expression of the broad phenomenon of color, an aspect of light that astronomers gradually harnessed to great advantage.

The multicolored character of light has been studied with man-made contrivances since at least the thirteenth century. Trying to imitate the action of a raindrop, Leonardo da Vinci placed a water-filled glass globe in his sunlit window and noted the array of colors cast onto the floor. In an appendix to his 1637 *Dioptrique*, Rene Descartes sketched the formation of an artificial rainbow by a triangular glass prism, a scheme more clearly depicted in Robert Boyle's treatise *Colours*, from 1664. Two years later, inspired by Boyle's work, Isaac Newton directed a sunbeam from a hole in the window shade into his darkened room and through a prism. On the opposite wall was projected a multicolored band, formed from overlapping images of the round aperture in the window shade. In his report to the Royal Society in London, Newton stated that, "Light it self is a *Heterogeneous mixture of differently*

In a darkened room, a prism decomposes a beam of sunlight into its constituent colors; a second prism repeats the process for a narrow section of the original spectrum.

refrangible Rays." Within this amorphous *spectrum*, as he called it, Newton identified precisely seven primary colors (violet, indigo, blue, green, yellow, orange, and red), a number without physical significance, but chosen to accord with the seven notes of the major scale in music. Again taking a lead from Boyle, he refocused the dispersed colors to restore the original white beam of sunlight.

During the mid-eighteenth century, Scottish scientist Thomas Melvill observed that a flame, when sprinkled with substances such as sea salt, saltpeter, or potash, yields a spectrum that is not continuous like the Sun's, rather a sequential array of discrete, colored bands. Half a century later, English physician William Hyde Wollaston closed his medical practice to become a scientist, funding his diverse research from the invention of a method that rendered platinum malleable. Wollaston developed an optical rig to gauge the refractive power of translucent materials, such as glass, amber, melted spermaceti, even the lens of an ox-eye. To illuminate his samples, he directed sunlight through a narrow slit, before dispersing it with a prism. The spectrum projected by a prism is a composite of contiguous or overlapping colored images of the luminous source, whether a candle flame, a glowing lightbulb filament, or—in both Newton's and Wollaston's setups—a sunlit aperture. A mere twentieth of an inch wide, Wollaston's slit-shaped aperture produced a sharper spectrum than Newton's original round aperture, whose projected colors bled into one another. (Newton mentioned his use of slit apertures in the 1704 treatise *Opticks*.)

Wollaston found that the Sun's apparently seamless spectrum is, in fact, interrupted at intervals by dark lines, that is, partial or complete absences of color. He (mistakenly) supposed that these lacunae are natural divisions among what he regarded as the solar rainbow's four primary colors, reduced from Newton's proposed seven. Wollaston next revisited terrestrial sources of illumination, clarifying Thomas Melvill's earlier observations. "By candlelight," he reported to London's Royal Society in 1802, "a different set of appearances may be distinguished. When a very narrow line of the blue light at the lower part of the flame is examined alone, in the same manner, through a prism, the spectrum, instead of appearing a series of lights of different hues contiguous, may be seen divided into 5 images, at a distance from each other. The 1st is broad red, terminated by a bright line of yellow; the 2nd and 3rd are both green; the 4th and 5th are blue, the last of which appears to correspond with the division of blue and violet in the solar spectrum."

In other words, the solar spectrum is a multihued field, riven by a series of dark lines, whereas the spectrum of a flame or an electric arc is a series of bright lines against an otherwise dark field. This apparent complementarity does not presume that the two types of spectra are mirror images of one another—that each dark line in the Sun's spectrum aligns with a bright line in the spectrum of a flame. Yet in the closing phrase of his description, Wollaston does cite one example of an evident coincidence in the spectral

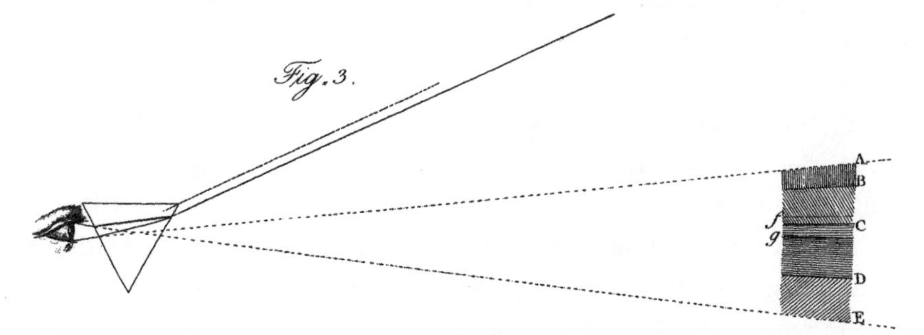

Solar spectral lines, as depicted by William Wollaston in 1802.

positions of a solar line and a terrestrial line. The relationship of dark lines in the Sun to their bright, laboratory counterparts would perplex astronomers, physicists, and chemists for decades to come.

From an instrumental perspective, the birth of solar and celestial spectroscopy dates to 1814, when master optician Joseph Fraunhofer magnified the Sun's spectrum with a small telescope. Adversity had dogged Fraunhofer for much of his youth. Orphaned at age eleven, he found himself bundled by his guardian into the back of a Munich-bound mail wagon, on his way to an apprenticeship with a glass cutter. The household proved to be a virtual prison, and the supposed apprenticeship a six-year term of domestic servitude. Books were forbidden, as well as a candle for his windowless room.

On July 21, 1801, when Joseph Fraunhofer was fourteen, the four-story building in which he worked collapsed. By chance, he had been standing beside a pile of crates, which arrested the fall of a beam that would have crushed him. Four hours later, he was extricated under the watchful eyes of Maximilian Joseph, Prince-Elector and eventual King of Bavaria. Fraunhofer found himself a sudden celebrity, his rise from the ruins a phoenix-like symbol of rebirth amid the nation's economic and political turmoil. That he was an orphan—indeed a fragile-looking one—only heightened the tale's flash. So disquieted was Maximilian by Fraunhofer's plight that he invited him for a private visit at his estate.

The royal palace at Nymphenburg, in the west of Munich, is a sprawling, red-roofed edifice surrounded by gardens, fountains, and expansive lawns. The unworldly Fraunhofer was surely rapt in wonder as he approached the central château, past the reflecting pool and the sculpted forms of Greek deities. His entrance into the building's Great Hall would have been equally jaw-dropping. Exuberant frescoes adorned the walls around him, while scantily clad gods, goddesses, and nymphs cavorted across the vaulted ceiling above his head.

Any unease Fraunhofer might have felt in the face of this Rococo-era excess was dispelled by the prince himself, a reform-minded leader who habitually engaged with commoners while strolling the streets of the city. The two sat for an amiable chat, during which Maximilian evidently perceived in this humble apprentice an aspirational energy to be nurtured. When their conversation was over, Maximilian presented his visitor with eight golden *Karolinen*—half a year's salary for a typical worker. Indeed, in a gesture that must have moved the orphaned Fraunhofer, the prince offered to help him as a father would a son.

Maximilian assigned one of Bavaria's most prominent citizens, the statesman and entrepreneur Joseph von Utzschneider, to see to Fraunhofer's education. Utzschneider would profit as much as his young charge from their arrangement, for Fraunhofer proved remarkably adept at science and mathematics. He joined Utzschneider's optical instruments firm in 1806 as an assistant, and by 1814 was running the entire production side of the company.

Primed by his readings on the theoretical foundations of optics, Fraunhofer found traditional glass- and lens-making methods to be positively medieval compared to the mathematical rigor in his books. Like generations of predecessors, opticians carved lenses intuitively, as would a sculptor, out of glass whose refractive properties were only approximately known. This haphazard process might produce just one acceptable lens out of every four or five completed. And the wider and thicker the lens, the greater the odds of a misshapen form or material flaw. Middling opticians maximized their chances of success by limiting production to small lenses. Yet in early 1800s Europe, the press for scientific and technological supremacy ran up against the limits of seat-of-the-pants optical fabrication. Finely crafted lenses were required for high-precision surveying instruments and astronomical telescopes.

To effectively compete against the entrenched English optics firms, Fraunhofer sought to replace—or at least supplement—the trial-and-error manufacturing model by mathematics-based design and rigorous testing. "In such a way," one biographer notes, "he fought the personal art and ability of artists like [English optician Jesse] Ramsden by means of the arms of applied science."

Joseph Fraunhofer.

The performance of a multielement achromatic lens

depends critically on the chemical compositions and shapes of its nested glass components. To optimize these qualities, the refractivity and dispersive power of various formulations of glass had to be determined by experiment. Refractivity—the degree to which a light ray is deflected within a piece of glass—is wavelength-dependent; thus, a prism casts a rainbow, while a lens generates ill-defined, color-fringed images. To enhance the former and suppress the latter, Fraunhofer stressed the need to quantify the optical parameters of glass over a wide range of wavelengths.

Fraunhofer's testing protocols hinged on the development of a reliable source of monochromatic light rays: "It would be of great importance to determine for every species of glass the dispersion of each separately coloured ray. But, since the different colours of the spectrum do not present any precise limits, the spectrum cannot be used for this purpose. . . . It was, however, absolutely necessary for me to have homogeneous light of each colour."

Having failed to generate monochromatic rays by passing lamplight through tinted glass or chemical filters, Fraunhofer instead constructed a prism-based apparatus to isolate particular colored lines in the spectral emission of a sodium lamp. Measuring the deviation of these lines with a surveyor's theodolite, he successfully deduced the refractive properties of the glass in the prism. He noted in particular a brilliant yellow line, unaware that Wollaston had reported its existence more than a decade earlier.

Out of curiosity, Fraunhofer applied his spectroscopic device to sunlight. Seeking the vivid yellow line he had previously spotted in the sodium-lamp spectrum, he was amazed to find what he termed a virtual infinitude of dark lines strewn among the colors. (Fraunhofer was unaware of Wollaston's prior observations of dark lines.) In the end, he tallied the spectral positions of 574 such features, some grayish and hair-fine, others black and comparatively broad. Fraunhofer labeled the most prominent lines A, B, C, D, and so on, from red to violet, with occasional lowercase letters interspersed among the list—designations that are still used today.

The bright yellow line cast by lamplight was missing here, its position taken up by a close-together pair of dark lines Fraunhofer would come to designate D. (Upon further magnification, D's bright, sodium-lamp analog proved to be a doublet as well.) The significance of this cosmic-terrestrial alignment was unclear to Fraunhofer, nor could he explain why the solar D lines were dark, whereas their laboratory counterparts were bright. However, on one critical issue, he did weigh in: the Fraunhofer lines, as they were soon to be called, originate in the Sun itself, and are neither optical artifacts of the spectroscope nor the result of selective absorption of sunlight within Earth's atmosphere.

By 1820, English craftsmen, who had been masters of the optics trade throughout the eighteenth century, found themselves lagging Fraunhofer in every facet of production: glassmaking, optical design and testing, and instrumental sophistication. To their chagrin, most English opticians were unable to make spectroscopes of sufficient grade to display all but the most prominent Fraunhofer lines, much less the nearly six hundred he had counted. (English makers were subject to a confiscatory excise tax on crown and flint glass, which crippled optical research and manufacture. A government-sponsored project to improve the quality of optical glass foundered, as did outright attempts to obtain relevant information through bribery. In 1824, John Herschel visited Fraunhofer in the hope of gleaning insight into his techniques, but was denied entry to the glass foundry and workshop.)

That Fraunhofer saw the solar dark lines at all, while previously unaware of their existence, is a testament to the resolving power of his spectroscope and his attentiveness as an observer. The lines' inconspicuousness is captured by a recollection from 1830 by English mathematician and computing pioneer Charles Babbage. Some years earlier, John Herschel had invited Babbage to his home to view the Fraunhofer lines. While setting up the spectroscope, Herschel remarked to Babbage:

> I will prepare the apparatus, and put you in such a position that [the lines] shall be visible, and yet you shall look for them and not find them: after which, while you remain in the same position, I will instruct you *how to see them*, and you shall see them, and not merely wonder you did not see them before, but you shall find it impossible to look at the spectrum without seeing them.

Babbage continues, "On looking as I was directed, notwithstanding the previous warning, I did not see them; and after some time I inquired how they might be seen, when the prediction of Mr. Herschel was completely fulfilled."

Fraunhofer adopted his eponymous lines as reference markers to assess the refractivity of glass samples. In principle, two identical samples should refract a particular line to the same position along a graduated scale. That is, the projected images of the lines should coincide; any relative displacement of the images indicates a difference in composition. Fraunhofer could thus alter the chemical recipes of his various glass mixtures—add, say, more or less lead oxide—to produce lenses of unprecedented homogeneity, clarity, and optical specification. In precisely matched pairs, these component lenses formed the world's best achromatic telescope objectives.

In the first halting steps toward celestial spectroscopy, Fraunhofer equipped a four-inch refractor telescope with a prism and viewed spectra of the Moon, planets, and several bright stars. He found that lunar and planetary spectra largely mimic the Sun's, suggesting that these bodies shine by reflected sunlight. Spectra of stars, on the other hand, are diverse. Although the overall line *patterns* are largely preserved from star to star, the comparative *prominence* of individual lines in the spectra of red stars like Betelgeuse differs from that of white stars like Sirius; and both, in turn, differ from the Sun's characteristic spectrum. Fraunhofer reports in his pioneering paper from 1817: "I have seen with certainty in the spectrum of Sirius three broad bands which appear to have no connection with those of sunlight. . . . In the spectra of other fixed stars of the first magnitude one can recognize bands, yet these stars, with respect to these bands, seem to differ among themselves."

Unlike his predecessors, Fraunhofer had taken a distinctly scientific approach to optical design and fabrication. He developed apt experiments, constructed sophisticated apparatus for measurement, and reported his findings in accepted scientific forums. Yet, at heart, Fraunhofer was an artisan. His observations were conducted in service to manufacturing imperatives, not to the explication of nature. He targeted stars, for instance, to ascertain whether spectral lines are an instrumental artifact of light passing through a slit; the spectra of stars could be obtained without a light-restricting slit (yet still showed spectral lines). Fraunhofer's working methods were never published, nor were outsiders permitted entry to his glassmaking or lens-grinding workshops. Having cast new light on the once-inaccessible realms of atoms and stars, Fraunhofer resumed his true calling: the making of precision optical instruments. He was equally innovative in this arena, presenting astronomers with an advanced generation of refractor telescopes. In his absence, the nascent field of stellar spectroscopy slumbered for almost fifty years.

After Fraunhofer, laboratory scientists in England and France studied the spectra of light emitted by substances incinerated in flames and electric arcs. In 1826, English photographic pioneer William Henry Fox Talbot evoked the freewheeling character of this frontier research:

> A cotton wick is soaked in a solution of salt, and when dried, placed in a spirit lamp. It gives an abundance of yellow light for a long time. A lamp with ten of these wicks gave a light little inferior to a wax candle; its effect upon all surrounding objects was very remarkable, especially upon such as were red, which became of different shades of brown and dull yellow. A scarlet poppy was changed to yellow, and the beautiful red flower of the *Lobelia fulgens* appeared entirely black.

Talbot was drawn to the flame's mesmeric yellow cast, which he ascribed to the combustion of the element sodium in the salt solution that infused the wick of his burner (although elsewhere he proposes both sulfur and "crystallized water" as the activating agent). A prism-view revealed the source of the distinctive color to be a brilliant yellow spectral line. To Talbot's bewilderment, the same yellow line showed up, like an uninvited guest, in the spectrum of *every* incandescent chemical—even those supposedly devoid of sodium. Why would elemental spectra, each one arising from a supposedly unique arrangement of matter, all share this particular spectral line?

Despite this irreconcilable state of affairs, Talbot painted a future where "a glance at the prismatic spectrum of a flame may show it to contain substances which it would otherwise require a laborious chemical analysis to detect." The ubiquitous yellow spectral line muddled the universal assumption that each chemical element possesses its own unique spectral pattern, and that spectroscopic classification of matter—perhaps even identification of new elements—might be feasible. It would be decades before these lofty aspirations were achieved, in large part due to impure test samples and lack of rigor in the laboratory. (Talbot himself frankly admitted as much when he confessed, "I am not much of a chemist, but sometimes amuse myself with experiments.")

Between 1820 and 1860, there was scant scholarly momentum behind the development of spectrochemical analysis; spectroscopic researchers were chiefly concerned with the production of monochromatic light for use in optical testing or photochemical experiments—or as ammunition in the hot debate over whether light is a particle or a wave. Those who did venture into the spectrochemical arena narrowed the scope of their investigations: they sought a spectroscopic means to differentiate between known, but chemically similar, substances. By the 1850s, one could distinguish the spectrum of, say, lithium from that of strontium or nitric-acid gas from bromine, but no trail was blazed toward Talbot's more expansive notion of spectral analysis.

Yet even as the vast inventory of chemical substances remained beyond reach of the spectroscope, the conundrum of the intruding yellow line was solved. A trio of British scientists—John Hall Gladstone, William Crookes, and William Swan—independently isolated the culprit: it was, as Talbot had suggested, the element sodium, in the form of sodium chloride—common salt. Such was spectroscopy's power to amplify the salience of matter that chemists had been flummoxed by nature's most mundane crumb. As nineteenth-century historian Agnes Clerke described it:

> [Salt] floats in the air; it flows with water; every grain of dust has its attendant particle; its absolute exclusion approaches the impossible. And withal, the light that it gives in burning is so intense and concentrated, that if a single grain be divided into 180 million parts, and one alone of such inconceivably minute fragments be present in a source of light, the spectroscope will show unmistakably its characteristic beam.

Given the omnipresence of sodium and its high reactivity in a flame, chemists realized that they had to painstakingly leach every trace of dissolved salt out of their samples before applying any spectrochemical process. John Hall Gladstone spoke for the chemical community when he concluded in 1857 that the analysis of flame spectra, "though doubtless very accurate in certain cases, is of limited and difficult application." Barely two years later, Robert Bunsen and Gustav Kirchhoff wrestled this limited and difficult application into submission and opened the door to cosmochemical analysis through light.

Chapter 15

Laboratories of Light

> [I]n order to examine the composition of luminous gas, we require, according to this method, only to see it; and it is evident that the same mode of analysis must be applicable to the atmospheres of the sun and the brighter fixed stars.
>
> —Gustav Kirchhoff and Robert Bunsen, "Chemical Analysis by Spectrum-observations," 1860

On November 15, 1859, Robert Bunsen wrote with evident excitement to his longtime colleague Henry Roscoe, in England: "At present, Kirchhoff and I are engaged in a common work that does not let us sleep. Kirchhoff has made a wonderful, entirely unexpected discovery in finding the cause of the dark lines in the solar spectrum, and he can increase them artificially in the sun's spectrum or produce them in a continuous spectrum and in exactly the same position as the corresponding Fraunhofer lines. Thus a means has been found to determine the composition of the sun and fixed stars with the same accuracy as we determine strontium chloride, etc., with our chemical reagents." As Bunsen laid out the particulars, Roscoe realized that he was privy to the unfolding of a scientific revolution.

Like other chemists of the day, Bunsen had convinced himself that color is key to the photochemical identification of elements—that each element, incinerated in a flame, blazes with a unique color that, if properly characterized, establishes the element's presence. But how to define that telltale tinge, how to distinguish this shade of yellow from that, how to describe a complex color formed from an amalgam of hues? For more than a year, Bunsen and a student from England, Rowlandson Cartmell, had exhaustively examined the colored emission of flames. Their goal was to construct a series of glass and chemical filters to mask colors released by known impurities such as sodium, thereby disclosing the *true* characteristic color of a sample. Gustav

Kirchhoff was brought in to gauge precisely which wavelengths of light were blocked by the various filters. How he accomplished this is undocumented, but he probably used a spectroscope to assess each filter's selective absorption of sunlight. Indeed, only a few months later, in mid-1859, Kirchhoff took up a parallel study of the optical properties of crystals. Although his measurement apparatus is not pictured, Kirchhoff's detailed description is of a sophisticated, high-magnification spectroscope.

Around the same time, Bunsen admitted in a letter to Henry Roscoe that he had squandered his time on color-filter chemical analysis. Yes, the method did work to a limited degree for select substances. But mulling over its expanded use, Bunsen recognized the folly of developing the requisite multitude of filters. Yet he clung to his faith that chemical elements could be characterized by their light.

Just as paint incorporates individual pigments, the light of incandescent materials is a mixture of discrete colors. However, the human eye maps images based on the spatial distribution of the *amalgamated* light from a source, and is poor at deducing the presence—much less the relative intensities—of specific color components. Thus, two glowing elements might differ markedly in their color constituents, offering a perceptible difference in the character of their emitted light. Another pair of elements might differ only slightly in their luminous makeup, presenting no discernible difference to the eye. Even an element that appeared orange in a flame might harbor subtle—and unseen—shades of blue or violet. In essence, Bunsen lacked the means to assay the pigments dissolved in the paint. And the obstacle, he came to realize, lay not in his familiar domain of chemistry, but in the realm of physics.

Perhaps it was during one of their late-summer ambles along Heidelberg's Philosopher's Walk that Bunsen asked his friend Kirchhoff for advice on his experimental conundrum: How does one *definitively* parse the colors within light? Kirchhoff would have had a ready answer; the apparatus he had only recently used to study crystals contained precisely the instrument Bunsen needed—a spectroscope. Kirchhoff reasoned that if each incandescent element emits light of a characteristic color, as Bunsen asserted, then the spectrum of that light must necessarily be unique to that element. The spectroscope would disperse the intertwined colors of the flame into an ordered regiment of spectral lines, whose pattern corresponds to a given element.

In the early autumn of 1859, Bunsen's and Kirchhoff's experimental goals converged: one's dream of practical photochemical analysis melded with the other's research into the optical properties of matter. The two scientists retreated to their respective laboratories: Bunsen to his gleaming

Chemical Institute, where he prepared highly purified samples of chemical compounds; Kirchhoff to a back room in the dilapidated physics building several blocks away, where he adapted his spectroscopic apparatus to the stringent needs of the new project.

The spectroscope's optical components were housed in a squat, trapezium-shaped enclosure, elevated a few inches above a wooden base. At the center of this homely "cigar box" was a prism-shaped bottle containing carbon disulphide, a fluid with almost twice the light-dispersing power of expensive flint glass. The box's side walls carried a pair of small, fixed telescopes: one to guide a sliver of incident light into the prism, the other to view the emergent spectrum. The first of these telescopes was inserted backwards: its objective lens sat inside the box near the prism, while its focus-end—normally an eyepiece, but here a brass-edged entry slit—protruded horizontally from the box. This telescope served as a collimator, taking incoming light rays, which diverge upon passage through a slit, and rendering them parallel before they reach the prism. (Developed in the 1830s by English optician William Simms, the collimator maximizes a spectroscope's resolving power—its ability to render fine details in a spectrum.) The second telescope, on the opposite side of the box, was fitted with a vertical sighting wire to assist in placing the various spectral features dead-center in the field of view.

Bunsen's chemical samples, each cradled in a tiny loop of platinum wire, were successively incinerated in a gas burner placed before the spectroscope's entry slit. Mounted on a swivel, the central prism could be rotated azimuthally to bring different parts of the spectrum into view. The swivel,

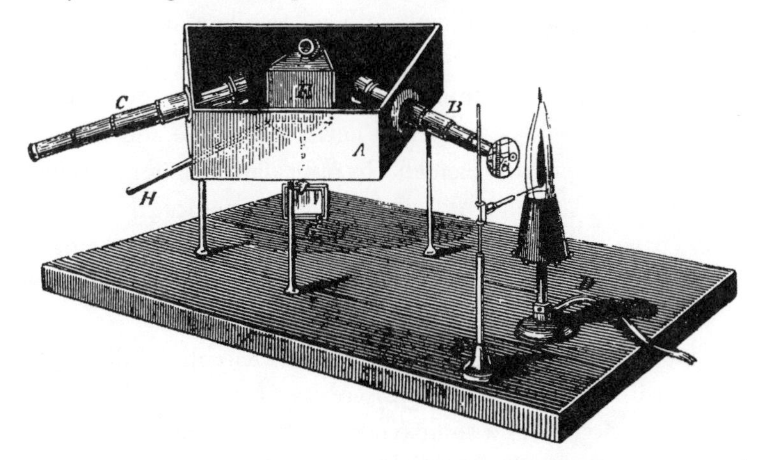

The original spectroscope used by Gustav Kirchhoff and Robert Bunsen to explore the spectra of chemical elements at Heidelberg in 1859.

in turn, held a vertical rod, which poked out the bottom of the enclosure and bore a small mirror. The mirror allowed Kirchhoff to sight, through an external telescope, the reflection of a finely graduated scale several feet away. By axial adjustment of the prism, any spectral line could be brought into coincidence with the vertical wire in the viewing eyepiece; the associated scale reading quantified the line's position.

By April 1860, Kirchhoff had subjected several of Bunsen's chemical specimens to spectroscopic analysis. The acuity of their home-brewed spectroscope was sufficient to reveal faint emission lines never before seen. Kirchhoff was easily able to distinguish among the spectral-line patterns of various elements, including sodium, lithium, potassium, barium, strontium, and calcium. He compared elemental spectra produced in a variety of flames and in electrical spark gaps to prove that line patterns are innate to matter itself and not affected by either the mode or temperature of incandescence.

Both men acknowledged that Bunsen's impeccable laboratory technique was crucial to their enterprise. Bunsen was vigilant about eliminating impurities—especially the ubiquitous sodium—that would have otherwise muddled the interpretation of the spectra. In one test, Bunsen burned several milligrams of salt in the corner of the laboratory opposite the spectroscope. Within a few minutes, sodium lines appeared in the spectrum of a previously featureless gas flame, the sodium particles having wafted across the room. Bunsen reported that the combustion of as little as one three-millionth of a milligram of sodium generates a perceptible line. (He added that airborne sodium might play a role, for good or for ill, in the spread of contagious disease.) In one stroke, Bunsen's test cautioned would-be spectrochemists against airborne contaminants, while affirming the exquisite sensitivity of the spectroscope itself.

That a seasoned "wet-chemist" like Bunsen would endorse spectroscopic analysis lent special credence to the new process. He and Kirchhoff predicted the discovery of new elements by optical means, "for if bodies should exist in nature so sparingly diffused that the analytical methods hitherto applicable have not succeeded in detecting, or separating them, it is very possible that their presence may be revealed by a simple examination of the spectra produced by flames."

Barely a year later, in June 1861, Bunsen and Kirchhoff fulfilled their own prediction with the spectroscopic discovery of two new elements: cesium and rubidium. To obtain a sufficient quantity of cesium for spectroscopic analysis, Bunsen had refined forty tons of mineral water from the springs at Dürkheim. The "mother-liquor," as he called it, was chemically stripped of known elements until the resulting concentrate displayed a pair of previously unseen blue emission lines. (The name cesium, or caesium, derives from the Latin *caesius*, for bluish gray.) A similar effort was mounted against 180 kilograms of the mineral lepidolite before the distinctive red line of rubidium rose to visibility. The dual discovery sent chemists scurrying to their laboratories to set up their own spectroscopes. Before decade's end, three more elements were discovered by spectroscopic means.

One evening, Bunsen recalled, he and Kirchhoff were working late in the laboratory, when they noticed a golden glow on the horizon. A fire was raging in the neighboring city of Mannheim, toward the northwest. Ever curious, they moved their spectroscope to the window and peered at the dispersed light. By now, both were intimately familiar with the spectral patterns of chemical elements. From the brilliant duo of green lines, which they had previously dubbed alpha and beta, they surmised that the fire had ignited compounds containing barium. And astride the barium pair, they recognized the distinctive eight-line fingerprint of strontium: six reds and an orange to one side, a lone blue to the other. If the chemical constituents of a blaze could be discerned at ten miles, Bunsen mused aloud to Kirchhoff, might not future spectroscopists just as reliably assay the remote stars?

As far as Kirchhoff was concerned, the future had already arrived. During the first weeks of his collaboration with Bunsen, he modified the spectroscopic apparatus to allow direct comparison of laboratory-generated spectra with that of the Sun. In a hastily composed announcement to the Berlin Academy, dated October 20, 1859, Kirchhoff alludes to his and Bunsen's unfolding success identifying chemical substances from their flame spectra. Yet he declines to elaborate, and shifts abruptly to the matter at hand: "I made some observations which disclose an unexpected explanation of the origin of Fraunhofer's lines, and authorize conclusion therefrom respecting the material constitution of the atmosphere of the sun, and perhaps also of the brighter fixed stars."

Using a clock-driven plane mirror (a "heliostat"), Kirchhoff had projected a sunbeam through a salt flame before it entered the spectroscope. Manipulating a sequential pair of calcite crystals, he increased or decreased the intensity of sunlight illuminating the flame. When the solar illumination was low—that is, when the flame was not backlit by the Sun—the lines of sodium appeared bright, as in a typical flame spectrum. However, when the solar backlighting was intensified, the bright lines transformed into dark lines. Kirchhoff reasoned that, while incandescent sodium always throws off its characteristic rays, these lines of emission were now seen in silhouette against the more brilliant Sun. Indeed, when Kirchhoff blocked the sunlight, the sodium lines reverted to their bright form. Moving the salt flame into and out of the solar beam, he found that the Sun's D lines were *darker* when the flame was interposed than when it was not. That is, the flame's radiance did not "fill in" the dark D lines, as he had expected, but reinforced the absorption of these wavelengths of light. Functionally, the flame was a far-flung extension of the solar atmosphere, both of them acting in concert on the light emanating from the Sun's interior. More surprising, an interposed lithium flame created a dark line in the Sun's spectrum where no such line had existed before; two superimposed luminous sources—the Sun plus the flame—somehow resulted in a selective *diminution* of light.

To reaffirm the evidence gleaned from nature, Kirchhoff decided to create Fraunhofer lines using an artificial sun. He acquired a Drummond lamp—the original "limelight" of theatrical fame—and swapped out the Bunsen burner for a less-intense alcohol flame. Not only does a Drummond lamp burn hotter than alcohol, its spectrum is virtually continuous like the Sun's. (The lamp's inevitable sodium-contaminant emission disappears after brief use.) As Kirchhoff interposed the salt-laced flame between the lamp and the spectroscope, he saw the bright D doublet go dark, as in the solar spectrum. He had artificially impressed the telltale sodium lines upon the spectrum of a lamp that itself contains no sodium. His conclusion was that sodium in a flame emits characteristic wavelengths, but also absorbs the same wavelengths from light passing through it. Kirchhoff asserts, in his 1859 announcement, "that the dark lines of the solar spectrum . . . exist in consequence of the presence, in the incandescent atmosphere of the sun, of those substances which in the spectrum of a flame produce bright lines in the same place." Fraunhofer's D lines, manifested in emission or absorption, mark the presence of sodium in a laboratory flame, in the Sun, or in a star.

In the months that followed, Kirchhoff developed a physical theory to account for the origin of spectral lines. Analogizing to the natural flow of heat from a region of high temperature toward that of low temperature, he deduced that a body with a propensity to emit light at a given wavelength must have an equal propensity to absorb light at that wavelength. Thus, the luminous emission of a hot, diffuse gas takes the form of bright lines when viewed directly, but dark lines when viewed against an incandescent body of higher temperature. (In fact, the bright-dark reversal of sodium lines was observed and similarly explained in 1849 by French physicist J. B. L. Foucault, who superimposed solar and electric-arc spectra. Kirchhoff was unaware of Foucault's work.)

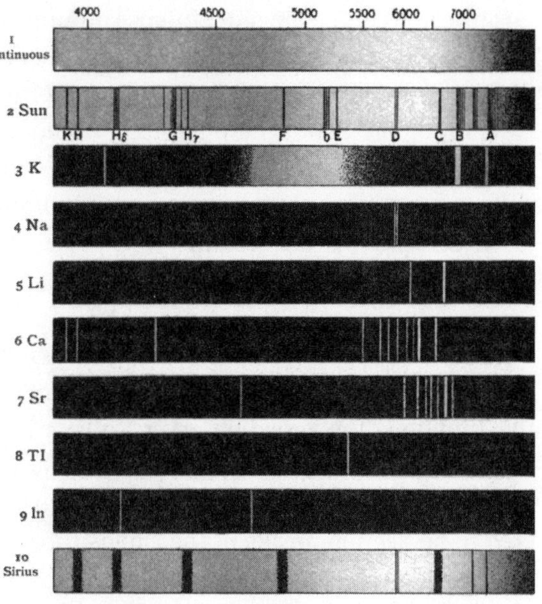

Spectra of the Sun, Sirius, and several chemical elements. Note the coincidence of the dark D-lines in the stellar spectra with their bright-line counterparts in the laboratory spectrum of sodium (Na).

Kirchhoff's radiation theory explains spectral-line formation to the degree that Freudian theory explains the psychological manifestations of brain function. They are functional models of phenomena whose root cause, in their respective eras, were unknowable. Thus, Kirchhoff says nothing about the atomic-level phenomenon that gives rise to the emission or absorption of light by matter, nor could he: the structure of the atom was as yet unknown. However, the prevailing idea that the atom is indivisible—that it represents the fundamental iota of matter—was shaken by the sheer number and variety of spectral lines associated with any given element. How could a unitary particle give off light at a multitude of discrete wavelengths? Scientists suspected that spectral lines reflected something about the form and behavior of the atom, but they were unable to decode the clues. In the absence of a realistic atomic model, which would not come until the twentieth century, physicists proposed scenarios of spectral-line production that typically involved complex vibrations of the atom.

Having placed spectral-line production on a theoretical footing—at least to his own satisfaction—Kirchhoff, in 1861, stated explicitly what he had only intimated in his prior communication to the Berlin Academy:

> The sun possesses an incandescent, gaseous atmosphere, which surrounds a solid nucleus having a still higher temperature. If we could see the spectrum of the solar atmosphere, we should see in it the bright bands characteristic of the metals contained in the

> atmosphere. . . . The more intense luminosity of the sun's solid body, however, does not permit the spectrum of its atmosphere to appear; it *reverses* it . . . so that instead of the bright lines which the spectrum of the atmosphere by itself would show, dark lines are produced. Thus, we do not see the spectrum of the solar atmosphere, but we see a negative image of it. This, however, serves equally well to determine with certainty the presence of those metals which occur in the sun's atmosphere.

Kirchhoff's spectroscopically founded assertion that the solar interior is *hotter* than its luminous envelope marked a break with decades of speculative tradition. The famed observer William Herschel had claimed that the visible Sun is actually a hot, opaque cloud-layer concealing a central world whose surface is temperate—and likely inhabited. Sunspots were considered to be fleeting breaks in the clouds through which one might glimpse the hospitable regions below. François Arago, director of the Paris Observatory and early champion of photography, assured an 1840s audience that "if anyone asked me if the Sun could be inhabited by a civilization like ours, I would not hesitate to say, yes." As late as 1854, English physicist David Brewster maintained that "we approach the question of the habitability of the sun, with the certain knowledge that the sun is not a red-hot globe, but that its nucleus is a solid opaque mass receiving very little light and heat from its luminous atmosphere."

English astronomer J. Norman Lockyer declared in 1881 that Kirchhoff's spectroscopic observations "at once destroyed, at a blow, the idea . . . that the sun was a cool habitable globe, with trees, and flowers, and vales, and everything such as we know of here. If the atmosphere were in a state of sufficient incandescence to give these [spectral-line] phenomena it was absolutely impossible that anything below that atmosphere should not be at the same time at a high temperature."

Kirchhoff's 1861 report presents the first evidence-based solar model fully informed by the maturing physics and chemistry of the mid-nineteenth century. Crude as it may read in retrospect, the model's underlying implication was profound: The spectroscope had effectively projected the scientist's critical faculties almost a hundred million miles, into the very body of the Sun. A "victory over space," trumpeted physicist Arthur Schuster in a review for *Popular Science Monthly*. Astronomers would never look at the Sun the same way: It was, henceforth, a physical body—a star—whose structure and luminance arise from, and are ever subject to, physical law. Nor could scientists fail to recognize the immense potential of spectroscopy, both in the cosmic realm and in the laboratory.

By early 1860, Bunsen and Kirchhoff had acquired a pair of improved spectroscopes from the firm of C. A. von Steinheil in Munich. The newer instrument featured a sequential quartet of flint-glass prisms, the latter trio each dispersing the spectrum transmitted by its predecessor. No other spectroscope in the world could display spectral features with comparable resolution. A small reflecting prism grafted to the entry slit allowed the spectra of two sources to be seen simultaneously in the eyepiece, one directly above the other. This feature was put to immediate use to better compare solar and laboratory-generated spectra. (By this time, Bunsen had provided Kirchhoff with samples of thirty-two elements that could be set alight in a flame or spark.)

Visiting from England, Henry Roscoe viewed the Sun's spectrum arrayed against that of the electrical discharge between a pair of iron electrodes energized by an induction coil:

> In the lower half of the field of the telescope were at least seventy brilliant iron lines of various colors, and of all degrees of intensity and of breadth; whilst in the upper half of the field, the solar spectrum, cut up, as it were, by hundreds of dark lines, exhibited its steady light. Situated *exactly* above each of the seventy bright iron lines was a dark solar line. These lines did not only coincide with a degree of sharpness and precision perfectly marvellous, but the intensity and breadth of each bright line was so accurately preserved in its dark representative, that the truth of the assertion that iron was contained in the sun, flashed upon the mind at once.

The presence of iron lines in the Sun's atmosphere electrified astronomers, whose prevailing theory of solar formation (later discredited) involved an accretion of iron from impacting meteors. Exercising his mathematical chops, Kirchhoff computed the odds of a chance alignment of the sixty most prominent iron lines in the Sun and the flame at less than one in a million trillion. In addition to sodium and iron, there were multiple line coincidences for barium, calcium, chromium, cobalt, copper, magnesium, nickel, and zinc. Not only did the positions of the solar lines match their laboratory counterparts, so too did the relative intensities—essentially, the widths of the lines. The lone line-coincidence for gold was judged insufficient to verify its presence in the Sun.

In the ensuing months, Kirchhoff began to draw an eight-foot-long map of the solar spectrum, diluting the India ink in his pen to render progressively fainter lines. (While visually appealing, color is redundant in such a map; a line's position is all astronomers need.) Suffering from severe eyestrain, Kirchhoff delegated the project's completion to an assistant. The published map served as the benchmark reference of spectral-line patterns for the next decade.

The scientific community quickly recognized the revolutionary dimensions of spectroscopic analysis as practiced by Bunsen and Kirchhoff. Barely two years after the Heidelberg duo's initial report, veteran and novice researchers in Europe and the United States were applying the process to a variety of projects, spurred on by indications that solar-surface activity influenced terrestrial climate and magnetism. At a gathering of London's Chemical Society on June 20, 1861, pioneering celestial photographer Warren De La Rue remarked that "if we were to go to the sun, and to bring away some portions of it and analyze them in our laboratories, we could not examine them more accurately." In giving astronomers the means to determine the intrinsic nature of celestial bodies, Bunsen and Kirchhoff's contribution was every bit as far-reaching as the introduction of the telescope itself two and a half centuries earlier. The commonality between laboratory and solar spectra extended the sense of cosmic unity already witnessed in Newtonian mechanics. Just as the orbital movements of planets, comets, and binary stars are governed by the same laws, the spectroscope showed that our Earth and Sun consist of like matter that interacts with light in identical fashion.

The accomplishments of Bunsen and Kirchhoff were much lauded, leaving the two unprepared for the tempest over priority that blew in from England. In 1861, William Crookes, founding editor of London's *Chemical News*, expressed his overtly nationalistic opinion that "our readers will feel an interest in knowing that many of the observations which are now being followed by Continental *savans*, have been investigated in a more or less perfect manner by English experimentalists." Crookes's barrage was followed by salvos from spectrum researcher William Allen Miller, who laid out a similar version of the history of spectral analysis at professional meetings in 1861 and 1862. A series of counter-lectures by Bunsen's protégé Henry Roscoe failed to still the outcry over the originality of the Heidelberg observations.

Bunsen and Kirchhoff affirmed that aspects of spectral analysis had been studied by others, all the way back to Joseph Fraunhofer, David Brewster, Henry Fox Talbot, and John Herschel. And that there were subsequent contributions on the subject by Crookes, Miller, George Stokes, William Swan, and John Hall Gladstone in England, Anders Ångström in Sweden, and Julius Plücker in Germany, among others. But they denied that their own research was in any way influenced by these works, much of which never came to their attention or were variously limited in scope.

Kirchhoff countered in November 1862 with a hard-hitting rebuttal, "Contributions Toward the History of Spectral Analysis and the Analysis of the Solar Atmosphere," which appeared in the venerable British scientific journal *Philosophical Magazine*. He parses the narratives of John Herschel and Henry Fox Talbot about their research on the spectra of colored flames. "In these expressions," Kirchhoff writes, "the idea of 'chemical analysis by spectrum-observations' is most clearly put forward. Other statements, however, of the same observers, occurring in the same memoirs . . . flatly contradict the above conclusions and place the foundations of this mode of analysis on most uncertain ground."

Crookes praised William Allen Miller's 1845 renderings of flame spectra, which he claimed were superior to those from Heidelberg, but Kirchhoff finds them to be utterly worthless for scientific use: "I have laid Prof. Miller's diagrams before several persons conversant with special spectra, requesting them to point out the drawing intended to represent the spectrum of strontium, barium, and calcium, respectively, and that in no instance have the right ones been selected." Kirchhoff likewise dismisses Crookes's endorsement of the explanation of the D-line reversal phenomenon posed by Miller. He suggests brusquely that Crookes "read Miller's words with some *slight* attention," so that he might realize how the scheme produces a result that is counter to what is observed.

Kirchhoff reminds readers that William Swan restricted his study of the D lines to hydrocarbon compounds like ether, paraffin, and turpentine; therefore, no conclusion can be drawn from that work regarding the general question of the uniqueness of elemental spectra. Anders Ångström fares no better in Kirchhoff's history: "It is seen that the proposition which forms the basis of the chemical analysis of the solar atmosphere floated before Ångström's mind, but only, indeed, in dim outline."

In a summary statement worthy of a judicial proceeding, Kirchhoff declares that no one "had clearly propounded this question [of the uniqueness of elemental spectra] before Bunsen and myself; and the chief aim of our common investigation was to decide this point. Experiments which were greatly varied, and were for the most part new, led us to the conclusion upon which the foundations of the 'chemical analysis by spectrum-observations' now rests."

The priority debate stoked by Crookes and Miller prompted an inevitable question: If English scientists were aware of the potential of spectral analysis as early as 1845, if not before, why hadn't they developed it themselves? The issue flared at an 1861 meeting of the leading lights of English chemistry, where an exasperated Edward Frankland (yet another Bunsen trainee) pointed an accusatory finger at one of the drama's leading players:

> I have recently read, with very great interest, the beautiful researches which Dr Miller made some sixteen years ago upon this very subject. It is really wonderful that sixteen years ago we had the real pith of the whole of this matter thrown before us, but up to the present time we have been unable to use it. I will not acquit Dr Miller for being partly to blame, for, perhaps, he himself ought to have shown us the practical way of employing the instrument at the time it was revealed to us.

Already before Bunsen and Kirchhoff's monumental work, the majority of scientists believed that spectral analysis could be applied—in principle, if not yet in practice—to delineate the makeup of terrestrial and solar substances. Many also trusted that every element or compound, when brought to incandescence, would prove to display a unique spectrum by which it might be identified. Yet assertions and imprecise observations are rickety bones upon which to hang a complete body of theory and practice. Until Bunsen and Kirchhoff, there was only the feeblest momentum toward the establishment of practical spectrochemical techniques. A sustained and vigorous effort was required before all doubts about the viability of the new process were erased.

Kirchhoff and Bunsen pursued a higher and more comprehensive level of experimental verification than any previous researcher. Their chemical samples were purer, their spectroscopic apparatus more refined, their results more compelling by virtue of an almost fanatical attention to detail. They originated an analytical technique that was independent of the underlying physics governing the interaction of matter and light. While theorists sorted out the latter, experimentalists could proceed with the spectrochemical analysis of radiating substances.

Perhaps the major advantage the Heidelberg scientists had over their cohorts was each other. The two worked both individually and in concert, as dictated by the research needs of the moment. Their respective specializations eased their collective experimental burden, relative to that faced by a lone physicist or chemist in the laboratory. In Bunsen, Kirchhoff had a ready source of reliable emission spectra by which to identify elements in the Sun. In Kirchhoff, Bunsen had the broad theoretical outlook and instrumental expertise of the physicist. And in their collaboration, we witness the escalatory give-and-take of a musical duo, stoking their mutual creative agency through each other's cues. Together, Bunsen and Kirchhoff comprised one of the most productive joint ventures in science.

Perhaps Crookes's cadre of scientists were cowed by the sheer weight and intricacy of work needed to fully develop a new spectrochemical process. In any event, such an all-out effort would have distracted them from their primary research goal, which focused on the physical more than the chemical aspects of spectra: What might these ordered arrays of spectral lines reveal about the nature of light, electricity, or thermodynamic processes?

Equally spurious are the English claims of priority in the photochemical analysis of the Sun; their achievements are closer to a proof of concept than development of practical techniques that could be implemented by others. Indeed, as late as 1860, David Brewster and John Hall Gladstone still questioned whether Fraunhofer lines arise in Earth's atmosphere or the Sun's. As Bunsen and Kirchhoff's tireless champion Henry Roscoe asserted in a letter to Gladstone on May 10, 1861, "The real importance of . . . [a] discovery is not to be measured by the first imperfect notices which have been made on the subject. . . . The discovery is really made when the true importance of these observations is shown, & when they are connected together in a scientific exact manner."

While the dispute over priority simmered, spectral analysis proceeded apace. The spectroscope quickly became an essential piece of equipment in virtually every physics and chemistry laboratory, and by century's end, in many observatories.

In 1877, Bunsen and Kirchhoff were awarded the Humphry Davy Medal of the Royal Society, an unambiguous sign that the issue of priority had been settled in most minds. Kirchhoff often related the story of a conversation with his banker, in which he mentioned how the spectroscope might, in principle, reveal the presence of gold in the Sun. The banker shot back, "What do I care for gold in the sun if I can not fetch it down here?" Shortly after he and Bunsen had received the Davy Medal—plus its weight in gold—Kirchhoff visited the banker and handed over his gleaming prize. "Look here," he said, "I have succeeded at last in fetching some gold from the sun."

Chapter 16

DECONSTRUCTING THE SUN

It is not an uncommon thing for the physicist to tread upon the ground which a chemist thinks belongs to him, and for the chemist to tread upon the ground of the physicist. Now we have the chemist occupying the ground of the astronomer.

—Warren De La Rue, comments to the Chemical Society, London, June 20, 1861

DURING THE 1860S, THE SPECTROSCOPE took up a central position in efforts by astronomers to probe the chemical and physical properties of the Sun. The first step had been taken by Bunsen and Kirchhoff, who established a working inventory of elemental lines in the solar spectrum. Although heartily adopted by solar astronomers, Kirchhoff's published spectral-line map had serious shortcomings. Each line's position is designated, not by its physical wavelength, as Fraunhofer had done, but by its millimeters of deviation along a scale with an arbitrary zero-point. Kirchhoff had to adjust the prisms to bring successive portions of the spectrum into view, imparting the scale readings with discontinuities. Furthermore, prismatic dispersion varies with the wavelength of light: On the Heidelberg map, a pair of lines in the red part of the spectrum spans a wavelength range different from that of a pair of equally spaced lines in the violet. The precise manner of dispersion being unique to Kirchhoff's apparatus, his map's numerical scale is better suited to line identification—akin to Fraunhofer's alphabetical labels—than to direct line-to-line comparisons against other's spectral maps.

In 1868, Swedish astronomer Anders Ångström released an extensive, wavelength-based solar-spectrum map. This was made possible by Ångström's use of a diffraction grating instead of a prism as his spectroscope's light-dispersing element. Introduced by Joseph Fraunhofer around 1820, a diffraction grating consists of a series of parallel, closely spaced, linear apertures through which light can pass. Fraunhofer had experimented with an array of wires stretched over the threads of a pair of bolts, then progressed to a sequence of slits peeled from gold film on a glass plate. (Gratings are also produced by cutting fine grooves into a transparent plate; a reflection grating can be made by grooving an opaque material.) Unlike its prism-based cousin, a grating spectroscope disperses colors uniformly with wavelength. Fraunhofer derived a straightforward formula that expresses the wavelength of a spectral line, depending on its deviation angle and the density of grooves in the grating.

Ångström's spectroscope had considerably more dispersive power than Kirchhoff's, and the number of visible lines rose accordingly to more than twelve hundred. Once again, each line was rendered in ink as it appeared to the eye, not only in its proper place, but with its observed breadth and gradation, from pale gray to coal black. Several dark bands on Kirchhoff's map resolved themselves into arrays of tightly packed lines. Among the solar elements added by Ångström were hydrogen (later proved to be the Sun's primary constituent), manganese, strontium, and titanium; he rejected Kirchhoff's claimed line coincidences for barium, copper, and zinc.

The ascendance of Ångström's visual map proved to be nearly as short-lived as its predecessor's. With the same nod toward objectivity, permanence, and efficiency that would propel the camera into the depiction of faint nebulae, photographic pioneers began to record the solar spectrum onto a sensitized plate. New York astronomer Lewis M. Rutherfurd developed a succession of telescope-mounted spectroscopes during the wet-collodion era of the 1860s. Kirchhoff himself is said to have remarked that had Rutherfurd's 1864 solar-spectrum photograph come sooner, it would have spared him a year's crossed-eyed toil. Rutherfurd's ultimate spectroscope, completed in 1877, featured a 1.7-inch-square diffraction grating of his own manufacture with an astonishing 17,296 rules per inch. The following year, Rutherfurd stunned attendees at a meeting of the Royal Astronomical Society when he displayed a ten-foot-wide mosaic of the solar spectrum, assembled from prints of twenty-eight negatives. Lord James Lindsay, the society's president, remarked, "I certainly have never seen any thing before so fine as this photograph."

Rutherfurd distributed his machine-ruled gratings, each signed and dated, to spectroscopic professionals worldwide. Their allotment was haphazard, as American solar observer Samuel Pierpont Langley describes in an 1879 letter to his English colleague Norman Lockyer:

> On reaching New York I called at Mr Rutherfurds to see about your grating. He is away but [his assistant] Chapman told me he (Daniel C. Chapman) had sent you one, two weeks ago which I suppose you have now got. I seized upon two large ones I found there and bore them off, that being the only way of getting them, as he has apparently promised so many people that he has no longer any very exact idea of the order of priority of claimants for the few he makes.

Of the more than fifty precision diffraction gratings Lewis Rutherfurd gave away during his lifetime, three went to his uptown colleague Henry Draper. In 1872, Draper set up a heliostat to reflect sunlight into a fourteen-foot-long grating spectroscope, the projected spectrum recorded on a series of wet-collodion plates. The spectroscopic camera, or spectrograph, proved to be a keener eye than a human's. In one section of Ångström's hand-drawn map of the solar spectrum where 118 spectral lines had been depicted, Draper's negative captured 293. Presenting his spectrum photograph to Britain's *Philosophical Magazine* in 1873, Draper assured the editors that the "picture is absolutely unretouched. It represents, therefore, the work of the sun itself, and is not a drawing either made or corrected by hand." So dense was the forest of spectral lines that London instrument maker John Browning wrote to Draper, "I am glad that you have stated so clearly that the plate is a perfectly untouched photograph, for I have not been able to get my friends to believe this in many instances."

A collective huzzah arose from the astronomical community with every release of a solar spectral-line map, the new map invariably eclipsing the old in clarity or extent. However, the iterative increase in the dispersive power of spectroscopes and spectrographs, each jump impressive in its own time, was still insufficient to resolve many questionable line coincidences. To make matters worse, systematic errors or discontinuities were discovered in the numerical scales of the maps, requiring complex corrections before they could be referenced for research. Without more definitive positional matches between solar lines and those generated in the laboratory, astronomers were unable to verify the presence of certain elements in the Sun.

Solar-spectrum mapping reached an essentially modern standard of accuracy during the 1880s with the invention of the concave reflection grating by Henry Rowland, a physics professor at Johns Hopkins University. Replacing the flat diffraction grating with one of the proper curvature eliminated the need for focusing lenses, producing a sharper, brighter, and higher-resolution spectrum. The omission of lenses from the optical path also allowed the spectrograph to reach into the ultraviolet part of the spectrum, wavelengths ordinarily absorbed by glass. And with the new dry-plate photographic process, segments of the solar spectrum could be recorded in mere minutes.

In 1888, Johns Hopkins published Rowland's map of the solar spectrum, featuring some twenty thousand absorption lines and an easy-to-read wavelength scale. The dispersion in wavelength was so high that the spectrum, pieced together from its segments, stretched forty feet. As to its merits relative to its predecessors, an advertising circular for the map crowed that the "superiority is so great there is no possibility of comparison." Rowland's map and his subsequent compilation of spectral-line wavelengths, intensities, and chemical identifications were an immediate hit among solar astronomers and laboratory spectroscopists, and would serve as the standard reference for researchers well into the twentieth century.

By the 1890s, line coincidences on the various solar maps had raised the number of chemical elements identified in the solar atmosphere to around forty. Among the more curious tales of spectroscopic discovery is that of helium, the second most abundant element in the universe. Astronomers had realized that by attaching a spectroscope to the eyepiece-end of a telescope, they could zoom in on specific solar features, such as sunspots or flares. During the solar eclipse of August 18, 1868, French astronomer Pierre-Jules-César Janssen trained his spectroscope on a flamelike outburst called a prominence, which protruded beyond the Sun's occulted disk. As expected, the prominence's spectral lines were bright instead of dark, because its diffuse gas appears against the blackness of space, not the blinding solar surface. However, the emission-line array did not fully mirror the ordinary Fraunhofer spectrum; evidently, the physical conditions in a prominence differ from those in the region underneath, where line absorption takes place. The prominence spectrum included species familiar from the laboratory, but also a delicate yellow line whose placement was almost, but not quite, coincident with the well-known D doublet of sodium. The new line, Janssen realized, had no match in the spectral patterns of known elements.

In the aftermath of the eclipse, Jules Janssen and Norman Lockyer independently found that the circumsolar spectrum, ordinarily effaced by the luminance of the Sun's disk, was, in fact, bright enough to observe in full daylight. Lockyer confirmed the presence of the yellow line in the Sun's chromosphere, the red-tinged atmospheric layer that surrounds the solar photosphere (whose boundary roughly demarcates the Sun's "surface").

When laboratory tests failed to support astronomers' initial suspicion that the line arose from some form of hydrogen, the mystery element was dubbed helium, after the Greek god of the Sun.

Lockyer was loathe to suggest that a new element had been found on the Sun before being isolated on Earth. At the time, a host of spectroscopically theorized substances, like jargonium, nigrium, coronium, and nebulium, were jostling for scientific acceptance. All proved to be fictitious or else manifestations of known elements. A similar cautionary tale erupted in the late 1870s with Henry Draper's notorious "discovery" of oxygen in the Sun. Draper's assertion that the telltale pattern of oxygen appeared in *emission* among the Fraunhofer absorption lines contradicted the inviolable rule of line formation: spectral lines generated in the photosphere must be dark, as the ambient matter absorbs light from the hotter regions below. The five-year-long controversy, in which Norman Lockyer wielded the opposing cudgel, was stilled only by Draper's death in 1882. (While Draper's bright lines proved illusory, three dark lines of oxygen were identified in the solar spectrum in 1896.)

Helium remained a notional element until 1895, when Scottish chemist William Ramsay incinerated a chunk of cleveite, a uranium ore, hoping to isolate a sample of argon, the gas he had codiscovered with Lord Rayleigh the previous year. Peering through a hand-held spectroscope, Ramsay applied an electrical discharge to the recovered gas. He was perplexed to see, in addition to the telltale spectral pattern of argon, a secondary array of unfamiliar lines. After further research, Ramsay announced to his wife, "There is argon in the gas; but there was a magnificent yellow line, brilliantly bright, not coincident with, but very close to the sodium yellow line. I was puzzled, but began to smell a rat. . . . Helium is the name given to a line in

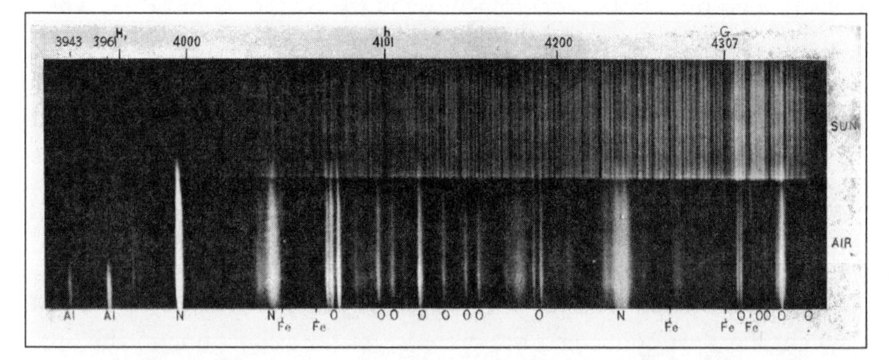

Henry Draper's 1876 photographic spectrum of the sun, depicting his spurious "solar emission lines" of oxygen.

the solar spectrum, known to belong to an element, but that element has hitherto been unknown on the earth." Within months, the terrestrial helium line was found to have a weak companion alongside, a feature that sent spectroscopic astronomers scurrying to their telescopes. The solar helium line proved to be double as well, proving beyond doubt that Ramsay's gas was the same as the gas resident in the Sun. From argon and helium, Ramsay went on to discover the noble gases krypton, neon, and xenon, and received the Nobel Prize in chemistry in 1904.

In addition to revealing the chemical composition of the Sun, the spectroscope confirmed the existence and physical extent of the Sun's light-absorbing envelope. In Kirchhoff's model, dark lines are impressed upon the otherwise continuous solar spectrum when light is selectively absorbed by chemical elements in the Sun's atmosphere. Were this absorbing layer viewed without the back-illumination of the Sun's interior, an emission-line version of the Fraunhofer absorption spectrum would appear instead. Astronomers realized that such a spectrum reversal might be glimpsed at the Sun's limb during an eclipse; in the brief interval before the Moon fully occults the Sun, the light of the solar disk is nearly absent and the incandescent photosphere might be seen unimpeded against the blackness of space.

In the campus newspaper, Dartmouth College astronomer Charles A. Young described the first-ever sighting of the aptly named "flash spectrum" during the December 22, 1870, solar eclipse expedition in Jerez, Spain:

> The slit of my spectroscope was placed tangential to the sun's limb, just at the base of the chromosphere, the 1474 line on the cross-wires. As the crescent of the sun (or decrescent, rather) grew narrower, this line, and the magnesium lines close by, as well as some others in the same neighborhood which I am accustomed to see bright in prominences, gradually increased in brilliancy, when suddenly, as the last ray of the solar photosphere was stopped out by the moon, the whole field of view was filled with countless bright lines—every single dark line of the ordinary spectrum, so far as I could judge in a moment, was reversed, and continued so for perhaps a second and a half, when they faded out, leaving only those I had at first been watching.

Young added in a subsequent report, "The phenomenon was so sudden, so unexpected, and so wonderfully beautiful as to force an involuntary exclamation." The brevity of the line reversal indicated that the Sun's light-absorbing stratum spans some five or six hundred miles, a mere one-thousandth of the solar radius. The existence of the flash spectrum confirmed Kirchhoff's rough model of solar structure, although he had never conceived that such a thin layer of gas would suffice to generate the profusion of Fraunhofer lines.

Spectroscopic studies of the Sun during the latter decades of the nineteenth century illustrate why astrophysics rose so slowly after its nominal birth in 1859 at the hands of Bunsen and Kirchhoff. Although spectroscopists were gratified by their success in fixing the chemical composition, and to a much lesser extent, the physical layout of the Sun, they were keenly aware that their instruments barely scratched its surface. The solar interior was the province of theoretical physicists, who had yet to formulate the mathematical means to probe the opaque depths of our central star. Yet even in the Sun's periphery, physical theory was sidetracked by preconceived notions about how matter should behave in the high-temperature, low-pressure, gravitationally governed solar environment. In retrospect, we know that at least two erstwhile assumptions about the solar atmosphere—that its temperature decreases radially outward and that its atoms occupy different atmospheric strata according to their weight—are simply wrong. (Temperatures rise steadily above the photosphere, and atoms distribute themselves freely throughout the gas.)

Even as the solar data accumulated, significant questions remained in the realm of spectroscopic practice: To what extent do temperature and pressure affect the creation and appearance of spectral lines? Which of the purported solar spectral lines arise instead from absorption of sunlight in Earth's atmosphere? To the perplexing pile of numerical and visual information were applied a host of speculative, often contradictory, physical theories. Benighted by their foundational ignorance, astronomers sparred uselessly in private correspondence, professional journals, and conferences over interpretations of one or another observation. Without a realistic model of the atom, for instance, the origin of spectral lines cannot be explained: How does an atom emit or absorb light in the doses and at the specific wavelengths it does? Why does the very same atom ignore light of other wavelengths? What, structurally, differentiates one element from another?

Superlatives abounded in the professional and public press about the evolving fusion of solar photography and solar spectroscopy. Yet the artifacts of this fertile coupling embodied, for the most part, technological advancements whose scientific value remained latent. The astrophysics of the Sun could not be teased out until the fundamental physics of both matter and energy marched into the twentieth century.

While solar researchers haggled over the nature of our nearest star, a handful of adventurous astronomers aimed their spectroscopes in the opposite direction, toward deep space. The challenge to see, much less photograph, a stellar spectrum using 1860s technology seemed ludicrous to most telescopic observers. Even the brightest star in the night sky is one ten-billionth the brilliance of the Sun. To steer a star's feeble glimmer into the guts of a spectroscope, disperse its aggregated wavelengths to the point of near invisibility, and then seek to extract any datum of scientific value strains the very definition of optimism. To seek such an outcome, not just for a star, but for a faint, diffuse nebula, enters the realm of delusion. Yet every uncharted realm draws its explorers. And the more distant that realm, the more intrepid the explorer.

Chapter 17

A Strange Cryptography

> [W]hen a molecule of hydrogen vibrates in the Dog-Star, the medium receives the impulses of these vibrations; and after carrying them in its immense bosom for three years, delivers them in due course, regular order, and full tale into the spectroscope of Mr. Huggins, at Tulse Hill.
>
> —James Clerk Maxwell, "On Action at a Distance," discourse at the Royal Institution, London, 1873

In 1855, thirty-year-old English silk merchant and amateur astronomer William Huggins sold the family business, moved with his elderly parents to the upscale London suburb of Lambeth, and observed the night sky the way it had been done since Galileo's time: peering into the eyepiece of a telescope and letting the cosmic light flood his retina.

Astronomy was the perfect hobby for Huggins, given his affinity for the sciences and his financial means. (It is said he dropped an interest in microscopy because of his distaste for dissection.) Huggins acquired his first telescope when he was eighteen. He replaced that instrument in 1853 with a pricey Dollond equatorial refractor of five-inches aperture. Like many neophytes, his observing runs were routine: peering at and sketching the Moon, planets, and the occasional comet. His drawings of Mars from April 1856 appear to be a hash of what he saw through the eyepiece plus what he had seen in existing illustrations. Although never formally trained in astronomy—what advanced academics he had came from private tutors—Huggins pursued his new avocation as an entrepreneur might a fledgling business: to gain one's market share of fame required hard work, self-belief, and persistence.

Huggins was elected in 1854 to the Royal Astronomical Society, where he enjoyed fellowship with veteran cosmic seekers like Warren De La Rue,

George B. Airy, and William "Eagle-eye" Dawes. (The Dawes' limit is still used as a gauge of a telescope's ability to resolve close-together double stars.) To his already substantial cottage on Lambeth's Upper Tulse Hill, Huggins added a twin-story, twelve-by-eighteen-foot observatory, whose iron columns raised the working floor above the surrounding trees. In 1858, on Dawes's recommendation, he swapped out his five-inch telescope for a top-quality, eight-inch Alvan Clark refractor. The telescope rested on a massive, pyramidal, brick pier, which was anchored in concrete at ground level. The original hemispherical dome, twelve feet wide at the base and sheathed in zinc, revolved atop three iron balls along a circular, iron channel. A doorway from the upper level of the house led straight into the observatory.

At the May 1856 meeting of the Royal Astronomical Society, Huggins presented a model and description of his new facility. He had fully entered the small army of independent amateur astronomers: men of means whose scientific aspirations superseded their more grounded ambitions in commerce or society. Often lacking the validation of an advanced degree, they delved into areas of research and invention spurned by their institutional counterparts. Over the next few years, Huggins received from these night-sky specialists a thorough education in the practical and theoretical aspects of astronomy. With time, he hoped he might garner the sort of acclaim bestowed upon his contemporary Richard Carrington, who cataloged the places of 3,735 stars from his private observatory at Redhill, Surrey. An 1858 encomium to Carrington captures the dynamism Huggins would come to share with his fellow amateur scientists: "Talent and zeal, untiring devotion and industry, and an unsparing but prudent application of private resources, have equally

William Huggins's observatory after its 1870 expansion. At front, Huggins's wife Margaret.

combined to produce the work in question. . . . The work was a labour of love; and the results are such as might be expected,—of unquestioned excellence, and a standing memorial of his ability and love of science."

In January 1862, having marked time in desultory studies of planets and double stars, Huggins attended a lecture at London's Pharmaceutical Society by his neighbor, William Allen Miller, Chair of Chemistry at King's College. It was here, presumably, that Huggins learned of the momentous discovery by German scientists Robert Bunsen and Gustav Kirchhoff: the perplexing array of dark lines in the solar spectrum—the Fraunhofer lines—might be used to deduce the chemical constituents of the Sun. No longer did the Sun's remoteness place it frustratingly out of reach of laboratory-type scrutiny; to the contrary, its relentless stream of light delivers its chemical signatures right to Earth.

Miller's lecture was nominally about the science of spectroscopy, but was also a retort to Bunsen and Kirchhoff's claim of priority in spectrochemical analysis. Thus, Huggins heard the subject's entire time line of development, from Newton's seventeenth-century experiments to present-day research in England and on the Continent. The saga would no doubt have inspired the autodidact Huggins, who privately sought validation by his peers. How might one establish one's own research credentials among the crowded roster of visual observers?

Decades later, in a much-read retrospective titled *The New Astronomy*, William Huggins described in heroic prose his response to the genesis tale of cosmochemistry:

> This news was to me like the coming upon a spring of water in a dry and thirsty land. Here at last presented itself the very order of work for which in an indefinite way I was looking—namely, to extend [Bunsen and Kirchhoff's] novel methods of research upon the sun to the other heavenly bodies. A feeling as of inspiration seized me: I felt as if I had it now in my power to lift a veil which had never before been lifted; as if a key had been put into my hands which would unlock the unknown mystery of the true nature of the heavenly bodies.

Whether this latter-day account portrays a genuine eureka moment or else the self-congratulatory embellishment of an aging man, Huggins did set his sights—and a spectroscope—on the stars. He stocked his household observatory with the trappings of the Victorian spectroscopist until it resembled a Frankenstein's laboratory: prisms, batteries, electrical spark

William Huggins in his observatory.

coils, Bunsen burners, and chemical powders and fluids. And in William Allen Miller, his neighbor, Huggins found a close-at-hand and knowledgeable (if initially hesitant) collaborator.

Like their contemporaries Bunsen and Kirchhoff, Huggins and Miller were a complementary pair, whose collective productivity far exceeded the sum of what either might have accomplished on his own. Miller was the spectroscopic expert of decades standing, who provided the materials and the know-how to carry laboratory practice into the observatory. Huggins was the astronomer and inveterate maker, whose stamina, manipulative skill, and visual acuity secured results. Lacking the deep state and institutional pockets available at Heidelberg, Huggins bankrolled every aspect of his outsized hobby.

From the start, in early 1862, Huggins and Miller set themselves an ambitious agenda for a technology so new and so tangled. The spectroscope had proven its worth in the energy-lavish domain of the Sun, but was virtually untried on the dim luminal inflow of stars. Like other researchers, Huggins and Miller wished to cross-check the elemental composition of the Sun against those of the stars, based on coincidences of spectral lines. That meant building a spectroscope with sufficient dispersive power to portray the full complement of stellar lines, but not so dispersive as to dilute the already-faint spectrum to the point of invisibility.

They also needed to implement an effective framework of measurement, one that would fix the positions of stellar lines relative to an ironclad fiducial standard that all astronomers could use. Analogous to Bunsen and Kirchhoff's solar spectroscope, the new apparatus had to permit simultaneous viewing of a star's spectrum alongside a calibrated spectrum generated in real time within the observatory. And these parallel spectra must be identically scaled in wavelength, so as to "enable the observer to determine with certainty the coincidence or noncoincidence of the bright lines of the elements with the dark lines in the light from the star."

The completed spectroscope employed two flint-glass prisms and was bolted securely to the eye-end of the Clark refractor. The stellar spectra were observed through an auxiliary telescope fitted with crosshairs. For comparison, an external mirror-and-prism arrangement deflected light into the spectroscope from an incandescent source in the observatory. The English climate being what it is, Huggins spent considerable "downtime" in his laboratory, applying a higher-dispersion, table-mounted spectroscope to the emissions of twenty-nine chemical elements provided by Miller. These spectra he laid out in charts for quick visual reference against the stellar line-arrays.

To maximize their precision of measurement, Huggins and Miller sought a real-time reference spectrum against which to gauge the positions of stellar spectral lines. The Fraunhofer lines could not be used, as the Sun was below the horizon during their telescopic observations. Instead, they settled on a readily available and virtually inexhaustible source of light: the incandescent air between a pair of electrodes. The discharge produced about a hundred widely spaced emission lines—a sufficient number of fiducial marks against which stellar lines could be measured. (By analogy, one might index the positions of Manhattan skyscrapers from a distance by holding up a ruler against the cityscape.) However, a downside to the process was that electricity occasionally jumped from the induction coil to the metal housing of the spectroscope, giving Huggins a rude shock.

Huggins and Miller understood the technical exigencies imposed by the nature of their work. Virtually every component of their spectroscopic apparatus was finely crafted and adjustable to minimize optical distortions; brass covers shielded sensitive components from dust, stray light, and air currents; even the microscopic flexure of the spectroscope in its various orientations was factored into every observation. Standing before their colleagues, many of whom were leery of new technology, the two men would have to justify every procedural step, every decimal point of precision, every scientific assertion. Huggins's fanatical attention to detail reflected equally the scientific imperative and the desire to avoid professional embarrassment. Technology was not his end game, as it appears to have been for Joseph Fraunhofer; it was the vehicle by which he might explicate nature, and simultaneously secure his credibility as a scientist.

Once again, the issue of priority attended the arc of scientific advancement in practical spectroscopy. The case against the originality of the solar and laboratory research published by Bunsen and Kirchhoff percolated in England. Meanwhile, Huggins and Miller found themselves competing with a host of researchers similarly inspired to apply a spectroscope to the stars: Giovanni Battista Donati in Florence, Pietro Angelo Secchi at the Collegio Romano, Lewis M. Rutherfurd in the United States, and even classical-astronomy stalwart George Biddell Airy at the Royal Greenwich Observatory. (Johann Lamont, director of the observatory at the Bavarian Academy of Sciences, had made a perfunctory study of stellar spectra in 1836.)

Both Huggins and Miller were predisposed to claim priority over their cohorts: Huggins, the nonacademic newcomer, keen to make his professional mark; Miller, the aggrieved laboratory veteran, disenfranchised by his rivals in Heidelberg (at least, in his own view). As Miller would have been loathe to acknowledge, credit for discovery occasionally sidesteps the first to reach the goal. The door to scientific fame is ever ajar for those who arrive later—those whose work surpasses the reigning threshold of completeness and exactitude, whose methodology and data instill trust among experts in the field, whose results sort out conflicting observations or theories.

Whether or not Huggins and Miller began their spectroscopic work in early 1862 as they claim, one point is certain: In January 1863, spurred by a development on the Continent, the pace of their project accelerated from a canter into a full-fledged gallop. The Royal Astronomical Society released a translation of a paper by Florentine astronomer Giovanni Battista Donati, describing the dark "striae" or lines he had observed as early as 1860 in the spectra of fifteen stars. Donati, whose eponymous comet of 1858 was the first ever to be photographed, had cobbled together a telescope using a fifteen-inch solar-burning lens from the city's Accademia del Cimento. (The great lens had been commissioned in 1690 by the Grand Duke of Tuscany. On a European tour in 1814, the celebrated chemist Humphry Davy and his assistant, a then-unknown Michael Faraday, had used the device to incinerate a diamond and infer its pure-carbon makeup.)

In Donati's setup, the focused stellar rays passed through a table-mounted prism, and the resulting spectrum was observed through a theodolite. A pair of metal indices—one fixed, the other movable by a micrometer screw—were superimposed onto the view-field, then adjusted to match the separation between any two spectral lines. To prepare for an evening's observation, Donati sighted the Sun, draping a cloth over the burning lens to diminish the intensity of the spectrum to a tolerable level. He rotated the prism to bring a selected solar line into coincidence with the fixed metal index. Leaving the prism in this state until nightfall, he pointed the telescope toward the target star, adjusted the micrometer screw to align the second metal index with a chosen spectral line, and recorded the separation between the indices. Measuring each line in this fashion, Donati constructed a crude drawing of the star's spectrum on the same scale as the Sun's.

Donati's results were problematic. Many of the stellar lines were significantly displaced from those Joseph Fraunhofer had identified decades earlier. In the light of the star Sirius, for example, a dark line that Fraunhofer had noted in the green portion of the spectrum appeared to Donati in the blue. Fraunhofer's famous D line of sodium was completely absent in Donati's spectral drawings of four of his fifteen stars. Donati could not account for the inconsistencies, except to suggest that Fraunhofer might have mistaken one line for another.

In the wake of Donati's paper, Huggins and Miller rushed a preliminary report to the Royal Society on February 19, 1863, outlining their spectroscopic studies of the stars Sirius, Betelgeuse, and Aldebaran. That very day, Huggins learned that Lewis M. Rutherfurd in New York had recently completed visual observations of the spectra of twenty-three stars, the results to be published in a forthcoming issue of the *American Journal of Science.* Huggins and Miller hurriedly appended a statement to their report informing the Royal Society that they had viewed the spectra of at least thirty stars over the previous twelve months; measurements of the lines of these stars were in progress.

A common complaint in this prephotographic era of stellar spectroscopy was the extreme difficulty sighting the ghostly spectral lines. The dispersed glimmer of a star is barely sufficient to stimulate the retina, much less ease measurement of lacunae strewn along a spectral-smudge of light. "On any but the finest nights," Huggins and Miller wrote, "the numerous and closely approximated fine lines of the stellar spectra are seen so fitfully that no observations of value can be made. It is from this cause especially that we have found the inquiry . . . more than usually toilsome; and indeed it has demanded a sacrifice of time very great when compared with the amount of information which we have been enabled to obtain."

By mid-1864, Huggins and Miller had endured enough. Instead of the anticipated inventory of fifty stars, they had obtained detailed results for only four, of which a mere two, Aldebaran and Betelgeuse, had complete spectral-line maps. The spectrum of low-lying Sirius, the brightest star in the sky, had been muddled by turbulence in the air. (The atmospheric roiling had also foiled attempts in January and March 1863 to produce a wet-collodion photograph of Sirius's spectrum.) Other stars had received only cursory treatment. Yet what seems a meager outcome relative to what had been expected is, upon reconsideration, a veritable triumph of industry. With the fickle weather, seasonal cycling of the constellations, limited nightly observing times, and all manner of methodological complexities, the new science of stellar spectroscopy had proved to be a protracted venture. Complete investigation of even a single star, Huggins and Miller realized, would take months, if not years. Their accumulated observations, incomplete as they were, permitted inferences about the chemical nature of stars. Faced with the daunting prospect of more nighttime labor, they decided to publish.

"On the Spectra of Some of the Fixed Stars," read by Miller to the Royal Society on May 26, 1864, led off with a generous nod toward Gustav Kirchhoff as the discoverer of the connection between the dark lines of the solar spectrum and the bright lines of terrestrial flames. (On this occasion, Miller evidently set aside his challenge to Kirchhoff's priority.) This tribute is followed by a stinging dismissal of Donati's spectroscopic work: "[T]he positions which he ascribes to the lines of the different spectra relatively to the solar spectrum do not accord with the results obtained either by Fraunhofer or by ourselves. As might have been anticipated from his well-known accuracy, we have not found any error in the positions of the lines indicated by Fraunhofer." The studies by Rutherfurd, Secchi, and Airy's staff at Greenwich are politely acknowledged, without comment. (A public critique by Huggins convinced Airy to suspend his spectroscopic efforts. Rutherfurd also closed out his spectral examination of stars, using the spectroscope instead to assess the optical properties of telescope objectives.)

Huggins and Miller's broad conclusion echoes that of Fraunhofer many decades earlier: In their essential properties, stars are but distant suns, each encompassing an array of familiar chemical elements. The spectrum of Aldebaran revealed the presence of sodium, magnesium, hydrogen, calcium, iron, bismuth, tellurium, antimony, and mercury. Betelgeuse showed evidence of sodium, magnesium, calcium, iron, and bismuth. There was no doubt about the correspondences between the stronger stellar and elemental lines. Both sets of lines had been viewed simultaneously in the eyepiece: one originating in the fiery atmosphere of a remote star, the other from an incandescent spark across the room.

The hydrogen lines in the spectra of white stars, such as Sirius, were visibly stronger than in yellow stars like the Sun, while lines of other elements were much weaker. Huggins and Miller suggest that the distinctive colors of stars, as well as their contrasting line patterns, might arise from differences in their chemical makeup. (Such variations would go unexplained until the twentieth

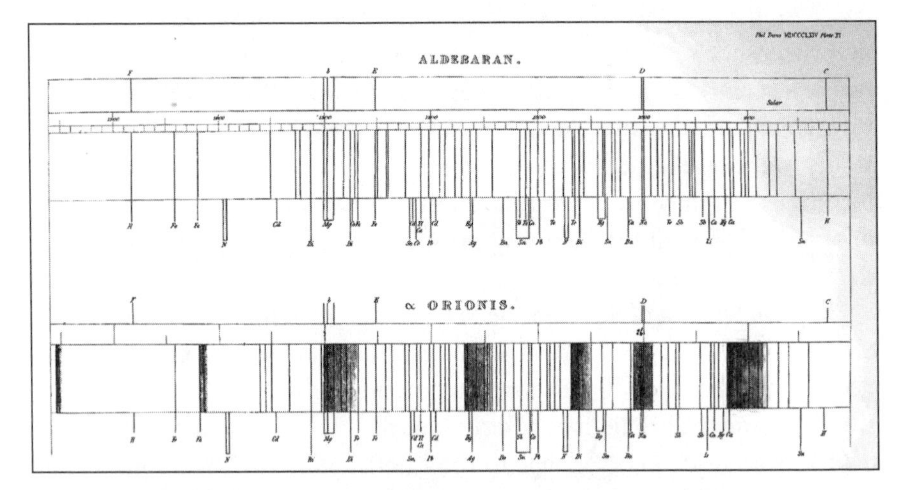

Drawings of the spectra of Aldebaran and Betelguese (α Orionis) by William Huggins, presented to the Royal Society on May 26, 1864.

century, when the physics of stars was elucidated. A star's color and spectrum are much more sensitive to its surface temperature than its composition.)

The symbolic core of the 1864 paper is its spectral-line maps of Aldebaran and Betelgeuse. Placed one above the other, each spans the width of three pages. Scores of fine lines crowd the graphical space, together as seemingly mundane as scratches in the dirt—until one notices the familiar Fraunhofer letters from the solar spectrum; and next, the assortment of italicized labels accompanying the stellar lines: H, for hydrogen; Hg, for mercury; Na, for sodium; Fe, for iron. Here, in what Huggins called the "strange cryptography of unraveled starlight," was science of the most profound order: visual confirmation of the chemical unity of the Sun and stars—and, by extension, of Earth and life.

In the early 1800s, William Herschel showed that Newton's laws of gravity and motion, which fix the trajectories of planets around the Sun, also govern the orbital movements of binary stars around one another. This mechanical unity of the cosmos led scientists to speculate whether a chemical unity reigns throughout space. With their limited, but compelling, spectroscopic evidence, Huggins and Miller infer "that the stars, while differing the one from the other in the kinds of matter of which they consist, are all constructed upon the same plan as our sun, and are composed of matter identical, at least in part, with the materials of our system. The differences which exist between the stars are of the *lower order*, of differences of *particular adaptation*, or special modification, and not differences of the *higher order* of distinct *plans of structure*."

Extending the theme of cosmic unity to its logical conclusion, they suggest that the stars are not just structurally analogous to the Sun, but are "surrounded by planets, which they by their attraction uphold, and by their radiation illuminate and energize. And if matter identical with that upon the earth exists in the stars, the same matter would also probably be present in the planets genetically connected with them, as is the case in our solar system. . . . On the whole we believe that . . . at least the brighter stars are, like our sun, upholding and energizing centres of systems of worlds adapted to be the abode of living beings."

With surprising swiftness, the spectroscope had demonstrated the chemical commonalities of Earth, the Sun, and the stars—and perhaps of life itself, wherever it might exist. Having grown accustomed to—or at least resigned to—the operational hurdles of celestial spectroscopy, Huggins wondered where next to exert this powerful diagnostic lever. Now working alone (Miller had returned to his own research), should he increase the inventory of stellar spectra? Or, having addressed the similitude among these spectra, should he now confront the tantalizing differences in the presence or the relative strengths of certain lines?

The idea of cosmic unity continued to stir Huggins's inquisitive instincts. To the now-seasoned observer, the spectroscopic frontier lay, not necessarily farther into the void, but fainter into the depths of visual perception. What sort of celestial object radiates the dimmest spectrum perceivable through an eight-inch telescope, situated to its detriment underneath the leaden skies of outer London? If the pinpoint blaze of a star is reduced by the spectroscope to a wan glow, Huggins reasoned, the greater challenge would be to hunt for lines within the enfeebled spectrum of a diffuse object: a nebula.

In 1864, little was known about nebulae beyond their ubiquity (some five thousand had been cataloged) and their visual appearance. And even that was an area of active disagreement. The stunning nebular images by Andrew Common and Isaac Roberts lay decades hence; the eye of the astronomer and the hand of the artist still reigned in the rendering of deep-space objects. The Royal Society had just published John Herschel's *General Catalogue of Nebulae and Clusters of Stars*, an expansion of his father William's eighteenth-century original. Descriptions were conveyed via numerical and alphabetical shorthand—*L.* for large, *rr.* for partially resolvable into stars, *bn.* for brightest toward the north side—with the rare triplet of exclamation points appended to objects whose splendor halts one's breath. By this time, Lord Rosse's six-foot-wide Leviathan reflector in Ireland had been applied to the study of nebulae for nearly two decades.

To the eye, some nebulae were large and asymmetrical, while others (dubbed planetary nebulae by William Herschel) were compact and round. Rosse's Leviathan revealed nebulae with a spiral shape, suggesting active coalescence of new solar systems, in the mode of Laplace's widely held nebular hypothesis. Regardless of their form, the intrinsic nature of these diffuse objects was a mystery: Were they clouds of incandescent gas, thus inherently irresolvable; or were they remote star systems, rendered indistinct by distance? Or might the nebular population encompass both of these celestial species?

Huggins believed that little of significance would come from further visual studies, no matter how large the telescope or acute the observer. But a spectroscope-equipped telescope was a different matter. “Prismatic analysis,” he offered, “if it could be successfully applied to objects so faint, seemed to be a method of observation specially suitable for determining whether any essential physical distinction separates the nebulae from the stars, either in the nature of the matter of which they are composed, or in the conditions under which they exist as sources of light."

Of the many diffuse targets visible from his observatory, Huggins gravitated toward the disk-like planetary nebulae. With their relatively high surface brightness, planetaries shine a pale bluish-green, a hue not often encountered among stars. Indeed, Huggins was seeking a class of nebulae more likely to consist of diffuse gas than unresolved stars. The latter would presumably emit a collective spectrum of the sort Huggins had already studied for the past two years. But the spectrum of a cosmic gas cloud had never been observed and might take on any conceivable form. It was here, Huggins surmised, that a greater discovery might be made.

On August 29, 1864, Huggins pointed his telescope—"armed with the spectrum apparatus," he notes in martial metaphor—toward the Cat's Eye, a bright planetary nebula in Draco discovered by William Herschel in 1786. The image in the eyepiece was confounding: "At first I suspected some derangement of the instrument had taken place; for no spectrum was seen, but only a short line of light perpendicular to the direction of dispersion. I then found that the light of this nebula, unlike any other extra-terrestrial light which had yet been subjected by me to prismatic analysis, was . . . monochromatic, and after passing through the prisms remains concentrated in a bright line."

Through Huggins's small telescope, the spectrum of the Cat's Eye appeared as a single, vivid stroke of green, never before seen in the laboratory or in the Sun. (Astronomers attributed the much-debated spectral line to a hypothetical element, nebulium; only in 1927 was it proved to originate in a rarified form of ionized oxygen.) Further examination of the nebular spectrum revealed a fainter emission in the blue, coincident with Fraunhofer's F line of hydrogen, then a third, positioned between the others.

Huggins's surprise stemmed not so much from the *presence* of nebular emission lines—low-density, incandescent gases exhibit such lines—but from their paucity. Planetary nebulae emitted almost all of their luminous energy in the blue-green portion of the color spectrum. The spectra of seven other planetary nebulae possessed at least one of the Cat's-Eye trio of emission lines. Huggins provided no explanation for this remarkable trait, but he did conclude that the spectra were not of stellar origin.

"It is obvious," Huggins reported to the Royal Society in 1864, "that [at least some nebulae] can no longer be regarded as aggregations of suns after the order to which our own sun and the fixed stars belong. . . . In place of an incandescent solid or liquid body transmitting light of all [colors] through an atmosphere which intercepts by absorption a certain number of them, such as our sun appears to be, we must probably regard these objects, or at least their photo-surfaces, as enormous masses of luminous gas or vapour. For it is alone from matter in the gaseous state that light consisting of certain definite [colors] only, as is the case with the light of these nebula, is known to be emitted."

To check his hypothesis, Huggins observed the spectra of several star clusters, as well as the great lenticular cloud in Andromeda. (Its spiral form had not yet been disclosed by the camera.) These objects gave off a continuous spectrum, indicative of diffused starlight, albeit too faint to manifest the characteristic absorption lines. The spectrum of the Orion Nebula, on the other hand, displayed the telltale emission lines of a gaseous body. But when Huggins aimed the spectroscope at stars embedded within the nebula, the gaseous and stellar spectra appeared together. The spectroscopic evidence was clear: nebulae are not a unitary class. Planetary nebulae are incandescent, gaseous objects, whereas at least some fraction of nonplanetaries are, in Agnes Clerke's lyrical phrase, "star clusters grown misty through excessive distance."

Huggins's theoretical assertions were more problematic, especially in hindsight. He took the extreme simplicity of nebular spectra to reflect their chemical makeup rather than physical conditions within the gas: The familiar stellar line patterns are missing because nebulae lack these elements. Thus, the accepted theory that element-rich stars condense out of more basic nebulous material is unsustainable. Where do all the stellar elements come from if not already present in the accreting gas? The spectra of nebulae, Huggins maintained, should be as crowded with bright lines as the stellar spectra are with dark lines. At the same time, he suggested that the Andromeda Nebula, despite its continuous spectrum, might be gaseous, only in a denser state than other nebulae. It would be many decades before the physics of nebulae, as well as the cosmic distance scale, were sufficiently refined to settle these issues. Evidently, William Huggins understood the risk of drawing conclusions from incomplete evidence. In his 1865 report to the Royal Society, he admitted that "science will be more advanced by the slow and laborious accumulation of facts, than by the easier feat of throwing off brilliant speculations."

By the mid-1860s, Rosse's six-foot Leviathan reflector had managed to resolve a number of nebulae into clusters of stars, or at least it appeared so to the observers. (Some of these "resolved" objects were subsequently proven to be gaseous, not stellar.) The nature of the other nebulae remained shrouded in mystery, with astronomers unsure about what their eyes were perceiving in the telescope's dim view-field. That shroud was partially lifted by a modest eight-inch refractor equipped with a spectroscope. Centuries after the power of the human eye had been compounded by the telescope, the act of seeing was again being transformed. Another optical breakthrough, as revolutionary as the first, endowed astronomers with a new, if rangebound, ability to analyze light. Huggins's spectroscopic study of the nebulae was the harbinger of the "New Astronomy," whose adherents would come to dominate the field.

Huggins's stellar and nebular work was immediately hailed by the scientific community. The Royal Society elected him a Fellow; he received awards from both it and the Royal Astronomical Society. The Astronomical Society's official history acknowledges that Huggins became the go-to guy for all things spectroscopic: "[H]e was frequently called upon to speak of the nature and real significance of the harvest of data that were being gathered in the new branch of astronomy to which he devoted his pioneering activities. His clear understanding, both of the power and also of the limitations of the new methods, did much to keep men's minds from jumping to hasty conclusions."

In the 1860s, spectroscopy was reckoned to be the critical breach in the once impenetrable barrier between terrestrial and celestial physics. The advance launched by Robert Bunsen and Gustav Kirchhoff and spurred on by William Huggins and others had penetrated with stunning rapidity into the unknown—indeed, what some had adjudged the unknowable. "Forward!" was Warren De La Rue's rallying cry to the Royal Astronomical Society in 1864, urging his colleagues to embrace the new technology so ably implemented by one of their own. The charge so ordered, but the battle lines unclear, spectroscopic astronomers wondered, "Which way is forward?"

Chapter 18

TRUMPETS AND TELESCOPES

This time was, indeed, one of strained expectation and of scientific exaltation for the astronomer, almost without parallel; for nearly every observation revealed a new fact, and almost every night's work was red-lettered by some discovery.

—William Huggins, *The New Astronomy*, 1897

ON TUESDAY, JUNE 3, 1845, IN A SCENE of Pythonesque strangeness, a locomotive sped through the Dutch countryside with a single flatcar, on which sat a trumpet player. The red-faced musician huffed desperately into his horn, trying to pierce the combined din of the steaming behemoth and the whipping wind. His prearranged tune was unadorned with tonal, rhythmic, or dynamic complexities; in fact, it was a single note, sustained as long as human lungs allow. Two other riders were intent on their colleague's performance despite the railcar's breakneck hurtle across the landscape. The trackside audience was similarly spare: at three places, a trio of listeners, heads cocked in unison toward the approaching thrum of music and machine. The men stood as close to the rails as they dared, having been told that a minimal distance from the travel path was mandatory. Meanwhile, the orchestrator of the surreal concert, Christoph Hendrik Diederik Buys-Ballot, a young meteorologist at Utrecht University, rode in the locomotive with the engineer, scribbling down the train's speed.

As the train approached at nearly forty miles an hour, the spectators along the track heard the crescendoing pulse of the steam engine. Then, from deep within the mechanical tempest, arose a faint whine: the bray of a trumpet. The instrument was audible for only a second or two before the locomotive whooshed by. But to the trained ears of the listeners, the preagreed note sounded perceptibly sharp. Now, in the train's recession down the track, again the trumpet's sound—only this time, unmistakably flat. In surreal splendor, the musical cavalcade swept past the other listeners' posts, where similar alterations of pitch were heard.

The train reversed in Maarssen, five miles from its embarkation point in Utrecht, and accelerated home. This time, a trumpeter at each of the ground stations sounded the standard note, while the flatcar riders strained to hear above the clatter. The result was the same as before: the pitch of the horn was sharp upon the train's approach and flat upon its recession.

Back in Utrecht, Buys-Ballot compiled the data from the day's experiment. He compared the attested degrees of sharpness or flatness to those derived from a mathematical formula based on the train's speed and the note's nominal frequency. The agreement between the mathematical predictions and the experimental outcomes was marginal at best. The designated listeners, a mix of concert-hall musicians and Buys-Ballot's friends, had complained of the background noise: they could barely hear the trumpet amid the clatter of the engine. The musicians responded that they could not play any louder without having to take more frequent breaths.

Holland's Interior Minister gave the go-ahead for a second trial, and Buys-Ballot reconvened his team later that week. This time, he had enlisted *two* players for the flatcar; one would continue to blow while the other took a breath. And he replaced the mellow trumpets with ear-piercing bugles. The results of this follow-up run were little better than they had been two days before. Yet Buys-Ballot was satisfied. He submitted his aptly titled report, "Acoustical Researches on the Dutch Railway," to the venerable scientific journal *Annalen der Physik und Chemie*. Buys-Ballot had pursued—and captured—the most alluring quarry in the intellectual hunt that is science: experimental validation of an idea.

Buys-Ballot's novel experiment had been driven by an equally novel treatise by Austrian mathematician Christian Doppler, who claimed in 1842 that the perceived frequency of a wave is altered by one's state of motion. The so-called Doppler effect

Christian Doppler.

was said to hold for any form of wave: ripples in a pond, ocean swells, sound, even light. (By the mid-1800s, there was ample evidence that light exhibits wave properties.) The key factor in the perceived frequency alteration is the relative motion between the wave source and the observer. Doppler provides a mathematical formula that quantifies the frequency changes under various circumstances: observer stationary and source moving, observer moving and source stationary, or both observer and source moving with respect to one another.

In Doppler's schema, waves from a steadily approaching source are compressed: as their frequency is increased, their wavelength is shortened. Waves from a steadily receding source are stretched: as their frequency is reduced, their wavelength is elongated. But if the source is stationary relative to the observer, no alteration of the waves is perceived.

Doppler next made a bold but ill-informed trespass into astronomy. At the time, no one knew the origin of stars' colors. Most astronomers presumed that color is an intrinsic property of a star, perhaps related to its temperature or some other physical condition of its matter. Doppler proposed instead that the perceived color of a star arises from its movement, specifically, that the line-of-sight, or radial component of the star's velocity through space, dictates its coloration in the telescope. Light waves from a star heading in Earth's general direction will be compressed; effectively, the star is chasing its own Earth-directed waves as it emits them. If, as Doppler believed, stars are intrinsically white, an approaching star will instead appear blue; its luminous energy is skewed toward shorter wavelengths of the spectrum. By the same token, a receding star will appear orange or red, its light shifted toward colors of longer wavelength.

The magnitude of the color shift, Doppler explained, depends on the velocity of the star relative to Earth. A recession velocity of around six hundred miles per second, for instance, would noticeably redden a white star to the eye. A line-of-sight velocity of sixty thousand miles per second would shift all of a star's luminous energy out of the optical range and render the star invisible. Doppler maintained, without evidence, that stars typically move at such breakneck speeds through space. (Modern studies indicate that stellar speeds rarely exceed one hundred miles per second relative to Earth.)

Doppler also ascribed the contrasting colors of many double-star systems to their orbital motion. Locked in a mutual gravitational embrace, the double-star member that is headed toward Earth might appear bluish, while its partner, headed away, might have an orange or red cast. Given that the orbital motion is periodic, Doppler predicted that double-star colors evolve over time. Indeed, because visual color estimation is so subjective and the 1840s stellar database was so sparse, Doppler did find spurious observational support for his assertions.

Only a handful of people were present when Christian Doppler presented his ideas to Prague's Royal Bohemian Society of Sciences in 1842. A year later, a summary appeared in *Annalen der Physik und Chemie*, and two years after that, the unabridged work, in pamphlet form, fell into the lap of Christoph Buys-Ballot. Like many who encountered the paper, Buys-Ballot was intrigued by the logical link between relative motion and wave frequency but suspicious of Doppler's astronomical inferences. As yet, there was no terrestrial means to test the Doppler principle using light waves. Not even a cannon could propel a light source rapidly enough to generate a measurable alteration of color and there was no astronomical route to verification. The radial velocity of an isolated star could not, as yet, be determined by any independent method. And although the laws of Newtonian mechanics do govern the orbital movements of binary stars, without prior knowledge of their distance, the velocities of the stars were incalculable.

Sound travels much slower than light; therefore it presents a more viable path toward a demonstration of the Doppler effect. Accelerating a sound source up to forty miles an hour, Doppler calculated, was sufficient to sharpen or flatten a musical note by a sensible degree. Christoph Buys-Ballot had only to look at Utrecht's new high-speed train to realize that he had the means to test Doppler's principle for acoustic waves. In no time, he convinced the railway's director and the Dutch Minister of the Interior to clear the tracks for his unusual musico-scientific project.

The first half of Buys-Ballot's 1845 report details his confirmation of the Doppler effect for sound waves, while the second half identifies a fatal flaw in Doppler's theory of star colors. The Sun had long been known to emit a significant amount of energy outside the bounds of the visible spectrum, in the form of infrared and ultraviolet light. Stars, too, are presumed to radiate such emissions. Therefore, any luminous energy Doppler-shifted *out of* a given region of the visible spectrum will be replaced by a comparable amount of infrared or ultraviolet energy shifted *into* that region. As a result, the perceived color of a star, regardless of its velocity, will be the same as if it were stationary relative to Earth.

Starting in 1846, Doppler published the first in a series of rejoinders to Buys-Ballot's argument, asserting, again without evidence, that the infrared and ultraviolet output of stars is negligible; hence, there is no replacement energy for emissions motion-shifted out of the visible spectrum. While the scientific community accepted the reality of the Doppler effect (train whistles were everywhere!), volleys over Doppler's star-color theory were exchanged in meeting halls and journal pages well into the 1850s. It didn't help that Doppler was hopelessly outdated in the experimental underpinnings of his own theory. Not only had he neglected the existence of solar ultraviolet and infrared radiation, he had adopted a long-discredited value of the speed of light from 1676.

When Doppler's collected works were published in 1907, Nobel Prize–winning physicist Hendrik A. Lorentz pronounced judgment on his controversial predecessor: "[I]n considering the importance of his principle and the productive use to which it has been put, we must include Doppler as one of the great men of science, although it seems to me, that neither his other work nor the manner in which he defended his theory against various objections and applied it to the colours of the stars, confer on him any claim to such an honorific title."

While astronomers were engaged in the Dopplerian debate over star colors, a little-known conjecture wafted in the background. In December 1848, French physicist Armand-Hippolyte-Louis Fizeau delivered a lecture at the Société Philomatique in Paris. (Fizeau and his collaborator Léon Foucault had taken the first successful daguerreotype of the Sun in 1845.) Unaware of Christian Doppler's prior work, Fizeau laid out the same mathematical formulations now identified with the name of his colleague in Prague. But Fizeau's astronomical assertions were rooted in hard observational facts. Instead of addressing the influence of motion on the overall colors of stars, Fizeau focused on the spectroscopic ramifications. Even at a small fraction of Doppler's breakneck stellar velocities, he found the Fraunhofer lines would be measurably shifted either redward or blueward, depending on the star's direction of motion. Significantly, the lines would retain intact, their diminishment of light merely displaced to a different wavelength. Measurement of the shift of a spectral line from its laboratory-standard wavelength allowed the computation of a star's radial velocity in space.

Since antiquity, astronomers have kept track of the positions of stars, both to provide a reference grid against which to measure the movements of planets and comets, and to track the movements of the stars themselves. These minuscule translations are seen in two dimensions, projected onto the plane of the night sky. If the distance to a star is known, its gradual creep through a constellation can be mathematically converted into actual velocity units, such as miles or kilometers per second. With Fizeau's proposal, the spectroscope promised to reveal a star's velocity in the third dimension, radially toward or away from Earth. Combining a star's planar and radial movements renders its full, three-dimensional motion through outer space. In sufficient number, stellar velocities are key to probing the dynamics of our galaxy. The measurement of radial velocities was one part of the new astrophysics with the potential to arouse the interest of classical astronomers.

Given their eventual impact on astronomy, it might seem incredible that Doppler's and Fizeau's findings were not more widely disseminated upon their release; technical journals did exist, but scientific exchange across disciplines and across national borders still relied heavily upon personal correspondence and word of mouth. (Fizeau's paper was not published until 1870.) It was not until the late 1850s that the famed Scottish mathematician James Clerk Maxwell encountered the Doppler–Fizeau effect in a retrospective volume on optical research. Maxwell had an abiding interest in the phenomenon of color, publishing a series of papers on the subject and conducting the first public demonstration of color photography at London's Royal Institution in 1861. (Maxwell obtained three black-and-white photographs of a tartan ribbon taken, respectively, through red, green, and blue filters, then projected the images through these filters onto a screen, where the magnified ribbon appeared in its true coloration.)

Maxwell was present in the audience on May 26, 1864, when William Huggins and William Allen Miller presented their paper, "On the Spectra of Some of the Fixed Stars," before the Royal Society. The discussion afterward turned to star colors. Huggins and Miller proposed that a star's color derives from a combination of chemical and physical conditions in its incandescent envelope, and is not imposed by external factors, such as the Doppler effect. They noted that in their limited sample, a star's color often relates to the distribution of its absorption lines: A reddish star like Betelgeuse features a profusion of lines in the green and the blue, whereas whitish Sirius displays a rather uniform obscuration of its colors. Of course, further observations were needed before Doppler's explanation of star colors could be ruled out.

At this point, Maxwell announced to his colleagues that there already exists a way, in principle, to test the action of the Doppler phenomenon on the colors of stars. With that, he introduced Fizeau's proposition that "if the colours were really tinged in consequence of the motion of the star or our earth, the lines in the spectrum of the star would not be coincident with the bands of the metal observed on the earth, which gives rise to them."

The shift of the Fraunhofer lines, Maxwell calculated, would be exceedingly small, but potentially measureable with a very high dispersion spectroscope.

There was only one person in the audience that evening with the expertise to run Maxwell's gauntlet: William Huggins. Over the next three years, while pursuing other projects, Huggins refined his telescope's clock drive and acquired a higher-dispersion spectroscope and a more precise micrometer. Despite the improved apparatus, he understood the operational challenge of discerning slight wavelength shifts of stellar spectral lines. Even as starlight is enfeebled by the dispersive action of the spectroscope's train of prisms, the entry slit must be reduced to the merest sliver, to render the lines as narrow as possible. But a razor-thin slit virtually extinguishes the continuous spectrum against which the dark lines are seen; at some point, the visual contrast between the lines and their backdrop becomes all but imperceptible.

Also, if individual line-shifts are to be gauged, it becomes imperative that the stellar and comparison spectra be properly aligned with respect to each other in the field of view. In Huggins's study of the chemical compositions of stars, precise spectral alignment was important, but not pivotal to the realization of the project. Each chemical substance generates a unique pattern of spectral lines; thus, Huggins could rely on a redundancy of line coincidences to support a given identification of a stellar element with a terrestrial element. Measuring a radial velocity inverts the experimental logic: instead of adopting a line coincidence as the sign of an elemental matchup, Huggins would have to assume that the nominal line coincidence has been disrupted by the star's motion; any wavelength disparity between the stellar and terrestrial lines is attributed to the Doppler effect.

Huggins made a series of radial velocity attempts between June 1867 and March 1868, acquiring his new thirteen-prism spectroscope midway through. He focused intensively on bright Sirius, comparing the position of its F line of hydrogen to one generated in an electrified discharge tube in the observatory. The observations were dogged by continual misalignment of the stellar and terrestrial spectra, requiring Huggins to develop a completely new means of diverting light from the discharge tube into the spectroscope.

On April 23, 1868, Huggins submitted his report to the Royal Society, concluding that the F line in the spectrum of Sirius is redshifted compared to its terrestrial counterpart: Sirius is speeding away from Earth at 29.4 miles per second. Huggins's result stood in stark contrast to that of Italian spectroscopic expert Angelo Secchi, who simultaneously reported no measurable line shift. In retrospect, the significance of Huggins's paper lies not in its quantitative result, which is indeed way off the mark (Sirius moves at six miles per second *toward* Earth), but in its symbolic place in the rise of astrophysical observation. It is equal parts instructional and aspirational. Huggins delineates the theoretical and mathematical foundations of stellar radial velocity measurement, quoting liberally from his correspondence with Maxwell.

The need for such review is evident from contemporary accounts of astronomers' lack of grounding in Doppler's and Fizeau's theories. Both Royal Astronomical Society president Charles Pritchard and Astronomer Royal George Airy sought assistance with these novel concepts. Evidently, Airy forgot Huggins's report, for he later wrote Cambridge mathematician George Stokes seeking an explanation of "das Doppelsche Princip," a term he kept encountering in German research papers. "It has something to do with change in the velocity of light but I see no clear description of it—can you help me?"

In his 1868 report, Huggins goes on to detail the procedure to measure stellar radial velocities. He documents his Herculean efforts to overcome a host of instrumental difficulties and justifies the exclusion of certain measurements that did not conform to the final answer. And, although he asserts the validity of his radial velocity of Sirius, there is a sense that he knows it matters little whether it is right or wrong. He sees the larger ramification for the future, that in pushing the limits of technology to their utmost, he has begun to define "a new method of research, which, transcending the wildest dreams of an earlier time, enables the astronomer to measure off directly in terrestrial units the invisible motions in the line of sight of the heavenly bodies."

In June 1872, now using a fifteen-inch Grubb refractor lent by the Royal Society, Huggins published a follow-up study of stellar radial velocities. Despite his own optimistic outlook on the results, the uncertainties of measurement proved to be nearly as large as the reported velocities themselves; bluntly put, the numbers were scientifically worthless. The upshot was clear: Spectroscopic determination of a star's radial motion cannot be accomplished by visual means. The key features of the stellar spectrum are simply too faint and too cramped—and the human eye too subjective—to yield trustworthy measurements through the eyepiece of a small telescope. Yet no sooner had this visual door slammed shut than the photographic door opened, if only a crack.

Just two months later, across the Atlantic and up the Hudson, Henry Draper recorded the first photograph of a star's spectrum with his new twenty-eight-inch reflector. Within the dime's-width spectral band of Vega, a million times more remote than the Sun, appeared several of Fraunhofer's absorption lines. Of course, there was no hope of gauging their Doppler shift on such a compressed scale, nor boosting the exposure times allowed by the crude wet-collodion process. Nevertheless, the future of stellar radial velocity measurement lay inchoate in that dusky image. Even as the visual study of stellar spectra exited the stage, its successor readied in the wings. The 1870s brought the essential chemical leap in photography—the gelatin dry-plate process—that would spark the auspicious merger of photographic and spectroscopic practice in astronomy.

Chapter 19

Burn This Note

> I want to tell Huggins how much you have done—for strictly speaking between ourselves I think he is afraid of you *now*.
>
> —U.S. Naval Observatory astronomer Edward S. Holden to Henry Draper, August 2, 1876

In 1875, with his colleague William Allen Miller dead now five years, William Huggins married Margaret Lindsay Murray, an Irish solicitor's daughter and an astronomy enthusiast since childhood. Twenty-four years his junior, Margaret swept into her husband's solitary household like an invigorating breeze. Serving initially as Tulse Hill's "scientific housemaid," as she wryly put it, she rose quickly to become William's full-fledged research collaborator. Within a year, Margaret's handwriting appears in the observatory notebooks, sometimes in a telltale first-person reference. Entries grow more detailed than before, and reveal her involvement in every facet of the work. She was the inexhaustible engine who kept Tulse Hill's demanding night work on track. Only many years later would she be listed as coauthor with William of their various research papers. For now, she accepted her unsung role, impelled by her own inquisitiveness to take up a vocation otherwise closed to women.

Margaret Lindsay Huggins.

In her recollections, Margaret opens a window onto the ritualistic tedium of astronomical observing, as well as the intellectual fire that drives one to endure this hard and sometimes hazardous line of work. She provides a glimpse of her ascension into the nuts-and-bolts domain of Victorian-era astronomy: "I had to teach myself what to do by degrees: at first I had my difficulties, but now my eyes are trained and are very sensitive. Also my hands respond very quickly and delicately to any sudden necessity. I can go and stand well at good heights on ladders and twist about well. (Astronomers need universal joints and vertebrae of India rubber.) . . . As I observe, I direct William as to what I need and he moves me bodily on my ladder, so that I am not disturbed more than is necessary."

The Tulse Hill research program, being astrophysical in nature, included a never-ending roll of daytime tasks to support the nighttime observations. In the household laboratory, for instance, the work is varied. "It may be photographic," Margaret relates, "in which case I should help in arranging instruments, keeping the light right, and so on, if we are working on the sun. If working electrically, I should work the batteries, fix electrodes, and be generally handy. I may take a turn mixing up chemicals, pounding, weighing, dissolving, boiling—in short be a jack of all trades. When needful I dust and wash up the laboratories, for no housemaid is allowed into those sacred precincts." Until William hired someone for the onerous task, Margaret took the cleaning of the laboratory's steam engine in stride: "One is interesting with a lump of engineer's waste in one hand and some nasty oily stuff in a can in the other."

In 1874, a year before Margaret's arrival at Tulse Hill, William Huggins had written to Henry Draper about the prospects for stellar spectrum photography. Having succeeded in 1872 where Huggins had failed, Draper was sanguine about his own efforts, but cautioned others: "I am very glad to learn that you think of continuing your former experiments in applying photography to stellar spectra. I have made some new trials in that direction with my silvered glass reflector of 28 inches aperture and find that I can get the great bands of Vega readily and even the spectrum of alpha Aquilae. It is a very difficult subject and requires so favorable a series of circumstances that a number of observers might work at it a long time before fine results were achieved."

For a veteran visual observer like William Huggins, the photographic art was a substantial deviation from past practice. By its chemical operation, it lacks the immediacy of peering into the eyepiece and seeing whether anything is there, whether proper focus has been achieved, whether magnification should be modified. In telescopic photography, the astronomer is repurposed as an auxiliary device, to direct and stabilize the light stream into the camera, flip the shutter, swap plates, and pray that the accruing image, when developed, is worth the investment of time put into its making.

If Huggins balked at pursuing Draper's "very difficult subject," his partnership with Margaret appears to have banished any doubts. With evident determination, Huggins set about to transform his spectroscopic study of stars from visual to photographic, exploiting his wife's longtime experience with the camera. (Photography was a rare, but growing, avocation of Victorian-era women.) The fifteen-inch Grubb refractor had been dismounted for an eighteen-inch reflector, also lent by the Royal Society, that was more suited to photography. Margaret reports that she "was occupied on all favorable days in testing and adjusting this photographic apparatus upon the solar spectrum: at the same time testing different photographic methods with a view to finding . . . the most sensitive and . . . the quickest method for star spectra." Both wet-plate and dry-plate technologies were tried, with the latter winning out by summer's end.

The photographic plates were one and a half inches long, with the spectrum itself crammed into a mere half-inch. Even so, the image definition sufficed to permit precise measurement of line positions under a microscope. Because of the lack of red-sensitivity of the photographic emulsion, only the violet and ultraviolet parts of the spectrum were recorded. The yellow D lines of sodium, for example, were literally out of the picture. (Such longer-wavelength spectral lines continued to be studied by eye.) Even with the telescope's improved clock drive, manual correction was necessary to capture the delicate features of the spectrum. The beam of starlight cast onto the spectroscope slit had to be monitored for the entire exposure—typically fifteen to thirty minutes—and the appropriate corrective jogs applied to the telescope.

Throughout 1876, William and Margaret Huggins tested and tinkered equally, adjusting their apparatus for optimal alignment, stability, sensitivity, and ease of use. Nevertheless, a wealth of instrumental obstacles and dearth of limpid nights conspired to impose a glacial pace of progress. By year's end, William Huggins was eager to assure the Royal Society that its telescope was being put to effective use. Although his results were preliminary at best, Huggins rushed a report on the spectra of Sirius and Vega to the Royal Society on December 6, 1876. Right up front, he hoists the priority flag, suggesting that the new project is a resumption of his failed photographic efforts with Miller in 1863. His concern over priority proved to be well founded.

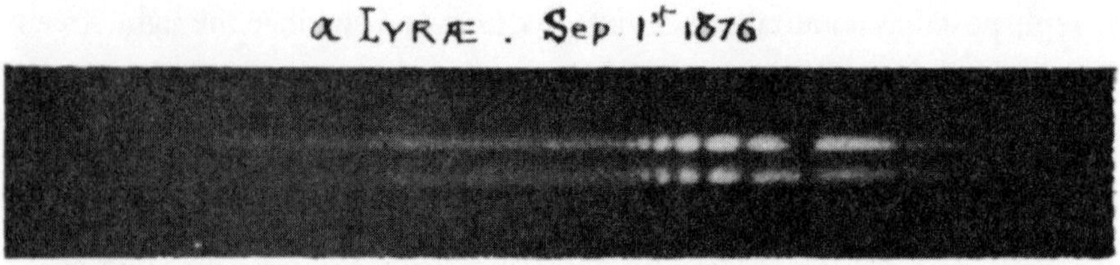

α Lyræ . Sep 1st 1876

Photographic spectrum of the star Vega (Alpha Lyrae) by William Huggins, September 1, 1876.

Just four months before their report to the Royal Society, Mr. and Mrs. Huggins received a visit from astronomer Edward S. Holden, sent by the United States government to assess the state of astronomical instrumentation in England. (Holden would later be appointed director of California's Lick Observatory.) "Huggins is very pleasant & everything about him is *thorough*," Holden informed his friend, Henry Draper, in New York, "his Obsy & working places are part of his house & every bit of apparatus in them works like a charm—smoothly and easily. He has a wife now, & she is devoted to him & science & altogether they seem to have work in them." To Holden, the high-tech equipment at Tulse Hill was of the sort in Henry Draper's observatory at Hastings-on-Hudson.

Since Draper's breakthrough picture of the spectrum of Vega in 1872, he and his wife Anna had practiced their photographic spectroscopy on a variety of celestial objects. In December 1876, coincident with Huggins's report to the Royal Society, Draper filed a progress note with the *American Journal of Science and Arts*. He confirms that, starting in October 1875, wet-collodion plates of Vega's spectrum were obtained through both his twenty-eight-inch reflector and twelve-inch refractor. A number of spectroscopes were tried, but none produced an image markedly better than Draper's original. Draper's 1876 report only hints at the overwhelming tide of frustration that rises from his observatory notes. "The research is difficult and consumes time," he complains, "because long exposures are necessary to impress the sensitive plate, and the atmosphere is rarely in the best condition. The image of a star or planet must be kept motionless for from ten to twenty minutes, and hence the driving clock of the telescope is severely taxed."

In the midst of his spectroscopic travails, Henry's father John counseled him from the sidelines: "You can either make star spectra, or examine different regions of the sun's surface, and have a paper ready for the Astronomical Society before Christmas. To do this you ought to come out every day on the 4 o'clock train, have dinner on your table at 5, have the carriage at your door at 6, get to Hastings at 6½ and work till 9½, then go home. Permit nothing whatever to interfere with this program and you will accomplish a great deal." With the publication of his report and with no evident route toward success, Henry Draper suspended his stellar spectrum studies. For the next two years, he was consumed by his controversial research on the alleged presence of oxygen in the Sun. Meanwhile, William and Margaret Huggins proceeded full-bore, gathering exposures of stellar spectra and hoping to get their results into print.

During the summer of 1879, while in England for a meeting of the Royal Astronomical Society, Henry and Anna Draper rode out to Lambeth to visit William and Margaret Huggins. After pleasantries, they spoke about their respective attempts to photograph the spectra of stars, as well as about the faster gelatin dry plates now commercially available in London. Touring the observatory, Draper asked where he might acquire some of Huggins's specialized prisms and lenses. Huggins arranged with his optical supplier that Draper return to the United States with the materials in hand.

"I was willing to show Draper everything," Huggins confided to his influential American colleague Charles A. Young on January 31, 1883, shortly after Draper's death. "He was greatly surprised at my spectra, and I told him the main points I had made out, and that my paper was in the course of preparation. I then offered to show him my special apparatus and arrangements; he immediately said 'I should like to see everything but I have given up star spectra and I do not intend to do any more, you need not hesitate to show me your apparatus. . . . I believed him to be a man of honour.'" Having conveyed his distress, Huggins instructed Young to "burn this note."

What sparked Huggins's ire was Draper's hurried resumption and publication of his stellar spectroscopy research. Draper presented his paper, "On Photographing the Spectra of the Stars and Planets," to the National Academy of Sciences on October 28, and Huggins his paper, "On the Photographic Spectra of Stars," to the Royal Society on December 11, 1879. Huggins firmly believed that Draper's uncannily rapid progress could only have come from his adoption of equipment and methods perfected at Tulse Hill. Margaret Huggins mailed a copy of her husband's 1883 letter to Edward Holden, a friend of both theirs and Draper's. "You cannot imagine the pain this Draper matter has caused us," she adds. "I was bitterly angry that my gentle noble-breasted husband should have been so used. . . . Who ever heard to any purpose of Dr. Drapers star results before his visit to Tulse Hill?"

Huggins's 1879 paper is at once educational, technological, and scientific, with a speculative coda on the physical ramifications of spectral line patterns. (In retrospect, one wishes Huggins had omitted this last.) He and Margaret adopted a bare-bones spectroscope with a single prism of Iceland spar, which was both more dispersive and more transparent to ultraviolet rays than either glass or quartz. Their primary targets were the so-called white stars: Sirius, Vega, Altair, Deneb, and Spica. The white-star spectra exhibited far fewer absorption lines than did the Sun; among the lines they had in common, some appeared more prominent in the star, others less.

The key feature that defines a white-star spectrum, declared Huggins, is its pattern of twelve ultrastrong lines, several of which were found to coincide with known lines of hydrogen. The spacing of these lines was intriguingly regular: as one proceeded from the blue toward the ultraviolet (toward shorter wavelengths), the lines grew progressively closer. Prior visual observations added two long-wavelength lines to this "grand rhythmical group," as Huggins described them. He wondered "whether these lines are not intimately connected with each other, and present the spectrum of one substance." (Indeed, the line array would come to be identified with hydrogen and its spacing quantified by a mathematical formula derived in 1885 by Swiss schoolteacher Johan Jacob Balmer. The unique pattern arises from the way light interacts with the hydrogen atom, whose structural regularity would be delineated by Neils Bohr in 1913.)

Huggins found that the orange star Arcturus was unlike white stars in its distinctly solar-type spectrum. Yet, unaccountably, its K line of calcium appeared much stronger than the Sun's. Huggins's published rendering of Arcturus's spectrum was so line-rich as to resemble a fine-toothed comb.

Not surprisingly, given its longer gestation, Huggins's paper is more far-reaching than Draper's. The latter reads like a tutorial and progress report rather than a completed scientific work: there is only qualitative examination of the spectral lines plus frequent mention of observations yet to be conducted. Draper's conclusions rest on a limited number of spectral plates of Vega, Arcturus, and Capella, obtained with his twelve-inch refractor

ARCTURUS. June 9th 1878

Two photographic exposures of the spectrum of the star Arcturus by William Huggins, June 9, 1878.

between August 6 and October 4, 1879, plus his prior plates of Vega and Altair. (The twenty-eight-inch reflector telescope was undergoing maintenance in 1879.) Like Huggins, Draper distinguishes the spectra of white stars from those of their yellow and orange counterparts. He also cites the prominent array of absorption lines in the spectrum of Vega; these he attributes to hydrogen, but makes no mention of their regular spacing.

Did Henry Draper seek to establish priority over William Huggins by publishing first? Or was his paper meant to inform American astronomers about the evolution and current state of stellar spectrum photography? By October 1879, Draper had reached the end of his observing season and would not return to Hastings full-time until the following June. To sit on his observational results, provisional as they were, until the following October must have seemed pointless. While he does not explicitly claim to be the inventor of stellar spectrum photography, he yields no ground to Huggins in his account of its development. Draper writes that his own "experiments and the preparations for them have extended over more than twelve years," that is, back to 1867 when he launched construction of the twenty-eight-inch telescope. He cites at length Huggins's abortive attempt at stellar spectrum photography in 1863. And he acknowledges his own recent inspection of the Tulse Hill observatory, thanking Huggins for his recommendation to use gelatin dry plates.

That Henry Draper leveraged his research through Huggins's willing aid is beyond doubt. Thus, his subsequent rush to publish before Huggins might strike modern sensibilities as a breach of collegial protocol, if not scientific ethics. In science, the publication date has always been regarded as the benchmark of priority in the case of *discovery*. Yet history has shown that credit for an evolving theory or field, such as stellar spectrum photography, often goes not to individuals who are first to publish, but to those who most convincingly establish the validity and worth of their results.

Not that this would have been of any comfort to William Huggins. He remained ever-vigilant to intrusion upon his scientific legacy, and was convinced that it had been undermined, at least in America, by Henry Draper. In his 1883 letter to Charles Young, Huggins suggests that "it may be possible for you after a time to bring about quietly without any direct reference to Dr. Draper a more truthful appreciation of my work on the photographic spectra of stars than exists in America if I may trust to popular prints." Charles Young took no side in the dispute, and Huggins continued to stew in private. The matter might have rested there had not Harvard University astronomer Edward C. Pickering swept up Henry Draper's banner.

A bulldog like William Huggins when it came to professional affairs, Edward Pickering sought to establish Harvard as the preeminent center for what he called the "physical side of astronomy." Pickering had toured the Harvard Observatory in 1861 when he was fifteen years old. Then-director George P. Bond treated the young astronomy enthusiast to views of the Moon, the Ring Nebula, and the quadruple star system Epsilon Lyrae, an experience Pickering set down in mind-numbing detail in his diary: "This star [Epsilon Lyrae] appears double with a very low power and even, as Mr. Bond said, it can almost be seen without a telescope, through *this* telescope it appeared to consist of 4 stars, by pairs, each pair appearing at some distance from the other . . ." The sentence tumbles on for another forty-five words before blessedly encountering a period.

Pickering received his bachelor's degree from Harvard in 1865, taught at the Massachusetts Institute of Technology (MIT) until 1877, then became Harvard Observatory's director for the next forty-two years. Having soon made his mark in the wholesale visual measurement of stellar brightness—the trademark-redolent "Harvard photometry"—Pickering embarked on its photographic equivalent. His aim of quick success suggested the strategic wisdom of adding stellar spectrum photography to the institution's research program. By the time William Huggins's 1883 letter of grievance was steaming its way across the Atlantic, Edward Pickering had already made his initial move. He contacted Henry Draper's widow, Anna, declaring his deep interest in her husband's spectroscopic research. He had spoken to Henry during the National Academy gala at their home last November, only days before Henry's death: "I urged upon him the importance of an early publication of his stellar spectra. . . . I called his attention to the similar work now in progress by Dr. Huggins, and that a delay might seriously diminish the value of the investigations."

With Henry gone, Pickering inquired, what was to become of his spectroscopic plates, his equipment? Anna Draper replied with noble resolve that she intended to take up where her husband had left off, "yet I feel so very incompetent for the task that my courage sometimes completely fails me—I understand Henry's plans and his manner of working, perhaps better than anyone else, but I could not get along without an assistant. . . . He wished to get the spectra of several of the winter stars, and this we intended this winter. It is so hard that he should be taken away just as he had arranged all of his affairs to have the time to do the work he really enjoyed, and in which he could have accomplished so much."

Pickering urged Anna to publish Henry's spectroscopic results, even if incomplete. The measuring engine at Harvard could be used to speedily determine the wavelengths of the spectral lines. In early February 1883, Anna traveled to Cambridge and handed Pickering twenty-one photographs of stellar spectra taken at Hastings. A month later, wavelengths in hand, Pickering sent Anna a draft of a paper that might be published under Henry's name. With an eye toward England, he asked permission to present the results at the April 11 meeting of the American Academy of Arts and Sciences: "It is not necessary that the paper should be completed as long as the observations are finished. The work will take the date from the time of presentation, and of course it is desirable that this should be as early as possible." Anna Draper agreed to the presentation, but held out for an extended written volume that would serve as a proper legacy to her husband. Henry Draper's *Researches on Astronomical Spectrum Photography* appeared in February 1884. The monograph contained seventy-eight stellar spectra and Pickering's wavelength measurements for twenty-one. At Anna's request, there was an introduction by Charles Young, excerpts from Henry's observing logs, and pictures of their observatory in Hastings. Copies of the book were distributed to a roster of prominent astronomers—including William Huggins.

Huggins dashed off a diatribe to Pickering on March 12, 1884, indicating that the published wavelengths were "*very wild indeed*. . . . There can be *no doubt whatever* that your [measurements are] quite *inaccurate* and increasingly so as the wavelengths get smaller. . . . It would be a great drawback to the advancement of science if your measures were allowed to pass without criticism on this point, but as it is obviously of the nature of a slip, I should be so glad if you would *yourself* look into the matter and yourself publish the correction as soon as you find out the cause."

Pickering forwarded the letter to Anna Draper. "Fortunately," he assures her, "Dr. Huggins' arguments, that results must be wrong if they do not agree with his, will not generally be regarded as conclusive unless supported by facts. . . . [W]e must conclude that American stars are more refractory or less refrangible than English ones." On the same day, he fired off a reply to Huggins, indicating that he would re-examine the Harvard data, but also scolding Huggins for failure to publish details of his own measurement process, as Pickering himself had: "Your publication [of 1880] does not enable the reader to verify your reductions, and he has no means of checking your results. I hope you will supply this deficiency by a fuller publication and thus remove what seems to me the weakest part in your very valuable contribution to the subject."

Undaunted, Huggins shot back in mid-April 1884, "I cannot tell whether you have been led astray by your formula or by measuring wrong lines, probably the photographs were very badly defined, or there may have been some shift in the apparatus. Some of our best men to whom I have shown your paper do not seem to think your method serious enough to make it worth while for me to take any notice of the paper. . . . I have written to you plainly indeed, because I felt you would take it as a mark of friendly feeling." This letter, too, Pickering forwarded to Anna Draper, who wrote with evident sympathy, "I felt very sorry that you should have been subjected to such an ungentlemanly attack, through your interest in Dr. Draper's work. If Dr. Huggins did not find the results of your measures agreed with his, there was no objection to his saying so if he had expressed himself in a more courteous manner."

The transatlantic volleys ceased at this point, neither man willing to concede error. (According to modern measurements, both Huggins and Pickering were off the mark.) Pickering continued to develop the celestial photography program at Harvard, despite a chronic shortage of funds. In 1885, he conducted tests of an objective prism spectroscope, which situates a large prism in front of the telescope's main lens. With such a configuration, the spectra of dozens or even hundreds of stars are captured at once on a single photographic plate. A five-minute exposure sufficed to record the spectrum of every star visible to the naked eye in a swath of sky ten degrees on a side. The spectra of forty stars in the Pleiades cluster were obtained in thirty-four minutes, a task that might take weeks for a conventional, single-star spectroscope. There was a price to the mass-collection process: the spectral images were tiny, each about half a millimeter long. (Henry Draper's spectra stretched some ten times that length.) Nevertheless, under microscopic scrutiny, Pickering could identify all the major absorption lines and contemplate the large-scale classification of the stellar species.

Anna Draper tried to continue her husband's work herself, but was unable to find a suitable assistant. On February 14, 1886, she gave Harvard $1,000 to support Pickering's stellar spectrum program, with the promise of more funds in the future. Pickering promptly mailed out a circular to astronomers and scientific institutions worldwide announcing the creation of the Henry Draper Memorial, whose initial object was to photograph the spectrum of every star visible to the unaided eye from Cambridge, Massachusetts. By summer's end, Pickering had taken a total of 224 plates depicting some six thousand spectral images of around three thousand stars. (The plates overlap in their sky coverage, so most stars appear twice.) The photographs were fine-grained enough that substantial enlargement preserved details seen in the minuscule originals. Anna Draper was thrilled with the quality of the Harvard spectra. With undisguised relish, she mused to Pickering, "I wonder what Mr. Huggins will say when he sees them."

Pickering knew what William Huggins would say. How could he fail to be impressed, even intimidated, by the profusion of crisp spectral images? But Pickering knew, too, that his competition lay not in the garden-based facilities of zealous amateurs, but in the well-endowed observatories of academic and governmental institutions. In a letter dated January 18, 1887, he alerted Anna Draper to the likelihood that Harvard's success would "induce other astronomers to undertake the same work." He offered a strategy to preempt competition by expanding the Draper project to fainter stars and to the Southern Hemisphere sky. Draper replied within the week, "I quite agree with you in feeling that I should like to appropriate the entire ground that is possible." She raised her financial stake in the project and subsequently transferred the Hastings Observatory eleven-inch refractor and twenty-eight-inch reflector to Harvard.

William Huggins learned of the Henry Draper Memorial in May 1887. Pickering's announcement had its intended effect. "I have just received a paper from Harvard Observatory," Huggins wrote to mathematician George Stokes at Cambridge University, "& there, through the large endowment of Mrs. Draper the photography of star spectra is to be carried on upon a magnificent scale. Three large instruments are to be kept at work all through the night by relays of photographers & photographs to be enlarged by special methods & measured by other men. The question is, is it worth my while to continue working in this direction now that it is being done under circumstances with which no zeal & perseverance on my part will enable me to be in an equal position."

Huggins pivoted his spectroscopic research to avoid overlap with the ongoing work at Harvard. He and Margaret concentrated on the ultraviolet part of the stellar spectrum, while pursuing further studies of the Orion Nebula and the problematic nebulium line. In 1889 came the first of many Tulse Hill publications listing Margaret as coauthor. She was awarded honorary membership in the Royal Astronomical Society in 1903. (Full membership was denied women until 1915.) William and Margaret Huggins will be remembered as pioneers, if not the parents, of celestial spectroscopy. Their scientific achievements, combined with those of others, established the foundation of stellar and nebular spectrum photography.

The first substantive result of Anna Draper's largesse was the publication in 1890 of the *Draper Catalogue of Stellar Spectra*, listing the position, brightness, and spectral type of 10,351 stars. By far the most extensive star compilation to date, it was the opening salvo in Pickering's plan to monopolize large-scale photographic mapping and spectroscopy of the stellar realm. To generate the catalog, Pickering had hired a photographer to take the plates, and Williamina Fleming, his former housemaid, to measure and classify the spectra. (Fleming, a Scottish schoolteacher, immigrated to Boston in 1878 with her husband, who abandoned her and their unborn child. Pickering hired Fleming as a domestic while she was pregnant, then offered her part-time clerical work at the observatory. In 1881, he added her to the permanent staff.)

Because the photographs captured more spectroscopic detail than visual observations, Pickering and Fleming developed an alphabetic spectral classification system to replace the 1860s-era, Roman-numeral scheme of Italian spectrum pioneer Angelo Secchi. (Secchi's I through IV became Harvard's A through N.) The system was revised during the 1890s by Annie J. Cannon, a Wellesley physics graduate hired to reduce the plates from Harvard's new high-altitude observatory at Arequipa, Peru. Cannon went on to lead a team of female assistants in the completion of the *Henry Draper Catalogue*, published in nine volumes between 1918 and 1924, and containing more than 225,000 stars. (This sort of low-wage work was virtually the only career option for women in astronomy.) As Pickering had hoped, Harvard's spectral classification system was officially adopted by the astronomical community; an expanded form is still in use today.

Williamina Fleming.

The goal of spectral classification is to seek commonalities among stars, to whittle down a vast and diverse population into a manageable number of species. And while discoveries were made during the inspection of spectra for the Draper catalog, these were incidental to the project's purpose: to provide astronomers with an essential stellar database they might use to explore the physical nature of stars. To

Annie J. Cannon.

Pickering, Cannon, Fleming, and the other spectroscopic analysts, the various spectral line patterns were nothing more than a visual means of comparison, of sorting stars into categories based on a particular facet of their light. The underlying physics, as yet unknown, was immaterial to the work. After eyeing thousands of spectra, Annie Cannon recognized an overall progression in the intensity of certain line patterns, and within that progression, she found subtler subprogressions. On that basis, she merged and shuffled Williamina Fleming's original alphabetical arrangement of spectral types into the seemingly random sequence O, B, A, F, G, K, and M. Each letter category was broken into subdivisions indicated by a numerical suffix 0 through 9, as in G2, the spectral type of the Sun. This purely taxonomic arrangement was later proven to parallel the range of stellar surface temperatures, from high to low. Secondary correlations with stellar luminosity were discovered in the early twentieth century, leading to the development of the Hertzsprung-Russell diagram, a key graphical tool used in exploring the physics and evolution of stars.

Pickering's entire enterprise hinged on the successful transition of stellar spectroscopy from a visual to a photographic art. That transition occurred with the introduction of the gelatin dry plate. Whereas spectral classification at the eyepiece requires an individual trained in the complexities—and inured to the discomforts—of telescopic observing, photographic classification can be conducted by workday office staff, whose expertise is channeled toward a single goal: spectral classification. Specimens are available for reinspection, communal consultation, and reproduction. English astronomer Normal Lockyer summed up the powerful synergy of the plate and the prism: "I do think we have in photography not only a tremendous ally of the spectroscope, but a part of the spectroscope itself. Spectroscopy, I think, has already arrived at such a point . . . in connection with the heavenly bodies, that it is almost useless unless the record is a photographic one."

Chapter 20

A Spectacle of Suns

> Astronomy paints its picture in the brighter colors. A star, regarded as a center of attraction, or as a reference point from which to measure celestial motions, awakens little enthusiasm in the popular mind; but a star regarded as a sun, pouring out floods of light and heat as a consequence of its own contraction, torn by conflicting currents and fiery eruptions, shrouded in absorbing vapors or perhaps in vast masses of flame, appeals at once to the popular imagination.
>
> —James E. Keeler, "The Importance of Astrophysical Research and the Relation of Astrophysics to Other Physical Sciences," 1897

THE ANALYSIS AND CLASSIFICATION of stellar spectra paved the way for scientists to determine the physical properties of stars. Yet it failed to rouse the interest of number-crunching classical astronomers, for whom stars remained points of light speckling a celestial vault. To this cohort of scientists, photography held promise as a vehicle of mass measurement of stellar positions, as evidenced by Lewis Rutherfurd's plates of star clusters and the vast *Carte du Ciel* mapping project. And in chronological photographs of star fields, taken across decades, lay the potential to detect the glacial drift of stars across the night sky, a clue to their actual movements through space.

As adept as they were at quantifying such two-dimensional metrics of position and motion, nineteenth-century astronomers found it hard to pin down the scale of the cosmic third dimension, radially away from Earth. The distances of a few dozen stars had been deduced from visual estimates of their annual parallax, the tiny shift in a star's position when viewed from opposite extremes of Earth's orbit. These generated a sparsely populated map of our stellar neighborhood. (Photography would swell the number of parallax measurements into the thousands during the early twentieth century.) Thus, the spectroscope was judged irrelevant to the work of classical astronomy—except for one prospect: Christian Doppler's supposition that stellar radial velocities might be gauged from spectral-line shifts.

By the mid-1880s, the visual spectroscopy of William Huggins, Angelo Secchi, and the Royal Greenwich Observatory had failed to produce a single reliable Doppler shift. No matter how steadfast the observer and well built the instrument, the errors of measurement were comparable to the stellar velocities being measured. Then, with stunning rapidity, the camera-enabled spectroscope, or spectrograph, accomplished what its eye-based predecessor could not. And in that success, even the most committed visual adherents came to see the potential of the spectroscope in their line of work.

Hermann Carl Vogel at Potsdam Astrophysical Observatory outside Berlin was well versed in the merits and pitfalls of the spectroscope. In the early 1870s, while directing a private observatory near Kiel, he tried unsuccessfully to discern the Doppler shifts of stellar spectral lines. However, in observing the limb of the Sun, he found that lines at the solar disk's eastern and the western edges were shifted in opposite fashion: the eastern limb, toward shorter wavelengths; the western limb, toward longer wavelengths. The Dopplerian conclusion is that luminous matter at the Sun's eastern limb is moving toward the observer, while the matter at the western limb is moving away from the observer. That is, the Sun rotates. Vogel's computed solar-rotation velocity agreed with the value derived from the movement of sunspots across the Sun's face. Furthermore, Vogel confirmed that the Sun does not rotate as a solid body: Its rotation rate varies with solar latitude, fastest at the equator, progressively slower toward the poles. At a time when scientists still debated whether the Doppler effect observed for sound waves applies as well to light waves, Vogel's solar observations provided definitive evidence that it does.

In 1887, Vogel embarked on a four-year-long program to determine the radial velocities of stars by photography. Having tried his hand at visual detection of Doppler shifts, he hoped that time exposures would smooth out the skittery spectral images seen by eye. Potsdam's eleven-inch, wooden tube refractor was a mule of a telescope: generally capable, but hardly suited to the rigors of ultraprecise work. Vogel knew that generating the required high-dispersion spectra demanded the utmost stability of the entire optical–mechanical system against expansion, contraction, and flexure. The frame of his dual-prism spectrograph was fashioned out of steel, with structural bracing inside and out. The camera, too, was made of steel, its plate holder of brass. The comparison spectrum was provided by a hydrogen-gas discharge tube located within the telescope itself; so placed, the starlight and the hydrogen light formed parallel rays. Any flexure of the instrument shifted the stellar and comparison spectra in unison. A tiny auxiliary scope allowed Vogel to sight the star's beam on the spectrograph slit, and keep it stationary for the duration of the hour-long exposures.

Vogel's brand of spectrographic work was far afield from either cosmochemical analysis or stellar spectral classification. Instead of photographing a broad swath of spectrum displaying dozens or hundreds of lines, he recorded only a narrow interval around each star's Fraunhofer G line, plus the proximate gamma line of hydrogen from his discharge tube. The separation of these lines was measured under a microscope, then compared to their separation in a photograph of the solar spectrum. The deviation of the star's G line from its solar position revealed the star's Doppler shift and, via a mathematical formula, its line-of-sight motion.

In 1892, Vogel published the radial velocities of fifty-one stars, streaking through space at up to thirty miles per second—more than one hundred thousand miles an hour—relative to Earth. The average measurement uncertainty was less than two miles per second, a tenfold improvement over that of Greenwich Observatory's long-running visual program. In a welcome confluence of the new and the old, Vogel's radial velocities for Arcturus, Aldebaran, and Betelgeuse were in close accord with those only recently measured by eye by James Keeler using Lick Observatory's thirty-six-inch refractor. The consistency between the photographic and visual results lent further credence to Vogel's work. That an eleven-inch refractor, fitted with a camera, matched the capability of the world's largest mountaintop telescope highlighted the potential of photography to revolutionize the practice of astronomical observation.

Vogel also released complete specifications of the apparatus and methods used at Potsdam, inspiring similar spectrographic programs in France, Russia, England, and the United States. The most far-reaching of these was at Lick Observatory, where William Wallace Campbell coupled a three-prism spectrograph to the giant refractor in 1896. (A one-time civil engineering student at the University of Michigan, Campbell changed his career path after reading Simon Newcomb's *Popular Astronomy*.) After surveying the Northern Hemisphere sky, Campbell secured funds to erect a telescope in Chile, extending the survey to the southern sky. Overall, more than fifteen thousand plates were taken by thirty-one observers and measured by fifty-eight assistants. By the 1920s, Campbell and his team had expanded Vogel's original page-long roster of radial velocities into a true, all-sky catalogue, containing entries for 2,771 stars. With this extensive compilation, twice as precise as Vogel's, Campbell quantified the Sun's progression through space relative to its stellar neighbors: about twelve miles per second toward the constellation Hercules.

William Wallace Campbell, pictured around 1890 with Lick Observatory's thirty-six-inch refractor, here fitted with a spectroscope by John Brashear.

Like his East Coast contemporary Edward Pickering, W. W. Campbell was an astronomical entrepreneur: he envisioned a large-scale research program, developed the methods, designed the instruments, raised the money, assembled a team, and managed the work. What Pickering had accomplished for stellar spectral classification with the Henry Draper project, Campbell had accomplished for stellar radial velocities with the Lick catalog. Their works provided expansive datasets against which astronomers could test theoretical models of stellar and galactic motion, as well as stellar energy production and evolution. Conceived in the nineteenth century and gestated in the twentieth, the catalogs stand as milestones in new modes of astronomical observation. Pickering, Vogel, and Campbell strove to lay the foundations upon which astronomers might build a more comprehensive picture of the physical universe. As incremental and seemingly prosaic their sort of work, they remained ever alert to the possibility of discovery.

Of the Big Dipper's seven stars, the most historic lies second from the end of the handle. Zeta Ursae Majoris, also known by its Arabic name, Mizar, forms a wide pair with dimmer Alcor, the dyad easily perceived by observers of antiquity. Mizar itself is a telescopic double star, the first ever photographed, by Harvard's George Bond in 1857. Its components, designated A and B, take at least five thousand years to circle each other. Mizar A is the 116,656th entry in the *Henry Draper Catalogue*, its spectrum an unremarkable smudge among the dozens that crowd the objective-prism photographs of that part of the sky. However, upon microscopic inspection of plates taken between 1887 and 1889, Edward Pickering found a startling anomaly that set apart Mizar A's spectrum from the rest: on some exposures, the Fraunhofer K line was double, while on others, it was single. And between these extremities of appearance, the K line was fuzzy, like an unresolved line-pair. Mizar A's numerous other lines were either too broad or too faint to exhibit duality.

To complete the spectral analysis of Mizar A, Pickering hired Antonia Maury, Henry Draper's niece and a Vassar graduate in mathematics and chemistry. From measurements of seventy spectrographic plates, Maury confirmed Pickering's hunch that the transition between single and double lines is periodic. Pickering concluded that, although it appears solitary in a telescope, Mizar A is a binary system. (English photographic pioneer William Henry Fox Talbot had proposed in 1871 that the orbital motion of a binary star might reveal itself in such periodic behavior of its spectral lines.)

Locked in a gravitational embrace, the stars of Mizar A swing alternately toward and away from Earth with clockwork regularity. "When one component is approaching the earth," Pickering explained, "all the lines in its spectrum will be moved toward the blue end, while all the lines in the spectrum of the other component will be moved by an equal amount in the opposite direction, if their masses are equal. Each line will thus be separated into two. When the motion becomes perpendicular to the line of sight [thus, no Doppler shift], the spectral lines recover their true wavelength and become single."

From the measured Doppler shifts and the time interval between successive line-doublings, Pickering computed the orbital velocity, combined mass, and miles of separation of Mizar A's components. On November 13, 1889, Pickering reported his results to a gathering of the National Academy of Sciences in Philadelphia. Within a month, he informed his patron, Anna Draper, that the star Beta Aurigae was also a spectroscopic binary. The crucial spectrum had been obtained during Draper's recent visit to the observatory, he reminded her, then added, "[I]s it not a sufficient argument in favor of your coming oftener?" (Pickering's initial findings have been updated: the stars of Mizar A trace out a highly elliptical orbit about the size of Mercury's every twenty and a half days. Mizar B and Alcor were each determined to be binary as well. With the recent discovery that Mizar and Alcor are gravitationally bound, the Mizar-Alcor "pair" is actually a sextuple star system.)

Unknown to Pickering, Potsdam's radial-velocity expert, Hermann Vogel, had simultaneously homed in on the spectrum of the most ill-omened luminary in the heavens: Algol, the so-called Demon Star. Algol shines prominently in the constellation Perseus, whose mythological namesake slew the dreaded Medusa. Within the array of stars that comprise Perseus, grasped in what is held to be his outstretched hand, is the Gorgon's serpent-covered head. And at the hub of that little starry spray lies Algol, the glinting eye of Medusa herself. Algol's gloomy reputation is said by some to stem from the variability of its light, which fades threefold, then swiftly recovers, every sixty-nine hours. Eighteenth-century observer John Goodricke speculated that the periodic dimming arises from eclipses of the star by an unseen orbital partner. However, more than a century of visual and photographic scrutiny revealed no such star. If Algol was a binary system, the fainter member was lost in the glare of its brighter companion.

In December 1889, one month after Pickering's report on Mizar, Vogel announced that Algol's spectral lines shift in synchrony with its fluctuating brightness. Before Algol dims, the lines are redshifted from their nominal positions: the star is receding from us. Upon its return to normal brightness, the lines shift blueward: the star is approaching. During the event itself, the lines appear in their proper places in the spectrum. Vogel concluded that Algol is indeed part of an eclipsing binary system, as Goodricke had suspected.

From the spectral data and the timing of the eclipses, Vogel estimated the diameter, mass, separation, and orbital velocity of Algol's component stars. Although his parameters, like Pickering's, proved to be erroneous, astronomers at the time were electrified by these twin discoveries. (Modern observations reveal Algol to be a luminous blue-white star, paired with a much dimmer orange star, about 2.9 and 0.8 times the Sun's mass, respectively. The main eclipse occurs when the companion passes in front of Algol, a secondary eclipse when their positions are switched.)

The photographic spectrum of the bright star Spica, in the constellation Virgo, also presented a cyclic wobbling of its spectral lines. Here, again, Vogel inferred a binary system, revolving about a common center of mass. Unlike Algol's telltale eclipses, Spica had displayed no prior sign of duplicity. In the "new" Spica, Lick Observatory's James Keeler saw the spectrograph's power to inflate a luminous speck in the night sky into a physical entity in faraway space:

> Translating the mathematical formulae of Professor Vogel into the ideas which they represent, a wonderful picture of stellar motion is presented to our mind, and one to which the whole visible universe, as revealed to us by our greatest telescopes, offers no parallel. The spectacle of two great suns like our own, revolving around each other in only four days at a distance no greater than that which separates the sixth satellite of Saturn from its primary, is one which the inadequacy of our optical powers will probably forbid us from actually beholding, but the indirect evidence that such extraordinary circumstances of motion exist, is so complete that we must admit their reality.

By now, astronomers were confident that a spectrum can reveal the elemental constituents of a star. The work of Pickering and Vogel further taught them that the spectrograph, in the periodic Doppler-dance of stellar spectral lines, can disclose once-invisible properties of a star. The advance of technology had birthed a new observational species: the spectroscopic binary. In the observation and computation of binary star orbits, the interests of classical, astrometric astronomers and their astrophysical counterparts neatly meshed. The latter provided the former with previously unobtainable data for the mathematical analysis of stellar motions. Every star might now be subject to spectroscopic surveillance. For if Spica, one of the brightest and most-observed stars in the night sky, had been unmasked as a spectroscopic binary, how many other hidden pairs awaited exposure by the spectrograph?

As the twentieth century approached, with the classical mode of astronomy on the wane, the very definition of a research observatory began to change. Where once the human eye was deemed the optimal receptor of telescopic light, an observatory without a camera and a spectrograph was increasingly regarded as incomplete. And as the march of technology rendered direct vision superfluous, if not antithetical, to the act of cosmic exploration, so too the many amateur astronomers who had helped foster that technology. Prior to 1900, amateurs had participated fully in the meetings and governance of scientific societies. Their lack of formal academic training was no hindrance to ascension in the field. Their research papers appeared in journals alongside those of professionals. They received honors in equal measure to their institutional colleagues. But when astrophysics accelerated into the new century, its amateur acolytes found themselves rudely left behind.

Chapter 21

THE CLOUD THAT WASN'T THERE

> The born astronomer . . . is moved to celestial knowledge by a passion which dominates his nature. He can no more avoid doing astronomical work, whether in the line of observations or research, than the poet can chain his Pegasus to earth.
>
> —Simon Newcomb, at the dedication of Yerkes Observatory, 1897.

SINCE THE MID-NINETEENTH CENTURY, amateur astronomers had been key participants, if not prime movers, in the development of observational astrophysics. Photography and spectroscopy would not have reached their heights as quickly as they did had not the path been blazed by a worldwide cadre of volunteer scientists and inventors. Few academic astronomers had been willing to hitch their professional prospects to these emerging technologies.

Into the 1890s, amateurs and professionals enjoyed equal status in the astronomical hierarchy, both in the United States and in Great Britain. They contributed to research and technological developments, published papers, participated in conferences, belonged to and even headed major scientific organizations, and received awards in like numbers. In Britain especially, the wealthy-amateur tradition remained strong, with "backyard" observatories comprising half of the nation's astronomical research facilities; with the exception of the Royal Observatories at Greenwich and the Cape of Good Hope, there was virtually no government or philanthropic support for astronomy.

In the 1890s, relations between academically trained astronomers and their amateur brethren changed. As the performance requirements of astrophysical observation became more stringent and the analysis of photographic and spectroscopic data more mathematical, institutional practitioners moved to redefine their field. Previously, anyone (more correctly, any man) who possessed sufficient skills was welcomed into the research fraternity. However, by the start of the twentieth century, entry required academic credentials, specialized graduate training, and absolute rigor in scientific thought and practice.

As the years progressed, the cost of research-grade telescopes, auxiliary equipment, and personnel skyrocketed beyond the means of wealthy amateur observers. Astrophysical observatories were no longer one-person, domestic operations, but complexes sited on remote mountaintops and attended by full-time staff. Though finding themselves marginalized in research, American amateurs remained a force in fostering public awareness of astronomy's wonder and value. Presumably, this active societal interest and its attendant media coverage gave even the most self-absorbed benefactors reason to consider astronomy as a route to secure their legacy.

A quiet tension between professional and amateur observers emerged as the cohort of academically trained astronomers prospered. Where once the two groups had shared a common platform, distinctions were being drawn between them in terms of academic credentials, mathematical acumen, and most devastatingly to the amateurs, the scientific merit of their work. The door to amateur participation began to ease shut when institutional astrophysicists revised their organizations and their journals to foster the professionalization of the field. Professional astronomers were expected to keep up with the published literature and research protocols in their area of specialization, and to understand the relevance of their work within the overall context of cosmical studies.

Research methods and results were subjected to unprecedented scrutiny. Astrophysical measurements were now held to the same rigorous standards as traditional astrometric quantities, the wavelength of a spectral line and the position of a star regarded with an equal eye to error. Artifacts of astrophysical observation were valued in proportion to their scientific worth, no longer their aesthetic appeal. In sum, the expectations of what constituted a publishable work were raised, as the gatekeepers of astrophysical journals reflected the ascendance of the academic astronomer.

Professional astronomers' concerns about their amateur colleagues were vented publicly at conferences and in the pages of research journals. In 1898, longtime amateur astronomer Frank McClean released two volumes of stellar-spectrum images obtained from his home in England and at the Cape of Good Hope. The work received a lukewarm review in the *Astrophysical Journal* from Yerkes Observatory's Edwin Frost, one of America's premier spectroscopists. Frost alludes to the many imperfections in

McClean's images, as well as to his omission of dates, exposure times, and even the barest description of his optical system. He characterizes the work as "pictorial and qualitative rather than metrical and quantitative," that is, more family photo album than scientific document. On McClean's discovery of oxygen lines in one star's spectrum, Frost withheld judgment—until the observation had been confirmed by a professional astronomer.

A nastier face-off erupted at the May 1895 meeting of the Royal Astronomical Society with a presentation by the preeminent celestial photographer Isaac Roberts. Entirely self-taught, Roberts had established his bona fides with an acclaimed series of images of celestial nebulae, most notably the giant spiral cloud in Andromeda. At the meeting, Roberts projected a lantern-slide of a three-hour exposure that displayed the nebulous surroundings of the star 15 Monocerotis. He followed this slide with an image of the same region by Lick Observatory's Edward Emerson Barnard.

A high-strung workaholic and passionate comet hunter—he spotted sixteen during his lifetime—Barnard made his name in 1892 with his discovery of Jupiter's fifth moon, Amalthea. He became an expert in wide-field celestial photography, using portrait-camera lenses to capture the magnificent landscape of the Milky Way. (The photographic prints for his 1927 sky atlas had to be remade after a stray bullet from a Chicago mobster's gun pierced the originals in the print shop.) Of his time with Barnard, astronomer Philip Fox recalled, "One could always tell how the night had been by his reaction from it. If clouds or bad seeing had marred the observing, his unconscious sighs were clearly audible. If the sky had been kind, his spirit was gay with song."

Although the pair of images offered by Isaac Roberts depicted identical regions of the night sky, they looked distinctly different. The stars in Roberts's photograph, taken through his twenty-inch telescope, were virtual pinpoints; those in Barnard's picture, taken with the low-magnification sky camera, appeared as bloated disks. Roberts acknowledged that he had enlarged Barnard's photograph fivefold to bring its scale up to that of his own. But he had done

Edward Emerson Barnard, circa 1885.

so to make a point. The nebulosity depicted in Barnard's picture, he asserted, was illusory; it stemmed from the poor spatial resolution of Barnard's wide-field camera, which captured broad swaths of sky at a highly pinched scale. Through his own twenty-inch reflector, he continued, the purported knots and swirls of nebulosity are seen for what they are: aggregations of faint, barely resolved stars. And scattered among these, a profusion of even dimmer stars, entirely beyond the light grasp of Barnard's minuscule instrument. The luminous veil around 15 Monocerotis, Roberts told his audience, awaits only a larger telescope to confirm its underlying stellar nature.

A salvo had been fired from an amateur astronomer's station south of London into the heart of a professional research institution in California. The return volley arrived with unsurprising swiftness in the December 1895 pages of the *Monthly Notices of the Royal Astronomical Society*, where Barnard points out the absurdity of Roberts's conclusions. Comparing a five-times enlargement of one photograph to an unenlarged version of another reveals nothing. Roberts's truncated field of view isolated the nebulosity from its overall visual context. Wide-field camera views are meant to be studied in the original format, with diaphanous nebulae standing out against their dark surround. "They are intended to be looked at and studied as *pictures* of the regions they show," Barnard writes, "and are not to be examined microscopically. It is unjust to use an enlargement such as Dr. Roberts used, because it necessarily puts these pictures at a disadvantage. My picture was simply spoiled by this, while Dr. Roberts' retained its original qualities, not being enlarged."

In his broadside, Barnard further points out a counterintuitive aspect of nebular observing that the amateur Roberts had failed to grasp. True, a large-aperture telescope has more light-gathering power than one with a small aperture, and will therefore reveal fainter stars. But unlike the light of a star, which is essentially point-like in the sky, the light of a diffuse nebula is spread out. The parameter that governs its visibility is its *surface brightness*, that is, its luminous energy per unit area of sky. For a nebula, telescope aperture is a mixed blessing: a large telescope renders an interstellar cloud in more detail, yet no brighter than a small telescope. The additional light that is collected is diluted in equal proportion by magnification, that is, the nebula's surface brightness remains the same. What brings a low-contrast nebula to visibility are a telescope's optical efficiency (the ratio of its focal length to its aperture), the clarity of the night sky, the photographic exposure time, and the plate sensitivity. Indeed, Barnard's wide-field camera was an ideal instrument to capture the species of extended, wispy nebulae that night-sky observers had long noted by eye.

To underline his dismay, Barnard included three high-quality prints of the 15 Monocerotis region, which highlight the cramped scale of Roberts's telescopic field of view. Examining these pictures, he tells the Royal Astronomical Society, "it will be seen that Dr. Roberts' reasoning is decidedly wrong. This diffused light is in nowise confined to the star areas. It will be also readily seen that it spreads over a large region where there are essentially no stars at all—even where Dr. Roberts' reflector can show no stars. That this is real diffused nebulosity there is no reason whatever to doubt."

Roberts refused to yield the point. A large-aperture telescope, he maintained, picks up fainter stars and therefore possesses an equal capacity to reveal fainter nebulae—which, as anyone can see, are simply not evident. To buttress his argument, Roberts mounted a five-inch, portrait-lens camera alongside his twenty-inch reflector and tried his own hand at wide-field photography. By 1896, the next battle erupted, this one regarding the presence of nebulosity surrounding the Pleiades star cluster. Here again, Barnard insisted on the reality of the extensive, diffuse corona, Roberts on its absence. Barnard dashed off an exasperated note to the Royal Astronomical Society. Three professional astronomers, he informs them, have taken unequivocal wide-field images of the supposedly fictitious nebulae around the Pleiades. Nowhere does he mention Roberts by name, but it is clear to whom he is speaking when he writes, "These nebulosities . . . have been amply verified (if such a verification were at all necessary). . . . It would therefore appear that a failure to show these remarkable features with an ordinary portrait lens and an exposure anything like 4 or 5 hours must be attributed to something else than their nonexistence."

The final exchange between Isaac Roberts and E. E. Barnard—and the one that most acutely highlights the escalating rigor demanded in astronomical observation—took place in 1903, this time in the pages of America's *Astrophysical Journal.* The subject of Roberts's article was William Herschel's venerable catalog of faint, extended nebulosity. Over the past six years, Roberts had photographed all fifty-two of Herschel's regions, yet confirmed nebulosity in only four; evidently, one of history's keenest visual observers had conjured the rest. So extraordinary was this conclusion—as well as Roberts's apparent ignorance of others' work in this area—that the editors of the *Astrophysical Journal* engaged Barnard, now at Yerkes Observatory, to write a companion piece to address the matter.

Feel the frustration flowing from his pen when Barnard asks his fellow astronomers whether it is not "a little unreasonable to suppose that Herschel, who made so few blunders compared with the wonderful and varied work that he accomplished, should be so palpably mistaken in forty-eight out of fifty-two observations of this kind." (In fact, working at the very limit of visual perception, Herschel had recorded a number of spurious celestial objects.) Barnard points out that Roberts's standard exposure time of ninety minutes is insufficient to show fainter interstellar clouds; his own camera shutter often remains open for four or five hours.

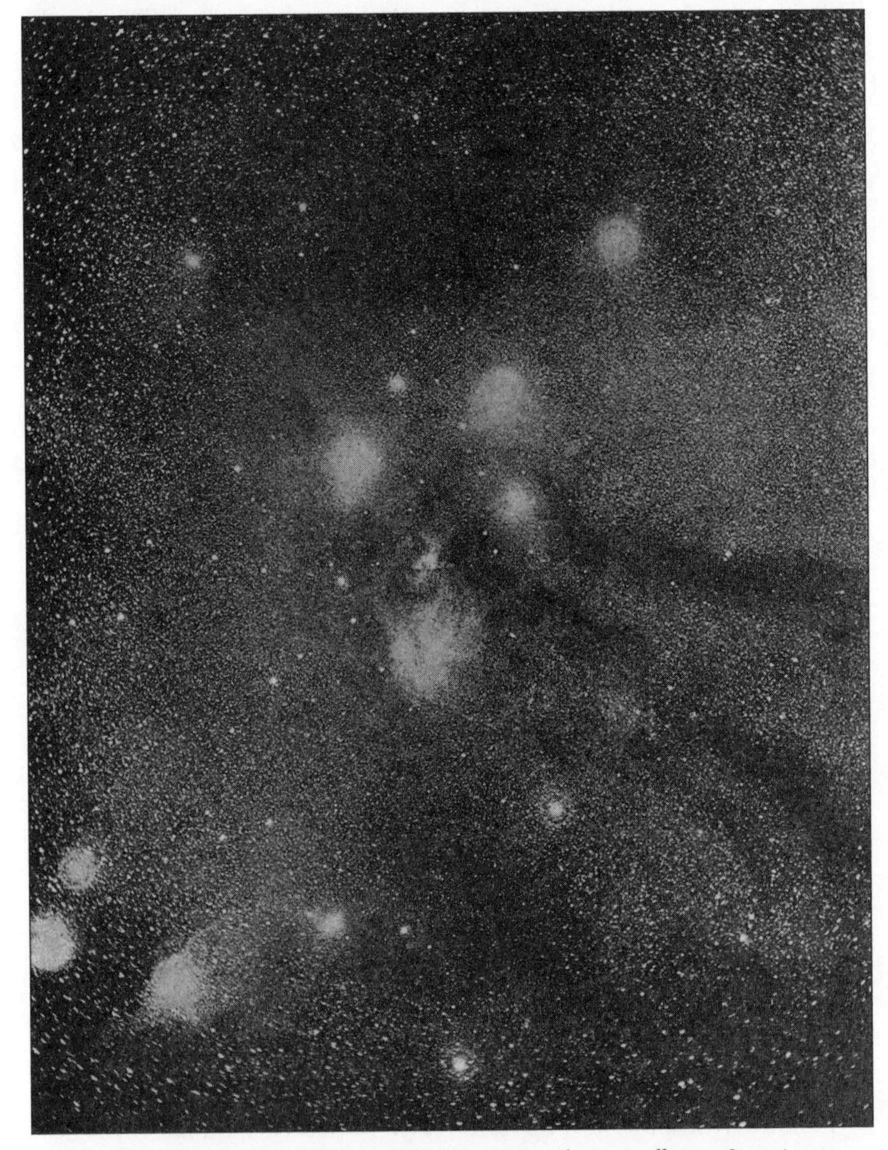

A seven-hour exposure of a nebulous region in the constellation Scorpius, taken by Edward Barnard on June 21 and 22, 1895.

As one might tutor a neophyte, Barnard returns to the fundamental differences between stellar and nebular photography. To demonstrate the irrelevance of telescope aperture on nebular visibility, he describes a picture he took of Herschel's region No. 27 in Orion through a cheap lantern lens, a mere 1.6-inches across. "Most of the great curved nebula is clearly shown," he writes about the ghostly arc, since dubbed Barnard's Loop, "especially the region described by Herschel. . . . There is therefore no question but that this nebulosity exists where Herschel saw it."

In closing, Barnard thrusts a verbal dagger at Roberts's competence, reminding readers that it was "with the same instruments described in his present paper that Dr. Roberts failed to get any traces of the exterior nebulosities of the *Pleiades*, which have been shown by four observers with four different instruments not only to exist, but to be not at all difficult objects."

A simultaneous assault on Roberts's disavowal of Herschel's nebulae swept in unexpectedly from the Continent. Attendees at the March 1903 meeting of the Royal Astronomical Society were treated to an exquisite, wide-field image of Barnard's Loop, taken by Heidelberg University astronomer Max Wolf through the observatory's new sixteen-inch photographic telescope. According to the minutes of the meeting, the picture "showed, besides the great Orion Nebula and the nebula around [the star] Zeta Orionis, considerable masses of detailed nebulosity joining these, and also connecting the great nebula with the head of the snake-like nebulosity (first photographed by Prof. Barnard), which winds through a large part of the constellation."

Where, fifteen years earlier, Isaac Roberts had astounded the audience with his own breakthrough image of the Andromeda Nebula, he was now confronted with an equally vivid portrait of a celestial object he claimed did not exist. At a stout sixteen-inches aperture, Wolf's telescope was neither the small toy nor child's lantern-lens that Roberts had called Barnard's diminutive cameras. And Wolf's exposure time of six hours and fifteen minutes eclipsed Roberts's own ninety-minute standard. (Wolf's pièce de résistance would come in 1911 with a twenty-five-hour exposure of the spectrum of the Andromeda Nebula accumulated over twenty nights. As one astronomical luminary put it, "Difficulty seems to have a peculiar attraction to him.")

Faced with seemingly incontrovertible evidence of the nebula's reality, Roberts grew defensive. Perhaps, he suggested, counter to their long experience, Wolf and Barnard had each targeted the wrong object. If not that, then the apparition on their plates might have arisen from atmospheric glare or optical imperfections—again unlikely, given the multiplicity of cameras and observing sites. Had Roberts accepted the verdict of his own eyes, he would have recognized (as most others already had) why extended nebulosities did not reveal themselves on his plates. A ninety-minute exposure under moisture-laden, semitransparent English skies is frequently insufficient to capture a low-contrast space cloud. Roberts had used three- and four-hour exposures with great success for his Andromeda plates. Yet he clung to his willful misconception about the essential role of surface brightness on nebular imaging: if he exposed long enough to show the faint stars William Herschel had seen, then Herschel's faint nebulae should likewise appear. Then in his seventies, Roberts might simply have lacked the energy to push his mammoth project to its required limits. (In fact, Barnard announced that he did find traces of nebulosity on Roberts's plates, a claim Roberts himself denied.)

Another reason Roberts's photographs differ from Wolf's was voiced at the 1903 meeting by Cambridge University astronomer Arthur Hinks. Echoing Barnard's standing criticism, Hinks reminded the audience that Roberts's plates span a mere two degrees, whereas Wolf's cover a virtually panoramic ten degrees. Extended nebulae overrun the borders of Roberts's plates, causing at most a perceptible brightening of the entire field. The same nebulae sit isolated within Wolf's pictures, their feathered edges made manifest against the blackness of the night sky.

The nebular controversy ended only with Isaac Roberts's death in 1904. At first glance, the conflict might be taken to epitomize the professional–amateur divide in astronomy at the turn of the century. But, as the self-taught Barnard might be considered something of an über-amateur himself, it is more properly seen as the consequence of rising scientific standards. In his passion to become an effective researcher, Barnard aligned his working methods and his critical judgment (if not his prickly personality) to those of university-trained astronomers. His knowledge base was largely empirical, but filtered by an acute sense of the scientific constraints one learns in the halls of academia.

When Barnard criticized Isaac Roberts for not keeping up with the technical literature, for shrugging off gaps in his ken of pertinent physics, for plowing ahead without regard to advancements in the field, he was enumerating the hallmarks of an unprofessional scientist. That Roberts was an amateur was not the point. A decade earlier, Barnard had taken umbrage at the mediocre observing skills of his boss, one-time Lick Observatory director Edward Holden, who tied up the thirty-six-inch refractor two nights a week with his own desultory observations of planets. Of course, Holden viewed Barnard's grumbling in a wholly different light, remarking to a colleague,

"I shall not be sorry to have him go—for in his place we shall get a professional—and I must say that Amateurs are hard to get along with!"

As the twentieth century proceeded, amateurs evaporated from the ranks of professional scientific societies. By 1909, they constituted 12 percent of the American Astronomical Society, but contributed only 1 percent of the research papers. Amateurs had already formed their own associations, where members enjoyed the social bond of shared purpose and wonderment, minus the overtly mathematical aspects of the subject. An amateur-led incarnation of the American Astronomical Society flourished in Brooklyn between 1883 and 1888. Although several local professionals signed on, the group was roundly condemned by Simon Newcomb, dean of American classical astronomy, for its arrogance in portraying itself as a national body.

In 1889, impressed by the joint efforts of Lick Observatory and the Pacific Coast Amateur Photographic Association during a recent solar eclipse, Edward Holden moved to found the Astronomical Society of the Pacific (ASP), the first national astronomical organization in the United States. The ASP's liberal membership requirements are stated in its recruiting circular of that year: "The new Society is designed to be popular in the best sense of the word. We wish to count in our membership every person on the Pacific Coast who takes a general interest in Astronomy, whether he has made special studies in this direction or not." By the end of its first year, the ASP had recruited 178 professionals and amateurs, and thrives today with an international membership base.

The British Astronomical Association (BAA) was established in 1890 by a circle of amateur observers who found the discourse at the Royal Astronomical Society too technical and the membership fees too high. The BAA organized itself into sections by interest and instrumental means, and offered members practical instruction in the art of observing. And unlike its professional counterpart, the new organization was open to women. (Margaret Huggins was among the original board members.)

The early twentieth century saw the formation of regional amateur astronomy clubs worldwide. Harvard's Edward Pickering collaborated with noted observer and author William Tyler Olcott in 1911 to establish the American Association of Variable Star Observers (AAVSO), a still-active group devoted to long-term monitoring of stellar light fluctuations. In Pickering's ever-practical view, an army of amateur astronomers under the supervision of professionals might render valuable (and low-cost) service in certain areas of cosmical studies. Women especially should be recruited, he advised in an 1882 guide to variable star observation:

> Many ladies are interested in astronomy and own telescopes, but with two or three noteworthy exceptions their contributions to the science have been almost nothing. Many of them have the time and inclination for such work, and especially among the graduates of women's colleges are many who have had abundant training to make excellent observers. As the work may be done at home, even from an open window, provided the room has the temperature of the outer air, there seems to be no reason why they should not thus make an advantageous use of their skill. . . . The criticism is often made by opponents of the higher education of women that, while they are capable of following others as far as men can, they originate almost nothing, so that human knowledge is not advanced by their work. This reproach would be well answered could we point to a long series of such observations as are detailed below, made by women observers.

Among the AAVSO's charter members were Helen Swartz, a mathematics teacher in Norwalk, Connecticut; Vassar astronomy instructor Helen Furness; and Mount Holyoke's Anne Sewall Young, who logged more than sixty-five hundred observations over three decades. The AAVSO amply carried out its mandate, yet most amateurs—women included—preferred to view the sky for their own enjoyment, and largely ceded the business of research to their institutional brethren.

The emergence of astrophysics during the 1890s was at once a revolution in scientific practice and in the social structure of its community of devotees. The exigencies of forefront science came to overwhelm the majority of self-taught amateurs who had helped usher in the field. The successors of these wealthy gentleman-scientists, constrained by resources and training, could do little more than regard with interest the colossus of scientific research accelerating past them. By the turn of the century, astrophysical observation and analysis had become a full-time job, typically requiring institutional support, a team of workers, and increasingly, a remote mountaintop telescope. With amateur astronomers relegated to the history but no longer the future of astrophysical research, professionals had to sort out what it meant to be an astronomer in the twentieth century.

Chapter 22

The Union of Two Astronomies

> The domains of the physical sciences are not, like the political divisions represented on a map, capable of being defined by boundary lines traced with mathematical precision. They pass into one another by imperceptible gradations, the unity of nature opposing itself to rigid systems of classification. . . . Such is the nature of the science . . . known as astrophysics.
>
> —James Keeler, address delivered at the dedication of the Yerkes Observatory, October 21, 1897

By the 1890s, the New Astronomy was no longer new. The term *astrophysics*, or its hyphenated variant *astro-physics*, had become part of the astronomical lexicon. Initially, this meant the application of photography or spectroscopy to garner data about the intrinsic properties of celestial objects. Only during the early twentieth century, when physical theory achieved a sufficient degree of sophistication, would this branch of study swell into an all-encompassing science. Indeed, astrophysics would come to overshadow the traditional study of celestial positions and motions, not the least for its flexibility of application. Astrometry was a diamond, hardened by a century of experience, its procedural facets shaped by precise strikes along fixed planes. Astrophysics was gold, malleable to the purpose, its form ever evolving under the imaginative guidance of observer and theorist.

Astrophysicists harbored an expansionist outlook toward the exploration of the universe, a recognition that the scientific frontier lay beyond the circumscribed realm of traditional astrometry. At the dedication of the University of Chicago's Yerkes Observatory in 1897, astrophysicist James Keeler acknowledged that "there may be some who view with disfavor the array of chemical, physical, and electrical appliances crowded around the modern telescope, and who look back to the observatory of the past as to a classic temple whose severe beauty had not yet been marred by modern trappings."

Keeler's remarks convey the dislocation felt by many veteran observers, who had long served as stewards of cosmic science. Their unease over the future of the field was captured by Agnes Clerke, doyenne of the history of astronomy during this tumultuous period. Clerke was a walking-talking fount of astronomical news, from whom top researchers learned what their colleagues—and their competitors—were up to. "The majestic elder astronomy," she writes in an 1888 article, "might, it was to be feared, suffer neglect through the predominant attractions of its younger, more versatile, and brilliant competitor; or its lofty standard of perfection might become lowered through the influence of workers more zealous than precise, recruited from every imaginable quarter, inventive, enthusiastic, indefatigable, but unused to the rigid requirements of mathematical accuracy."

Typical of Clerke's zealous workers was radial-velocity expert W. W. Campbell, who conveys the allure of astrophysics in an 1894 letter to Keeler: "I regret having to give some time to the *old* work, & want to get out of it entirely. In spectroscopy one can cut loose from traditions and roam as free as he likes. . . . There is so much waiting to be done, & the nights are few and short."

Campbell was far from alone in longing to free himself from the bonds of traditional observation. The coming decades would bring a sea change in the career aspirations of rising astronomers. Allegheny Observatory's Frank Schlesinger, a pioneer in the photographic measurement of stellar positions and parallaxes, lamented in the 1920s that "the superior attractiveness of astrophysical work has nearly monopolized the services of our most promising young men."

The bifurcation of professional astronomy between the old and the new was cemented in the United States with the launch of the *Astrophysical Journal* in 1895. The magazine billed itself as an international review of spectroscopy and astronomical physics. Astrophysicists who had formerly published their results in observatory annals, the *American Journal of Science*, *Monthly Notices of the Royal Astronomical Society*, or Germany's *Astronomische Nachrichten*, now had an outlet devoted to their particular mode of research. (The venerable *Astronomical Journal*, founded by Benjamin Gould in 1849, specialized in mathematical and astrometric papers.)

At the same time, astronomers at Harvard, Lick, and other major American institutions realized that they had equaled, if not surpassed, their European brethren in the breadth and vitality of their research. Buoyed by the nation's growing contributions to the global enterprise, Simon Newcomb and University of Chicago astrophysicist George Ellery Hale sought to establish a national professional organization. American astronomy at the time was institution-centric, and the effort to unify within a professional body was viewed with some suspicion. But Hale's organizational genius and Newcomb's prestige together convinced the dispersed community of researchers that a national forum for the exchange of ideas was merited.

The Astronomical and Astrophysical Society of America held its first meeting in 1899 at Yerkes Observatory in Williams Bay, Wisconsin, with some fifty of its 113 charter members in attendance. Despite its aspirational name, the new society was far from a national body: Only 62 percent of United States professional astronomers joined, and over one-third of these worked at Harvard, Lick, the University of Chicago, and federal agencies. Many astronomers eyed the yearly meetings in distant cities as a needless impingement upon their time and their independence. However, under the energetic presidency of Edward Pickering, membership rose steadily. By 1910, 80 percent of American astronomers had joined, and forty-two institutions were represented.

In a fifty-year retrospective, University of Wisconsin researcher Joel Stebbins highlighted the critical social role of the new organization: "Astronomers must have been a group of lonely individuals in the [1880s and 1890s] when they had little opportunity to get together. Those doing research were pretty far apart with few personal contacts." In time, solo astronomers gravitated to the society's periodic forums and the opportunity to interact with colleagues. The gatherings hatched research collaborations and more than occasionally dampened professional animosities. Stebbins recalled a case of two individuals "who were at sword's points before they came to a meeting; but it was arranged for them to sit next to each other at the dinner; and at the end it was gratifying to see them shaking hands and saying how glad they were to have had the opportunity to talk things over."

Another issue that sparked dissension was membership eligibility: Should the society impose a restrictive standard of academic credentials and advanced research training or open up its ranks to amateur observers and armchair aficionados? The society's bylaws allowed the nomination of "any person deemed capable of preparing an acceptable paper upon some subject of astronomy, astrophysics, or related branch of physics," to which one applicant—a lawyer—inquired, acceptable to whom? Might he himself, the lawyer posed, submit a paper acceptable to his janitor on how to start a fire? (The society agreed that, in this case, the applicant's letter was sufficient proof of eligibility.) Others suggested wryly that "a good-sized check on a bank might be considered an acceptable paper on astronomy."

Unsurprisingly, traditional astronomers vied with astrophysicists for supremacy within the fledgling organization. Cofounder Hale even proposed that the presidency of the society alternate between the two factions to ensure that neither gained the upper hand. (The suggestion was never enacted.) In 1914, recognizing that astrophysics had become an inseparable part of the entire astronomical enterprise, the membership settled on coequal status and renamed themselves the American Astronomical Society. Once distinct, next intermingled, the terms astronomy and astrophysics were henceforth synonymous.

The initial perception of astrophysics as a "soft" science faded as its technical underpinnings grew more complex. Nor could old-school astronomers dismiss the scientific fruits of its investigations. Agnes Clerke foresaw the eventual acceptance of astrophysics as a full-fledged science, as well as its positive influence on its astrometric stablemate. "The new astronomy," she writes, "has submitted to bear the yoke of the old. The old astronomy has adopted the new methods, and is even now anxiously fitting them to its own sublime purposes. It has enlarged its boundaries without departing one iota from its principles. By an effort that shows it to be still young and elastic, it has seized the key of the situation, and now stands hopeful and dominant before the world. This union of the two astronomies has long been in remote preparation. . . . Circumstances concurred to bring it about just at the right moment."

Not every celestial observer aspires to be the next Galileo or Herschel. To secure one's reputation by revealing new worlds is a fetching vision that tends to dim with experience. For many nineteenth-century astronomers, the lure of discovery was set aside for the quiet, sustained pursuit of more workmanlike goals: mapping the night sky, classifying stellar spectra, counting sunspots, gauging double-star separations, refining the mechanics of the telescope. In this regard, amateur astronomer William Huggins and his professional counterpart Edward Pickering were in sync. Both relished the intricate, snail's-pace tasks attendant to the acquisition of astronomical knowledge. Pickering defined the peculiar joy of the night-sky observer in his journal in March 1900: "Although the work is the most acute and absolutely monotonous, I find it very fascinating, and always watch the next morning for the progress made in the completion of the work." Huggins expressed the astronomer's ethic with a veteran's shrug-shouldered succinctness: "Life is work, and work is life."

As alike as their scientific proclivities might have been, Huggins and Pickering diverged in their manner of execution. The contrast derived only partly from choice; in significant measure, it was imposed upon each of them by political and cultural differences in their respective scientific communities. Through the 1890s, while its European competitors lagged, the United States experienced an unprecedented rise in astronomical research capacity. And most of this expansion stemmed from an increasing academic and institutional focus on astrophysics.

The growth of astrophysical observation in America paralleled that of universities and colleges featuring specialized departments of study. As centers of higher education proliferated, so too the number of small- and moderate-aperture telescopes fitted for visual, photographic, or spectroscopic research. By the turn of the century, a campus observatory had become almost as common a fixture as an instructional chemistry laboratory, a stunning reversal of their paucity at midcentury. American universities were free to pursue virtually any avenue of research for which funding might be secured from individuals or foundations. Unlike tangled government bureaucracies, university presidents might fast-track an astrophysical initiative that promised to confer prestige upon their institution.

Factory-style research programs, such as those at Harvard and Lick Observatories, stimulated growth in the population of astronomers. The academically trained astronomer—effectively, the factory boss—directed specialized, low-wage employees in their various mundane tasks. One team might handle nighttime photography and chemical processing, the other daytime measurement and initial analysis. Adept workers might be assigned their own projects, altogether a multipronged research effort that multiplied the capacity of a single scientist. Freed of such time-consuming duties, the lead astronomer completed the analysis, drew conclusions, and presented the results—as his own—to the world. This assembly-line model emulated that of the Royal Greenwich Observatory's astrometric program, only applied to astrophysical observation.

Where the 1840s had been a virtual desert for astronomical research in the United States, the latter half-century saw widespread blooming of the landscape. By 1886, professional astronomers in America far outnumbered their European colleagues: 128 versus eighty-one in Germany, seventy in Great Britain, and sixty-three in France. The United States likewise led Germany, Great Britain, and France in the count of research observatories: forty, twenty-six, thirty-two, and sixteen, respectively. Fully three-quarters of American observatories at the time were based at universities and colleges, versus only 34 percent in Germany, 25 percent in Great Britain, and zero in France. (Almost all French observatories were government-run.) By the early twentieth century, the number of American astronomers and observatories had each more than doubled from their 1886 levels. As today, the road to advancement for these academics lay not in undergraduate teaching but in the tangible scholarly products they themselves had been taught to generate: original research, publications, and advanced-degree graduates trained to conduct independent scientific investigation.

Astronomy had long held a place in the minds of the nineteenth-century American public. "Of the sciences in America," writes contemporary historian John Lankford, "astronomy has the deepest cultural roots. . . . Public support for astronomy in the pre-Civil War era rested on its spiritual and cultural value, not on any material contributions astronomers might make to American life. Apparently Americans valued astronomy as a way of achieving cultural transcendence."

Cosmic exploration was the frequent subject of newspaper articles and popular-level books. Although science did stimulate the brain, reporters also latched onto the intellectual blood-sport of institutional politics and rivalries, such as Edward Holden's troubled tenure at Lick Observatory. Amateur astronomy societies played a significant role in promoting astronomy on a local and regional basis, through lectures and hands-on instruction in celestial observing. Monthlies like the *Sidereal Messenger* and its successor, *Popular Astronomy*, delivered the latest astronomical developments to the masses.

Driving the engine of astrophysical research in the United States during the late nineteenth century was the availability of money. Entrepreneurial astronomers solicited the wealthy, promising to burnish their social standing, as well as their legacy, through association with the cause of science. In Harvard's spectral classification program, spectroscopic pioneer William Huggins rightly perceived an institutional juggernaut elbowing him aside. To his potential donors, Edward Pickering offered an almost irresistible blend of self-confidence, graciousness, and persistence. Anna Draper gave $235,000 to Harvard during her lifetime and bequeathed another $150,000 upon her death. In 1878, Boston construction baron Uriah Atherton Boyden declined a request to pitch in $500 to the observatory's fundraiser, only to have the trustees of his estate bestow $230,000 in his name after his death.

Catherine Wolfe Bruce, daughter of a real-estate investor, answered Pickering's call for $50,000 to place a photographic telescope in Peru.

By the 1890s, Harvard's astronomy budget was second in the nation only to the U.S. Naval Observatory. This at a time when William and Margaret Huggins supported their research through personal savings and rents on a handful of properties. (Pickering referred to Harvard's finances as a "kind of wealthy pauperism," as its fourfold increase in endowment during the 1890s accompanied a fivefold increase in research expenses.)

Pickering's growing empire was but one wave in a tide of change that was about to engulf astronomical practice in the United States, as resource-rich academic centers and mountaintop observatories displaced citizen-astronomers operating out of their backyards. Business moguls with no scientific bent increasingly fueled the work of university-trained specialists. In 1888, California land speculator James Lick abandoned his pyramid-building scheme and gave $700,000 toward construction of a mountaintop telescope to search for life on the Moon. The thirty-six-inch Lick refractor failed to spot lunar creatures, but did find a fifth moon of Jupiter and plenty of double stars. James Lick's body was interred inside the telescope's massive support pier, turning it into a working monument to the man and to the power of money.

The thirty-six-inch refractor of Lick Observatory, as depicted in the December 1, 1888, issue of Knowledge *magazine.*

Andrew Carnegie was impressed by Lick's philanthropic largesse, even though self-aggrandizement had been its primary motivation. "If any millionaire be interested in the ennobling study of astronomy . . ." Carnegie advised fellow plutocrats in 1889, "here is an example which could well be followed, for the progress made in astronomical instruments and appliances is so great and continuous that every few years a new telescope might be judiciously given to one of the observatories upon this continent, the last being always the largest and best, and certain to carry further and further the knowledge of the universe and our relation to it here upon the earth."

Among the benefactors in Carnegie's model were Harvard-educated polymath Percival Lowell, scion of the venerable Boston family, who founded his eponymous research institution outside Flagstaff, Arizona, in 1894; streetcar magnate Charles Tyson Yerkes, who donated $300,000 the following year toward a forty-inch refractor and observatory for the University of Chicago; and William Thaw, Jr., who underwrote construction of Allegheny Observatory's thirty-inch photographic refractor in 1912. Andrew Carnegie followed his own advice in 1902 when he established a foundation to support astronomical facilities and research.

Europe provides a stark contrast to the ascension of astrophysics in America. Large-scale scientific philanthropy was relatively rare: Private donations brought a thirty-inch refractor to the Nice Observatory in France in 1886, and the Simeis astro-photographic station in the Crimea to Russia's state-operated Pulkova Observatory in 1912. As in the United States, government support went to mature fields—astrometry and celestial mechanics—not to astrophysics. But in America, private donors chose overwhelmingly to support astrophysical projects, endowing the New Astronomy with the facilities and career opportunities it needed to flourish.

With few exceptions, observational astrophysics languished in Europe. Government ministries allocated resources overwhelmingly to astrometric projects, such as the ill-fated *Carte du Ciel*. Little was left for institutional initiatives in observational astrophysics. The result was the application of more refined apparatus to *traditional* areas of observation. Given the funding constraints and cultural predilections, European astrophysicists achieved their greatest successes in theoretical rather than observational studies. As the opportunity gap widened between the United States and Europe, would-be astrophysicists flocked to American observatories for advanced training. Many never went back.

Europe was further hobbled by its ruinous embrace of the *Carte du Ciel* sky-mapping initiative, which calcified development of astronomical apparatus and methods. "The great irony of the project," historian John Lankford notes, "is that the international committee directing the *Carte* froze instrumentation and research design at the very time astronomical photography was developing exponentially. . . . In terms of individual careers and the direction of astronomical research, astrometry as represented by the *Carte* virtually defined European astronomy after 1890."

The project's international governing body ignored the implications of Edward Pickering's announcement that Harvard's Bruce telescope would survey the night sky in two thousand wide-field plates instead of the *Carte*'s eighty-eight thousand. At Oxford University, for example, it took five years to obtain the assigned plates, ten years for the five-person staff to measure the star positions, and another five years to publish the data in printed format. Other observatories dropped out, either strapped for cash or hungrily eyeing more exciting avenues of astrophysical research. No U.S. institution participated in the *Carte* project, leaving them free to pursue research directions closed to their European counterparts. In a 1943 review, longtime director of the University of Chicago's Yerkes Observatory, Otto Struve, pronounced that large-scale astrometric observing projects were responsible for "virtually killing the ambitions and scientific aspirations of hundreds of the younger astronomers in Europe."

Astrophysical observation, once the province of dedicated empiricists, had turned into a burgeoning, science-based enterprise. Time-exposure photography, for many decades the poor cousin of the human eye, had come to portray cosmic vistas only scarcely imagined by its inventors. The camera turned the familiar Milky Way into an expansive river of sky-borne starlight, phosphorescent clouds, and mysterious coves of darkness. The once-blurry Andromeda Nebula gleamed on the photographic plate in a majestic spiral, drawing gasps from even the most sober-minded astronomers. And dotting the heavens, as faint as the camera's glassy eye could see, thousands of tiny Andromeda-like swirls, each begging for close-up inspection.

Spectroscopy had likewise advanced from its primitive visual embodiment into a sensitive, photography-fortified probe of cosmic chemistry and stellar movement. Nebulae that appeared featureless to the eye sorted themselves spectroscopically into gaseous or starry species. Straightforward in concept, yet arduous in practice, the rendering of spectral lines joined the workaday routines of observatories around the world.

The further potential of these high-technology tools hinged on improvements in the observatory's core instrument. The telescope had undergone its own changes over the decades, having grown more mechanically and optically refined, and occasionally somewhat larger. Instruments once envisioned only in astronomers' daydreams were now being fabricated by optical specialists in America and Europe. Even so, astronomers' efforts to explore the farther reaches of space were frustrated by the dimness of the celestial objects there. The dribble of photons from these far-flung stars and nebulae was insufficient for a modest-sized telescope to enable the working of the camera or the spectrograph. Telescopes had to be wider, to funnel more light to the astrophysical instruments. Indeed, the mantra of the observational astronomer, then as now, is "more light."

"Light is all-important," James Keeler told the audience at the Yerkes Observatory dedication in 1897, "and while much can doubtless be accomplished with small telescopes, there is probably nothing that cannot be done better with large ones." But as astronomers contemplated the telescopic behemoths of the future, Lord Rosse's six-foot-wide, metal-mirror Leviathan lay fallow and rusting in the Irish countryside. Its abandoned carcass recalled the technological imperatives and social milieu of a bygone age, when local muscle, ropes, and pulleys worked such an instrument. By the dawn of the twentieth century, the accumulated astrophysical database had raised expectations for the succeeding stage of cosmic observation. To validate or refute rival astrophysical hypotheses—indeed, to generate such hypotheses in the first place—demanded photographs and spectra of much higher quality than the best of the day.

A twentieth-century Leviathan telescope would echo its predecessor only in terms of aperture and essential form. The new instrument would be free from vibration and flexure, no matter its orientation. It would be mounted so as to access any part of the sky, from horizon to zenith. Driven by machine, not men, it would nevertheless be so finely balanced as to move under a finger's pressure. It would track the diurnal movement of celestial objects doggedly from dusk to dawn, if need be. And it would be sited in the pellucid air of a mountain peak.

The path to the future of astrophysical observation had already been blazed by James Keeler's unlikely resurrection of Andrew Common's telescope atop Lick Observatory's Mount Hamilton. If a clunky thirty-six-inch reflector can morph into a premier vehicle of cosmic discovery, what might a larger and more sophisticated instrument reveal about the universe?

Part III

Money, Mirrors, and Madness

The doors of the observatory are never closed, and at almost any hour of the day or night someone can be found busy in observation or investigation. Indeed the energy, enthusiasm and earnestness of purpose of the Director are reflected throughout the entire institution; and the spirit of investigation seems to saturate the rare air about the summit of the mountain.

—Clarence A. Chant, Lick Observatory, 1907

Chapter 23

Mr. Hale of Chicago

> [George Ellery Hale was] slight in figure, agile in movement, of high-strung nervous temperament, over-flowing with formulae, technical facts and figures, theoretical speculations, almost ad infintum. His mind seems made of some stellar substance which radiates astronomical information as a stove sheds heat.
>
> —Reporter at the dedication of the Yerkes Observatory, 1897

In 1882, fourteen-year-old George Ellery Hale noticed a peculiar boxlike shed sitting in the backyard of a house at the corner of 36th Street and Vincennes Avenue, not far from his own home on South Drexel in the Chicago suburb of Kenwood. "A queer man lives nights in that cheese box," a friend confided, "and tells fortunes by the stars." Hale put aside such dunderheaded notions; he was already a seasoned amateur scientist and he knew an observatory when he saw one, even one as down-home as this. He introduced himself to mustachioed, cigar-chomping Sherburne Wesley Burnham, a court reporter by day and double-star observer by night. The taciturn Burnham, whose hair, one reporter claimed, "seemed bent on going where the wind listeth," took a liking to his spirited guest and became his mentor in the art of telescopic observation.

Sherburne Wesley Burnham.

Such was Sherburne Burnham's reputation as an observer that he had been chosen by the Lick Trust a few years earlier to check the suitability of Mount Hamilton in California for their new thirty-six-inch refractor. Having lugged his own six-inch Alvan Clark refractor up the mountainside, Burnham assessed the atmospheric conditions for two months, discovering forty-two double stars in the process. Many of the closer pairs proved to be a challenge to separate even in bigger telescopes, demonstrating both the clarity of the sky and Burnham's discriminating eye. Some were so close together so as to appear as a single star, their duplex nature revealed only in a subtle elongation of the luminous disk. Burnham's avocation would culminate with the publication in 1906 of his two-volume *General Catalogue of Double Stars*, containing measurement data on 13,655 double and multiple systems.

George Hale returned frequently to Burnham's "cheese box" to learn the ins and outs of astronomical observing. Burnham accompanied him on regular visits to the nearby Dearborn Observatory, where his young charge could peek through its eighteen-and-a-half-inch Clark refractor and receive guidance from George W. Hough, another double-star expert. Hale admired both men's persistence in tallying the particulars of double stars, but he harbored higher aspirations than repetitive measurement. The new science of astrophysics called to him. "I was born an experimentalist," Hale recollected decades later, "and I was bound to find the way of combining physics and chemistry with astronomy."

Hale's scientific passions were amply indulged by his father, William, who had made a fortune selling elevators during Chicago's postfire building boom. "George always wanted things yesterday," William recalled of his son's fierce impatience over the experiment of the moment. In contrast, George's mother, Mary, a brooding semiinvalid, feared that her high-strung, illness-prone child would exhaust his creative fire in one adolescent blaze. (George characterized his mother as "nervously organized.") William Hale tried to lure George into the elevator business, taking him to downtown building sites and enrolling him in a machine-shop course at the Chicago Manual Training School. But George's scientific muse would not be fettered.

Mary Hale gave over her upstairs dressing room to serve as the family's home laboratory. George and his younger siblings, Martha and Will, Jr., each had a work station with a Bunsen burner, batteries, galvanometers, and a host of self-made accessories. Soon arrived a microscope and camera, and eventually a lathe powered by a clamorous steam engine nicknamed "the demon." Among the more attractive experiments were chemical reactions that evolved combustible hydrogen or oxygen: pouring hydrochloric acid on zinc, mixing potassium chlorate and manganese oxide, decomposing water by an electrical spark. "Our delights were enhanced by frightening but delicious explosions," Hale remembered, "the sound of which sometimes pierced the carefully closed door and reached the floor below."

Hale designed and built his first telescope: a single-lens refractor that suffered all the optical and mechanical ills of a neophyte's creation. His astronomical mentor, Burnham, regarded the instrument with the contempt it deserved, and told him about a secondhand, four-inch refractor being offered by a local astronomer. The instrument, like Burnham's own, was manufactured by Alvan Clark & Sons, the nation's foremost telescope maker. George pleaded with his father for the money, citing the fact that a rare transit of Venus was less than a month away. William Hale took the matter under advisement. One evening, he arrived home with the Clark telescope in the carriage. On December 6, 1882, with Martha and Will shivering beside him, George Hale timed the passage of Venus across the Sun's face.

Hale laid a foundation for the telescope on the roof of the house, and from there viewed the beginner's roster of celestial objects. By way of experiment, he rigged a plate holder to the eyepiece-end of the instrument and managed an acceptable photograph of a partial solar eclipse. But Hale's billowing interest at the time was spectroscopy. Having read of Fraunhofer, Bunsen, and Kirchhoff, his curiosity was drawn to the wondrous devices that had opened up the cosmos to chemical investigation. "Even now," Hale would write in 1933, "I cannot think without excitement of my first faint perception of the possibilities of the spectroscope and my first glimpse of the pathway then suggested for me. No other research can surpass in interest and importance that of interpreting the mysteries concealed in these [spectral] lines."

Following directions in *Cassell's Book of Sports and Pastimes*, Hale assembled a pair of small spectroscopes for the upstairs laboratory. For one, he fashioned a prism out of a piece of glass "borrowed" from the household chandelier; for the other, he used a hollow prism filled with foul-smelling carbon disulphide. ("The odor of the disulphide abides with me after a lapse of fifty years.") Using Norman Lockyer's *Studies in Spectrum Analysis* as a guide, Hale studied a variety of flame and spark spectra in his laboratory. With a further monetary infusion from his father, he purchased a spectrometer for his rooftop refractor, swapping out the mediocre prism for a diffraction grating from Pittsburgh instrument maker John Brashear. The grating, Brashear informed Hale, was ruled by the noted physicist Henry A. Rowland of Johns Hopkins University in Baltimore.

Hale's observing notebook for 1886 features lists of Fraunhofer lines he had detected in the solar spectrum. So thrilled was Hale with the performance of the grating that he hopped a train to Pittsburgh and appeared at Brashear's workshop, introducing himself as "Mr. Hale of Chicago." Brashear was taken aback at the sight of his customer, surely not yet twenty, standing in the doorway. From their correspondence, he had assumed that Hale was a middle-aged man and a seasoned expert in astrophysics. A subsequent meeting with Henry Rowland, during which Hale hoped to discuss some fine points of diffraction gratings, elicited the huffy response that "any grating of mine should be good enough for an infant."

In 1886, Hale headed off to college at the Massachusetts Institute of Technology (then called Boston Tech). Shouldering a full course load in physics, chemistry, and mathematics, he hopped a horse cart to Harvard College Observatory every Saturday to serve as a photographic assistant to Edward Pickering. Hale valued the classics—his mother's influence—but referred them to science whenever possible: his English professor received not only the assigned compositions on Scott and Milton, but essays on celestial photography and diffraction gratings. Classmates remember Hale as an academic workhorse, genial but intense, ever en route to his next engagement. "To say that I have been busy is to put it mildly," Hale confessed to a friend, "for I have done nothing but hustle around at the top of my speed from morning till night."

Hale found the formalized, collegiate curriculum stifling and spent considerable time at the library reading books and journals of interest. He looked forward to summers when he could satisfy his experimental cravings. By the time he returned from his first year in Boston, Hale's family had moved to a fashionable mansion on Drexel Boulevard. The attic was set aside for George's new spectrograph, which he had purchased on a family trip to London, and its window-mounted heliostat (sun-tracking mirror). When this space proved inadequate, William Hale purchased the lot next door and erected a brick building to serve

George Ellery Hale.

his son's mushrooming research aspirations. In his first major publication, dated October 24, 1890, George Hale listed his institutional affiliation as "Kenwood Physical Observatory, Chicago."

Reading of the nascent science that would become astrophysics, Hale was intrigued by Janssen's and Lockyer's independent discovery in 1868 that the emission spectrum of solar prominences is bright enough to be seen outside of eclipse. Princeton's Charles A. Young subsequently found that if the spectroscope is centered on a particular emission line (say, the red alpha-line of hydrogen) and the spectroscope slit is widened, a single-color image of the prominence presents itself to the eye. In Young's popular book, *The Sun*, Hale drank in the stirring description of the sight:

> The red portion of the spectrum will stretch athwart the field of view like a scarlet ribbon, with a darkish band across it, and in that band will appear the prominences, like scarlet clouds—so like our own terrestrial clouds, indeed, in form and texture, that the resemblance is quite startling: one might almost think he was looking out through a partly opened door upon a sunset sky, except that there is no variety or contrast of color; all the cloudlets are of the same pure scarlet hue.

Young goes on to detail the technical facets of solar observation, then concludes in a coda that aroused Hale's curiosity:

> Setting the spectroscope upon this latter line [in the violet part of the spectrum] and attaching a small camera to the eye-piece, it is even possible to photograph a bright protuberance; but the light is so feeble, the image so small, the time of exposure needed so long, and the requisite accuracy of motion in the clock-work which drives the telescope so difficult of attainment, that thus far no pictures of any real value have been obtained in this manner.

Indeed, if the eye can so plainly see these solar outbursts outside of eclipse, why cannot the camera? The multiple impediments cited by Young—faintness, smallness, photographic insensitivity, and mechanical imprecision—were together a veritable siren song to Hale. Here was a complex challenge in astrophysical observation, one that required thoughtful, persistent, yet flexible attack.

In July 1889, Hale found himself riding the cable car along Chicago's Cottage Grove Avenue. As the coach rumbled along, he caught sight of a white picket fence lining the side of the road. The pickets reminded him of lines in a spectrum, the car's forward movement causing his fixed eye to scan across the lattice of vertical strips. In that instant, Hale envisioned the spectroscopic innovation that would permit the photography of solar prominences.

Hale's "spectroheliograph" would have two slits instead of the usual one. The entry slit would transmit a thin slice of the Sun's image-disk, as one might crop a portrait to depict only the nose and the features above and below it. The exit slit, situated after the light's blended colors have been dispersed by the grating, would isolate from the spectrum a solitary wavelength, which in turn strikes a photographic plate behind the slit. Were the two slits to remain stationary while the telescope tracked the Sun, the resultant picture would be a monochromatic rendering of the original slice of the solar disk. But were the slits moved in synchrony at the proper rate, a one-color scan of the Sun's surface is created.

Point the device at a prominence on the Sun's limb, and the slits' harmonized movement effectively creates a picture of the outburst: The height of the targeted spectral line lengthens and shrinks with the prominence's contours, while the line is simultaneously shifted to unexposed areas of the plate. (The same might be accomplished with stationary slits and a moving plate.) The width of each slit must be properly adjusted: too narrow and there is insufficient light to form the image, too wide and the ambient solar light fogs the plate.

In November 1889, Hale received permission from Edward Pickering to try his spectroheliograph on Harvard's fifteen-inch refractor, but the weight of the device threatened to snap the telescope's wooden tube. Instead, Hale affixed the apparatus to the observatory's twelve-inch, fixed, horizontal refractor, which receives starlight via an articulated, eighteen-inch plane mirror. However, the arrangement failed. "Not only is a large amount of diffuse light sent into the spectroscope," Hale complained, "but the distortion of the mirror by the sun's heat soon changes a prominence into a shapeless mass when the diffuse light does not render it entirely invisible." Frustrated by the lack of suitable equipment, Hale graduated from MIT in 1890, then sat down to design a telescope and spectrograph customized to the task at hand.

Once again, George Hale availed himself of his father's generosity. This time it brought to his Kenwood Observatory a gleaming twelve-inch

The twelve-inch refractor telescope and spectroscope in George Hale's Kenwood Observatory, Plate II from the August 1891 issue of the Sidereal Messenger.

Brashear refractor, premium Warner and Swasey mount, and ten-foot-focal-length spectrograph with an advanced Rowland grating. Rigid bracing effectively united the telescope and spectrograph into a singe, flexure-free unit. The building's new extension, topped by a twenty-six-foot dome and flanked by a reception room, library, and spectroscopic laboratory, was dedicated on June 15, 1891, before a hundred civic and scientific luminaries. At once, Hale's center became the nation's best-equipped private astrophysical observatory, a worthy successor to those of Lewis Rutherfurd and Henry Draper.

Even as the dignitaries gathered, Hale could announce that his spectroheliograph had yielded spectacular results. By utilizing the brilliant violet

George Hale's Kenwood Observatory, Plate I from the August 1891 issue of the Sidereal Messenger.

emission lines of calcium—the color at which 1890s-era dry plates were most sensitive—exposures of two or three minutes captured panoramic portraits of the Sun as it had never been depicted before. The Sun's seemingly placid surface, as portrayed in conventional white-light pictures, had turned into a roiling inferno of turbulent gas. In Kenwood Observatory's first year of operation, Hale and his photographic assistant, Ferdinand Ellerman, produced fifteen-hundred photographs of solar features, both prominences fringing the Sun's periphery and sunspots and bright regions—"faculae"—seen against the solar disk itself. (Reprising their childhood roles, George's siblings Will and Martha obtained an excellent prominence photograph themselves while their big brother was traveling in Europe.)

News of Hale's innovation spread quickly, as his spectroheliograms appeared in journals worldwide. In June 1891, Hale's initial report was read before the Royal Society in London and he was elected a Fellow of the Royal Astronomical Society. The British Association for the Advancement of Science feted him at its August meeting. "I am treated like a *Grand-Duke*!" he wrote a friend. The burst of acclaim also brought to Hale's doorstep the galvanic president of the new University of Chicago, William Rainey Harper. As driven to accomplishment as Hale himself, Harper had graduated from college at age fourteen, received his doctoral degree from Yale at eighteen, and now at age thirty-five, led an institution generously funded by John D. Rockefeller. Harper's visit was not unexpected, nor was its intent. Edward Pickering had already written to Hale that Harper had been asking about him. A friend from the Astronomical Society of the Pacific forwarded a similar inquiry, remarking, "The enclosed looks like business."

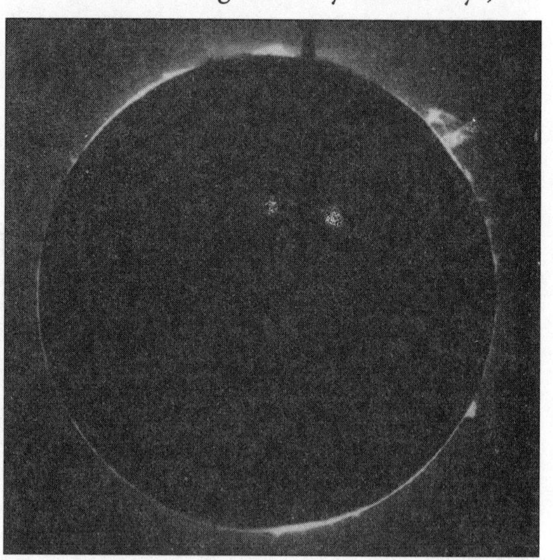

Solar activity captured by George Hale's spectroheliograph at the Kenwood Observatory in 1892. A circular metal plate within the instrument blocked the glare of the sun's surface.

Harper tried to lure Hale to Chicago's faculty, if Kenwood Observatory came along. Hale refused, having learned that Harper was more interested in acquiring the facility and its advanced equipment than in Hale himself. But within two years, the calculus had changed. The fledgling university had demonstrated its commitment to original research: among others, the great experimentalist Albert A. Michelson had been brought on to lead the physics department. Hale also knew that the planned successor to Kenwood's twelve-inch refractor lay well beyond his father's means to provide. The sight of the giant Lick refractor during his honeymoon in the summer of 1890 had inspired a vision of an even larger telescope devoted to astrophysical observation. To realize that dream, he needed an institutional affiliation, as well as William Harper's connections. To the independent-minded Hale, an academic post was but a means to an end: "I would not consider the thing for a moment were it not for the prospect of some day getting the use of a big telescope to carry out some of my pet schemes."

In 1892, George Hale was appointed associate professor of Astral Physics at the University of Chicago. His contract stipulated that, if his first year proved agreeable and if President Harper established a quarter-million-dollar research and maintenance fund, Hale would transfer ownership of the Kenwood Observatory to the institution.

While at a meeting of the American Association for the Advancement of Science, in Rochester, New York, in August 1892, Hale overheard a conversation that set his course for the next five years. Alvan Graham Clark, son of the founder of the great American optical house and the family raconteur, divulged that a pair of pristine, forty-inch glass disks from France lay in storage at the Cambridge workshop. The University of Southern California had ordered the castings in an effort to surpass the University of California's thirty-six-inch refractor at Lick Observatory. However, the regional economy soured and the prime donor withdrew his pledge of support, leaving the record-breaking disks up for grabs. Now the American telescope maker owed the French glassmaker $16,000, with no means to pay.

No doubt George Hale made several quick mental calculations. The light grasp of a forty-inch telescope is fully 23 percent greater than that of a thirty-six-inch. Its resolving power would render visible a quarter at a distance of three hundred miles. And the Sun's image, a measly two inches across in the Kenwood refractor, would swell to almost seven inches. But the telescope, while useful in itself, would form the observational hub of a world-class astrophysical laboratory.

After declaring his urgent interest to Alvan Graham Clark, Hale hopped the next train back to Chicago and urged President Harper to strike the deal that would bring the world's largest telescope to the University of Chicago. In a single, timely bound, Hale told him, the institution would vault to the top of the scientific establishment. The estimate for the telescope, mount, accessory equipment, and building: $300,000.

Harper had allocated the university's initial endowment to the establishment and staffing of the various academic departments. A massive expenditure toward the sort of highly specialized research facility that Hale envisioned would have to come from a new funding source. As distasteful as it must have been, Harper scheduled a meeting on October 2, 1892, with streetcar magnate Charles T. Yerkes. A "strange combination of guile and glamour," according to one observer, Yerkes wasn't called the "Boodler" for nothing. Once jailed in Philadelphia for embezzlement and notorious in Chicago for all manner of business chicanery, Yerkes had previously promised, then abruptly withdrew, his support for a new biology building. He received Hale's pitch enthusiastically, swooning over the size and power of the proposed telescope, as well as having his name affixed to the landmark instrument. Cost be damned, his goal was to "lick the Lick."

So began George Hale's five-year struggle to keep his new observatory on the rails. Almost immediately, Yerkes retreated from his original agreement to fund the entire project down to its last nut, bolt, and brick. Hale, he claimed, had misconstrued the terms: he was financing only the telescope itself, not the building, astrophysical laboratories, or associated equipment. Having no sense of the price associated with optical and mechanical refinement, he accused Hale of ginning up costs and creating a scientific edifice where a modest facility would do. (In fact, it was Yerkes, not Hale, who hiked the architectural splendor of the building.) Only after repeated entreaties and overt appeals to vanity did Hale coax Charles Yerkes to fund the observatory to its completion.

The Yerkes Observatory rose on the shores of Lake Geneva in Williams Bay, Wisconsin, a once-dark exurb eighty miles northwest of Chicago. (Hale favored a California mountaintop, but Charles Yerkes refused.) The stately, Romanesque-style building is shaped like a Latin cross, with the ninety-foot dome of the big refractor anchoring its base, and a pair of smaller domes and a meridian-telescope room at the remaining cardinal points. Laboratories, offices, optical and machine shops, a library, and a lecture hall occupy the structure connecting the observatories. On May 19, 1897, the precious lenses arrived at Williams Bay station in a private Pullman car, accompanied by Alvan Graham Clark and his shop foreman Carl A. Lundin. Within a day, installation was complete.

The Yerkes Observatory in the 1890s.
The large dome at right houses the forty-inch refractor.

The half-ton achromatic objective, pairing a two-and-a-half-inch-thick crown-glass lens with a one-and-a-half-inch-thick flint-glass lens, sits at the upper end of a sixty-three-foot-long sheet metal tube. The telescope and its movable parts weigh in at twenty tons, yet the tube is moved by the gentle impulse of a hand or an electric motor. The entire assembly rests atop a cast iron column, forty-three feet high, which itself is secured to a brick-and-concrete base. A thirty-seven-ton, counterpoised floor conveys the observer to the eyepiece, whose elevation varies by twenty-three feet depending on the telescope's orientation. (The floor collapsed when a support cable slipped from its pulley a few nights after observations had commenced. No one was hurt and the telescope was unharmed.)

The forty-inch refractor of the Yerkes Observatory, photographed in 1925.

On the night of May 21, 1897, after a thorough viewing of the planet Jupiter, George Hale, Edward Barnard (whom Hale had hired away from Lick), and observatory assistant Ferdinand Ellerman swung the new telescope toward deep space. The Ring Nebula in Lyra and the Dumbbell Nebula in Vulpecula shimmered with remarkable clarity, and the globular star cluster in Hercules seemed more populous than ever. Without question, Barnard pronounced, the celestial scenes surpassed those of the Lick refractor. Several days later, Hale reassured the project's mercurial benefactor that his "telescope is not only the largest but also the most powerful in the world."

While the observatory struggled to prosper—Charles Yerkes provided no operating funds—Hale's enthusiasm and heroic fund-raising efforts buoyed the disposition of the staff. Those who worked alongside him describe an almost palpable creative energy that transcended the muted confines of scientific investigation. Astronomer Frederick Hanley Seares noted that Hale's "logical formulation was shaped by the imaginative powers of a temperament essentially artistic. Problems thus presented became something engaging, to be undertaken as adventures of the spirit; and to Hale they were indeed such adventures, entered upon with a kind of joyous gaiety."

At Yerkes, Edward Barnard, now professor of Practical Astronomy, tracked the movements of planetary satellites and continued his wide-field photography of the Milky Way. Dartmouth College spectroscopist Edwin B. Frost was brought on to measure stellar radial velocities using a precision spectrograph funded by the noted astronomical benefactress, Catherine Wolfe Bruce. Hale's longtime friend Sherburne Burnham opted to keep his job at the courts, but arrived at Williams Bay station nearly every Saturday afternoon with a box of cigars under his arm and a case of Burgundy in waiting. For the next two nights, he measured double stars with the forty-inch, then returned to Chicago on the early Monday train. Walter S. Adams, Frost's graduate student at the time, joked that on the morning after Burnham's shift, "a fairly accurate analysis of his observing schedule could be made from the intricate trail of tobacco ashes on the floor of the dome."

From 1903 to 1905, Columbia University graduate Frank Schlesinger used the big refractor to conduct an exhaustive study of the instrumental and atmospheric factors that affect the photographic measurement of stellar parallax. The methods he derived at Yerkes were adopted at other observatories and became the basis for the *General Catalog of Stellar Parallaxes*, published in 1924. George Willis Ritchey, the staff optician, utilized special filters and photographic plates to effectively negate the telescope's chromatic aberration and turn the instrument into a formidable astronomical camera. Double stars that eagle-eyed Burnham had seen only under prime conditions were now imaged on a regular basis. The populous cores of globular star clusters were recorded in crisp definition. A half-second, high-magnification exposure on October 12, 1900, of the lunar crater Theophilus projected a vertiginous sensation of depth.

Secluded Williams Bay conveyed a sometimes serene, sometimes oppressive sense of isolation. Groceries were brought in daily by train, fresh meat twice a week if ordered ahead. Distant echoes of outside events wafted in, muted both by the observatory's remoteness and the astronomers' overriding focus on their scientific work. "This is a very peaceful region of the world," Hale said, "where we occasionally hear faint rumors of disturbance, but are not much concerned with such doings ourselves." A sense of exclusivity infused the premises, a selfless nobility of mission enveloped the staff. One astronomer, in a half-serious expression of Yerkes's special status, had his students and academic friends utter the password "syzygy" (an alignment of celestial bodies) before being admitted to the sanctum of the observatory.

During the warmer months, Williams Bay was a rural paradise, offering golf, sailing, hiking, and the simple pleasure of unhurried contemplation. The summer mansions of wealthy Chicagoans lined the shores of Lake Geneva, whose waters hosted a zigzag flotilla of steamers and small yachts. Hale and his family lived in a gracious house overlooking the lake; staff members' families resided in smaller homes on the observatory grounds or nearby. Bachelors and visitors lodged at a local rooming house. Transportation, when not by foot, was by horse-drawn carriage, the rutted roads an obstacle course to bicycles. As winter closed in, the village became an outpost, with blistering winds, chest-high snow, and howling wolves. Hale's wife and children spent the winter months in a rented apartment in Chicago, seeing him only on weekends.

Wintertime observing was especially arduous, with subfreezing temperatures and towering snowdrifts between the living quarters and observatory. Taking up his shift at 1:30 a.m., staff astronomer Storrs Barrett donned "two sets of underwear and one pair of pants, one shirt, two coats, two pairs of socks, a collar (no tie), a neckscarf, a sealskin cap, a fur overcoat, a pair of shoes, a pair of arctics and two pairs of mittens"—only to have clouds roll in by the time he trudged over to the telescope. Edward Barnard was described by visiting Irish astronomer Robert Ball as a "moving cylinder of fur coats . . . running as briskly up and down as if he were playing football. Indeed, he had to be well clad, for that night [December 17, 1901] he worked from five in the evening till six in the morning."

Despite the pressing burdens of the directorship, constant tug of family, and recurrent insomnia and headaches, George Hale carried on his research. In the solar arena, the forty-inch was fixed up with a jumbo,

Astronomer George Van Biesbroeck inspecting the lens of the Yerkes forty-inch refractor in 1928.

Yerkes-built spectroheliograph that provided new insights into the stratification and movement of gas in the Sun's upper atmosphere. Hale also delved into the extrasolar realm, obtaining the first-ever photographs of the spectra of faint red stars, a considerable achievement given the relative insensitivity of dry plates to these long wavelengths. By 1904, he had received the Janssen Medal from the Paris Academy of Sciences, the Rumford Prize from the American Academy of Arts & Sciences, the Henry Draper Medal from the National Academy of Sciences, and the Gold Medal of the Royal Astronomical Society.

Although secure in his scientific reputation, if not in his own physical and emotional health, George Hale could not quell the inner fire that drove him to greater accomplishment. He knew all along that the Yerkes refractor was the child of opportunity: The glass disks became available just when he needed a successor to his paltry, twelve-inch home telescope. In Hale's dawning estimation, the focal length of the forty-inch was too short for the brand of solar research he contemplated, and its largest-in-the-world aperture too small to capture the faint spectroscopic signatures of stars and nebulae at the fringe of the known universe. As grand as the Yerkes refractor had once seemed, it had been overswept by the observational challenges facing twentieth-century astrophysicists.

Even as the foundation was being laid for the Yerkes Observatory, Hale conceived plans for a much larger instrument, this time a reflector. A mirror-based telescope reflects light from its silvery objective element; the light never enters the glass substrate and thus is relatively undiminished when it reaches the eye or the camera. The reflected rays converge to a common focus regardless of wavelength, mitigating the chromatic aberration that afflicts refractors. (As early astrophysics advocate Agnes Clerke put it, "None of the beams they collect are thrown away in colour-fringes, obnoxious in themselves and a waste of the chief object of the astro-physicist's greed—light.") And unlike an objective lens, which is supported solely along its circumference, an astronomical mirror nestles securely in a cradle, braced against gravity's distorting influence. To a degree, the only limit on the size of a reflector telescope lies in the mechanics of the making—and in the aspirations of the maker.

In 1894, fueled by his son's latest vision—a telescope with *twice* the light-collecting area of the Yerkes refractor—William Hale ordered a sixty-inch mirror blank from the Saint-Gobain foundry in France. The eight-inch-thick, one-ton disk arrived two years later. But after grinding and polishing, it languished in the Yerkes basement while Hale sought funds to mount it. More than a decade later, the great mirror rose from its crypt in Wisconsin and, in 1908, found itself facing skyward on an airy mountaintop in California. The era of modern megareflectors was about to begin.

Chapter 24

The Universe in the Mirror

It has often been asserted by prominent writers on the subject that a large reflector is necessarily inferior to a large refractor in many vital points, such as permanence of optical qualities, freedom from injurious flexure of optical parts, permanence of adjustments or collimation of optical parts, rigidity and stability of the mounting, and convenience in use. . . . Nothing could be further from the truth.

—George W. Ritchey, "The Two-Foot Reflecting Telescope of the Yerkes Observatory," 1901

As the nineteenth century drew to a close, the near-universe was coming into focus, yielding to the allied technologies of the New Astronomy: the camera and the spectrograph. Practitioners of the maturing science of observational astrophysics sought to redefine astronomical research by demonstrating to their visual counterparts the potential of these new devices for cosmic analysis and discovery. Already they had teased out the chemical recipes of the Sun and several bright stars and nebulae, accumulated essential data about stellar motions through space, and imaged celestial structures that had eluded sharp-eyed observers and their exquisite telescopes. But what of the deeper reaches of space beyond the ragged borders of the Milky Way? Here, in the realm of the mysterious spiral nebulae, the camera and the spectrograph faltered because of the sheer dimness of the cosmic glow reaching Earth. To render these technologies effective with such miserly bits of light required a radical transformation—a scaling up—of that most fundamental astronomical instrument: the telescope.

The premier telescopes of the late 1800s, both in the United States and abroad, were refractors. A centuries-long evolution had made these instruments synonymous with optical and mechanical refinement. Yet their upsizing raised concerns about the deleterious effects of increased aperture:

A wider lens is necessarily thicker and heavier; it more substantially absorbs incident photons and might sag under its own weight. By contrast, an ideal reflector telescope transmits light relatively undiminished and is free of the refractor's chromatic aberration. With proper support from behind, even the bulkiest mirror might retain its critical concave form as the telescope swings from one position to the next.

As in celestial photography and spectroscopy, the development of reflector telescopes during the nineteenth century was driven in large measure by amateur astronomers, who took the conceptual and methodological risks their professional counterparts would not. Only toward the end of that period, once cornered by the limits of their existing instruments, did institutional astronomers fully comprehend the scientific imperative of larger aperture. Their introduction of sizable, yet refined, instruments paralleled an irrevocable change in the mode of astronomical research: from the lone observer operating a modest-sized, out-of-pocket telescope to the salaried, institutional team wrestling with a million-dollar machine on a remote mountaintop.

Isaac Newton is credited with the first functional reflector telescope, having developed in 1668 the hallmark design that now bears his name. (A small plane mirror deflects the gathered light through an eyepiece at the side of the tube.) His diminutive instrument—just over six inches long and one-and-a-third inches in aperture—magnified about thirty-five times. The mirror was formed of bell metal, or speculum metal, an exceedingly brittle alloy of copper and tin, to which Newton added a dash of arsenic for increased whiteness; still, it reflected only 16 percent of incident light and tarnished easily. Newton's telescope reportedly imaged the Moon about as distinctly as a typical, small-aperture refractor of his day.

In 1672, Guillaume Cassegrain in France produced an instrument with a secondary mirror that reflected light back through a hole in the center of the primary mirror, instead of out the side of the tube as in a Newtonian. (Englishman James Gregory had experimented unsuccessfully with a variant of this design several years before Newton constructed his telescope.) In the 1700s, English telescope makers John Hadley and James Short each conducted a brisk business selling metal-mirror telescopes with apertures up to eighteen inches—and wildly inflated magnification claims.

A one-man crusade for greater aperture began in 1781 with the discovery of the planet Uranus by German émigré William Herschel, a conductor and music teacher in Bath, England. Herschel sought to develop large telescopes with which he could survey the night sky for unseen planets, stars, and nebulae. Although high-precision refractors were ascendant at the time, he found reflectors to be more easily and economically scalable in size.

Disks of ever-larger diameter were cast from molten metal poured into molds of loam, charcoal, or compressed horse dung; each disk was then laboriously abraded with pulverized emery and rouge until its cross section had a parabolic shape. (Only a parabolic mirror converges all incoming light to a single focus.) Between 1773 and 1795, Herschel produced some 430 telescope mirrors, most sold for profit, the best kept for his own use.

Herschel's involvement in astronomy was all-consuming. His sister Caroline, who became an astronomer in her own right, complained in one diary entry that almost every room of the house had been turned into a workshop. She read to her brother and placed morsels of food in his mouth while he polished mirrors. On most clear nights, Caroline served as his amanuensis, sitting next to an open, second-story window while William called out his observations from atop a ladder beside the telescope.

William Herschel, depicted in an 1807 engraving of a painting by J. Russell.

In 1783, Herschel completed what would become his bread-and-butter research instrument for the next three decades: a reflector, nineteen inches in aperture and twenty feet long, suspended within a rotatable wooden frame. Herschel's next project, completed in 1789, was unprecedented: an instrument with a mirror forty-eight inches in diameter. American writer Oliver Wendell Holmes, father of the U.S. Supreme Court justice, recalled his first glimpse of Herschel's telescope: “It was a mighty bewilderment of slanted masts, spars and ladders and ropes from the midst of which a vast tube . . . lifted its mighty muzzle defiantly towards the sky.”

William Herschel's forty-eight-inch-aperture reflector telescope, completed in 1789 at Slough, England. Commonly known as the forty-foot telescope for its focal length.

Despite its yawning aperture, the forty-eight-inch never lived up to its potential. The massive metal mirror warped under its own weight and took hours to settle into equilibrium with the cool night air. On all but a few occasions, Herschel sacrificed size for the utility of his mainstay nineteen-inch telescope. Either way, he was in the enviable position of being able to see celestial objects invisible to the rest of the world's astronomers. The potential of large-aperture reflectors was underscored by Herschel's prodigious record of astronomical accomplishments, including descriptions of thousands of deep-sky objects, discovery of new moons around Saturn and Uranus, and from tedious star counts in different parts of the sky, the determination that our Milky Way system is shaped like a disk. Nevertheless, the troublesome forty-eight-inch telescope shouted a warning to any future enthusiast who sought to surpass it: this was a path reserved for those with single-minded devotion—and deep pockets.

Lord Rosse's seventy-two-inch Leviathan reflector, constructed in Ireland during the 1840s, was a worthy successor to Herschel's optical giants. But its utility was compromised by its restrictive support walls, which hemmed in its view of the night sky. The engineering challenge remained: how to make a massive telescope's mechanism of movement more conducive to the demands of astronomical observation. (The Leviathan was restored in the 1990s with a modern, aluminum-coated glass mirror.)

In 1845, inspired by Herschel's and Rosse's examples, William Lassell, a wealthy brewer in Liverpool, built the world's largest free-swinging, equatorially mounted telescope, incorporating a twenty-four-inch speculum-metal mirror. (An equatorial mount tracks the diurnal movement of celestial objects by rotating the telescope around one axis instead of two.) The 370-pound mirror, like that of Rosse's Leviathan, had been shaped by a steam-driven device that mimicked the epicyclic grinding-strokes of the artisan. Lassell incorporated a multipoint mirror support system, brainchild of noted Dublin optician Thomas Grubb, which prevented distortion of the reflector as its orientation changed. This model, adapted in various forms, promoted the growth of astronomical mirrors during the early twentieth century.

From his aptly named private observatory, Starfield, just outside Liverpool, Lassell discovered Neptune's satellite Triton in 1846, and two years later, the satellite Hyperion circling Saturn. (William Bond observed Hyperion at the same time through Harvard's fifteen-inch refractor.) At every turn, Lassell was hamstrung by Starfield's routinely poor observing conditions. His friend, the famed double-star observer William Rutter Dawes, dubbed the troubled site "Cloudfield." Lassell's frustration with the English climate permeates his scientific correspondence and foretells the sentiments of astronomers like George Hale and James Keeler in the 1890s: "I was never more struck with the conviction how necessary a pure tranquil sky is to the just performance of a very large telescope."

In 1852, Lassell took the dramatic step of disassembling his twenty-four-inch telescope and shipping it—and himself—to the more astronomically friendly climate of Malta. Over the next year and a half, he delighted in the clear night sky (and British naval protection) of his Mediterranean outpost. "The nights are as remarkable for their tranquility as transparency," he enthused to fellow astronomer Warren De La Rue in November 1852, "and I have not encountered one of those nights so frequent at Starfield, on which I opened the Observatory to close it in disgust." Lassell focused his studies on planets and their satellites, often taking advantage of the clear, steady air to ramp up his telescopic magnification to more than a thousand.

Lassell returned to England in 1854, relocated to a new estate farther from Liverpool's smudgy skies, and commenced work on his next project: a forty-eight-inch reflector. The new telescope saw "first light" at Liverpool in 1859 and under Maltese skies in 1861. Twin metal mirrors were cast, each weighing a ton, one placed in service while the other was repolished. This instrument, too, featured an equatorial mount, its iron-lattice tube jutting skyward between beefy conical tines. Like a star-struck Juliet, Lassell peered into his lofty eyepiece from the balcony of a rail-borne tower. On the ground below, an assistant turned a hand-crank to slew the telescope's eight-ton bulk in a manner sufficient for visual observing.

Ungainly as its mechanics might have been, the instrument's extraordinary light-grasp was evident from the start: Lassell likened the bright star Sirius to an incandescent diamond and the Orion Nebula's hazy swirls to luminous masses of wool. He tracked planetary satellites and cataloged six hundred previously unseen nebulae before leaving Malta for England in 1865, where he resumed observations with his more wieldy twenty-four-inch telescope. Dismantled, the forty-eight-inch lay in storage until shortly before Lassell's death in 1880, when it was sold for scrap. "I was not without a pang or two," he informed the Royal Astronomical Society, "on hearing the heavy blows of the sledge-hammers necessary to overcome the firmness of the alloy."

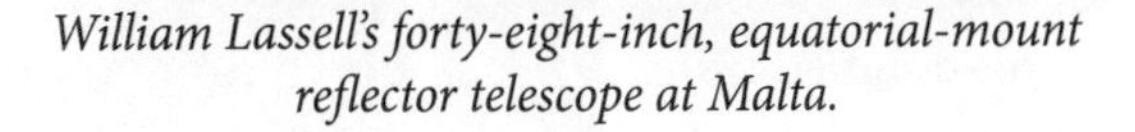

William Lassell's forty-eight-inch, equatorial-mount reflector telescope at Malta.

William Lassell was a stalwart observer who gleaned factual information about the cosmos, duly reported it, then let others speculate as to its meaning. He possessed neither the broad scientific ambitions nor the research résumé of his antecedent William Herschel. Yet his forty-eight-inch instrument surpassed Herschel's, Rosse's, and every other sizable reflector up to that time in functionality and mechanical sophistication. Although primitive compared to its sleek refractor cousins, it was the forerunner of a coming generation of large-aperture, fully movable reflector telescopes. Except for one glaring deficiency: its weighty metal mirror.

The era of jumbo metal mirrors came to a crashing end in the 1870s, following a pledge by the citizens of gold-enriched Victoria, Australia, to build a Southern Hemisphere telescope of supreme optical power. The planning committee politely rejected William Lassell's offer of his twenty-four-inch reflector (too small) and subsequently his forty-eight-inch (too cumbersome) and also a proposal from Joseph Fraunhofer's successor, George Merz, in Munich, for a thirty-inch precision refractor (too costly—and, well, not British enough). Instead the committee accepted a bid from Thomas Grubb's firm in Dublin for a more modestly priced, forty-eight-inch metal-mirror instrument of the Cassegrain design. This optical arrangement, invented in 1672, employs a convex secondary reflector that channels light through a central hole in the concave primary reflector, and from there through an eyepiece. As a result, the elevated viewing position of a Newtonian telescope drops conveniently to the base of a Cassegrain's tube.

The Great Melbourne Telescope, as it was known, was put into service in 1869 and became one of the biggest flops in the annals of instrumental astronomy. The mounting was unstable, the tube vibrated in the wind (the telescope was unsheltered in use), and the tarnish-prone metal mirror required frequent repolishing, a delicate process for which no one in Australia had been trained. Nor had the civil-service astronomers the experience or commitment to properly operate the eight-ton opti-mechanical beast.

The telescope's primary mission was to search for physical changes in Southern Hemisphere nebulae cataloged from South Africa during the 1830s by William Herschel's son John. The project was doomed from the beginning, given the inherent differences in human perception and artistry and in the telescopes themselves. The effective focal length of the Melbourne telescope was fully forty-one times its aperture, endowing it with a Lilliputian field of view that hampered the visual study of nebulae. (A shorter focal length would have increased the diameter and weight of the secondary mirror at the tube's mouth.)

The Great Melbourne Telescope in an imagined setting.

In terms of popular

sentiment, the death blow came in the autumn of 1877. Despite repeated attempts, the Melbourne observers could not see the recently discovered satellites of Mars, even though the pair were clearly visible in telescopes in the United States and England. (The imperfectly shaped mirror bled light from the Martian disk into an obscuring halo.) Even in the realm of deep-sky observation, where aperture reigns, the Great Melbourne Telescope was about to be eclipsed. In 1883, Andrew Common displayed his dry-plate photographs of the Orion Nebula, to which the Royal Astronomical Society added its imprimatur by awarding him its Gold Medal. Astrophotography was the coming wave, and the Melbourne reflector was ill adapted to the new technology. (A consortium of government and civic organizations is restoring the Great Melbourne Telescope for public viewing—this time, with a modern glass mirror.)

In retrospect, Yerkes Observatory optical designer George Ritchey judged the failure of the Melbourne reflector "to have been one of the greatest calamities in the history of instrumental astronomy, for by destroying confidence in the usefulness of great reflecting telescopes it has hindered the development of this type of instrument, so wonderfully efficient in photographic and spectroscopic work, for nearly a third of a century." Ritchey may have overstated the case, as there were plenty of other design or execution stumbles that reinforced negative perceptions about mirror-based telescopes. Reflector telescopes would indeed grow, nurtured as so often in science by technological innovation. But with the Melbourne debacle, the era of metal mirrors was over.

Even Isaac Newton knew the pitfalls of speculum-metal reflectors, having fabricated several of his own during the late seventeenth century. Much better to endow a lightweight, concaved piece of glass with a reflective surface. Newton deposited tin foil onto glass by immersing them in mercury for several days. But the resulting reflection was best seen through the backside of the glass, reintroducing the light absorption and chromatic aberration a first-surface mirror was supposed to solve. Thin meniscus mirrors and other optical correctives were tried with minimal success over the ensuing decades.

In 1856, driven by an effort to remove caustic mercury from the manufacture of looking-glasses, German organic chemist Justus von Liebig found that a solution of silver nitrate, caustic potash, ammonia, and sugar deposits a reflective silver film onto a glass plate. Shortly afterward, physicist Léon Foucault in Paris and optical designer Carl August von Steinheil in Munich independently applied the method to the creation of silvered-glass telescopes, ranging in aperture from four inches to thirteen inches. To reflector-telescope critics, Foucault cited the distinct visual separation of the close binary star Gamma^2 Andromedae in his thirteen-inch instrument.

Frustrated by the haphazardness of grinding glass disks to their proper concavity, Foucault developed his ultrasensitive knife-edge test, in which an illuminating beam reveals a mirror's surface contours to within a fraction of the wavelength of visible light. Individual bumps could be identified and then polished away until the curvature of the glass was uniform. Astronomical mirrors could now be vetted before leaving the workshop. The pinnacle of Foucault's telescope-making activity came in 1864 with a thirty-one-and-a-half-inch silvered-glass reflector for the Marseilles Observatory. Work on a government-funded, forty-seven-inch reflector was effectively halted after Foucault's death in 1868.

Unlike the secretive professional opticians, Foucault published complete reports of his mirror-making methods. In 1857, he addressed the British Association in Dublin on the merits of the silvered-glass reflector. Glass is both lighter and less brittle than speculum metal, and silver is more reflective. Resilvering a glass mirror is trivial compared to the restoration of a speculum disk, which entails a complete regrinding of its curvature. The initial reception to the newfangled technology was tepid: Following an excursion to see Lord Rosse's speculum-metal Leviathan—a "monstrosity," in the French scientist's opinion—Foucault complained to a Parisian colleague, "For the English, mine does not exist." Nevertheless, how-to articles on silvered-glass reflectors began to appear in English journals in 1859 and quickly leaped the Atlantic to America.

In the ensuing decades, a cadre of optical craftsmen and amateur astronomers stoked the development of silvered-glass telescopes, primarily for their high efficiency and low-cost path to increased aperture. Compared to the sleek, unitized refractor, the contraption-like reflector must have appealed to the astronomical tinkerer, much as a mechanically minded musician might be drawn to the banjo for its array of adjustable elements. In 1867, London optician John Browning published an advertising pamphlet, *A Plea for Reflectors*, in which he extolled the virtues of silvered-glass instruments. Browning assured his patrons that the days of tarnished metal mirrors were gone: "After some five years' constant experience in the use of these telescopes, I can assert that nearly all that has to be done, is to carefully *let the silver coating alone.* Several of my friends have mirrors which have been in use for two or three years, and they are almost unchanged in appearance, quite so in performance."

Given the cheapness of the glass and the relative ease of fabrication, reflectors enjoyed a huge cost advantage over refractors (and still do). A Browning ten-inch reflector, complete with equatorial mount and clock drive, could be had for about £200, a fraction of the price of the same-size achromatic refractor. And whereas Lick Observatory paid $50,000 just for the objective lens of its thirty-six-inch refractor, its thirty-six-inch Brashear reflector came in at half the amount—including the dome, mount, and spectrograph.

John Browning's contemporary George Calver likewise launched production of silvered-glass reflectors. Most notable was Calver's thirty-six-inch telescope, with which Andrew Common produced his breakthrough 1883 photograph of the Orion Nebula and James Keeler his revelatory plates of spiral nebulae. The 1870s saw Henry Draper's parallel endeavor to advance large-aperture reflectors in the United States. The ability of Draper's twenty-eight-inch telescope to record the first stellar spectrum in 1872 suggested the reflector as a viable research instrument. Still there was considerable pushback from the astronomical community. Oxford University astronomer Herbert Hall Turner spoke for many when he commented that the "reflector is so seriously influenced at times by air currents and changes of temperature as to be an instrument of moods and Dr. Common has accordingly compared it, somewhat ungallantly, to the female sex."

The reflector-versus-refractor debate continued into the 1890s with now some dozen lens-based telescopes exceeding two feet in aperture. Having completed the largest-in-the-world Yerkes refractor, Massachusetts optician Alvin Graham Clark boasted to George Hale about the feasibility of a seventy-two-inch lens. Hale dismissed the idea. Such a behemoth would be impossibly long and its thick glass matrix would severely winnow precious cosmic photons. For the astrophysicist seeking to photograph a spectrum, the refractor's chromatic aberration was especially problematic. "No combination of lenses yet devised," Hale advised his colleagues, "can compare with a paraboloidal mirror in the capacity to unite in a single focal plane all wave-lengths."

The Yerkes forty-inch, George Hale rightly suspected, marked the historical zenith in the aperture of refractor telescopes. If astronomers were to probe farther and fainter, they had no choice except to muster the resources to enlarge mirror-based instruments. Yet these next-generation reflectors could share none of the quirky, ad hoc construction of their predecessors; they would have to equal—indeed, surpass—the level of refinement long associated with their refracting counterparts.

While his sixty-inch mirror-blank lay fallow in the Yerkes basement, lacking funds for completion, Hale turned his attention to a more modest instrument only recent installed under the southeast dome. George Ritchey's twenty-four-inch reflector was an engineering masterpiece, the product of five years of labor by an unapologetic perfectionist. The low-slung mount, fabricated in the Yerkes workshop where Ritchey reigned, was designed to shoulder its burden with stoic mechanical aplomb. Its internal clock drive was accurate enough that a star centered in a high-power eyepiece drifted less than a hundredth of a millimeter over four hours. The tube—a cylindrical octet of steel bars bonded to cast-aluminum rings—was judged so rigid that the planned supplementary bracing was scrapped. The mirror had been painstakingly hollowed to a deep curvature, converging light far more efficiently than a conventional long-focus instrument. In an hour's exposure, Ritchey's reflector recorded stars beyond the reach of the renowned forty-inch refractor down the hall.

George Ritchey.

In a 1901 report, which included an arresting photograph of the Andromeda Nebula, Ritchey affirmed that his telescope was a test-bed for the future sixty-inch. The lessons learned would be applied to its eventual design and construction, virtually assuring its success. This was a telescope that needed to be built. And if the point was missed, Ritchey's paper included handsome cross-sections of the sixty-inch reflector, fully mounted and pointed skyward, under an expansive dome.

Although eager to progress, both Hale and Ritchey were aware that their envisioned leap in telescopic aperture and sophistication might be nullified by the mercurial climate of the Great Lakes region. When they looked upward from Williams Bay, Wisconsin, even on the crispest nights, their mind's eye saw a turbulent gauntlet of air that contorted light rays en route to the ground. The windswept outskirts of Chicago was not the proper site for a mega-telescope. "It would be interesting to think of the photographic results which could be obtained with a properly mounted great reflector in such a climate and in such atmospheric conditions as prevail in easily accessible parts of our own country," Ritchey mused, then added, "notably in California."

Chapter 25

Threads to a Web

> The stars looked like jewels on black velvet. The sky was rich and dark, and every star was a glowing, living point of light.
>
> —George Ellery Hale at the summit of Mount Wilson, 1912

Thursday morning, June 25, 1903, found George Hale on the back of a burro ascending the fogbound flank of Mount Wilson, surely the most erudite burden ever borne by the ragged beast. On a second burro rode Lick Observatory's William Campbell, who reassured the pessimistic Chicagoan that altitude would carry them above the mists that collared the hills outside Pasadena. Hale was unconvinced. The landscape was veiled in white, and a glance overhead revealed only a diffuse notion of the Sun's presence.

The two-foot-wide trail had been given the wishful designation "Mount Wilson Toll Road" long before its expansion merited the name. Hale was here only nominally on behalf of the Carnegie Institution, which had commissioned him, Campbell, and Lick's William Hussey to seek favorable sites in southern California for a new solar observatory. In truth, Hale had traveled west for his own interests. The climatic environment of the Yerkes Observatory oppressed him. A cloudless Lakes-region sky that, to the untrained eye, looked perfectly transparent, Hale saw as an enervating soup of air molecules that drowned cosmic photons before they reached the telescope. For all its fame, the forty-inch refractor was hobbled by its placement outside Chicago. Since its inception, Hale had pictured the instrument atop a mountain, like its kin at Lick Observatory. And what of his sixty-inch reflector, lying fallow, a mighty eye lacking a body? If provision ever arrived for its completion, what sense did it make to erect such a responsive telescope under the attenuated rays of a growing metropolis?

Hale had learned of Mount Wilson's pellucid skies as a college student in 1889, when a scouting team from Harvard pronounced the mile-high perch the best astronomical observing site in the nation. (However, they cautioned against the rattlesnakes.) Mount Wilson, Hale had been told, was cloudless some three-hundred days a year, and the fog that blanketed the valley rarely crept to the summit. Even the warm Pacific breeze tuckered itself out before reaching the mountain, leaving the air at the peak clear and steady. Now, fourteen years later, on the back of a burro, Hale was about to enter a world astronomers spoke of with almost mystical awe.

As Hale's animal plodded up a ridge, there was a perceptible brightening. Then, as though a curtain had been thrown aside, the fog evaporated to reveal a breathtaking vista. Mountain ranges stood in sharp relief against a cerulean sky. Along their slopes, a primitive landscape of rocky outcrops, canyons, trees, and thickets. Westward in the distance, the Pacific and, at the horizon, the silvery band of Catalina Island. Hale and Campbell arrived at the summit during the afternoon and joined Hussey that evening to view the heavens with a telescope. As the Harvard expedition had long ago concluded, the gain in altitude and the absolute clarity of the air rendered stars as brilliant specks.

Hale rose early to catch the sunrise, then scouted the rolling highland of pine stands and grassy meadows. At intervals, he climbed a tree to assess solar observing conditions above the ground; one notebook entry has him shimmying sixty-eight feet skyward on a yellow pine. Freshwater springs spewed a sufficient amount of water for a sizable community of workers. With each turn of the trail, each scrub-covered flat, each bulging outcrop, the institute of his dreams grew clearer: a solar telescope, the sixty-inch reflector, workshops, offices, lodgings—all appeared through his mind's eye in their proper places. One top-of-the-world promontory Hale dubbed Monastery Point, after the holy houses of the Levant. Here he envisioned a stone-walled aerie for himself and his fellow acolytes of astrophysics.

The prospect of a mountaintop observatory had long captivated George Hale, but a recent development raised his simmering hopes to a fever pitch. On January 13, 1902, Hale opened the *Chicago Tribune* to find that Andrew Carnegie had donated ten million dollars to establish a research institute of unprecedented magnitude and breadth. The organization's mandate, in the words of the reporter, was to "discover the exceptional man in every department of study . . . and enable him, by financial aid, to make the work for which he seemed especially designed his life-work." Hale had no doubt that he was such a man.

Eleven days later, Hale sent the Carnegie Institution a proposal for creation of a comprehensive astrophysical research center on Mount Wilson, including a solar observatory and a facility for the sixty-inch telescope. Estimated startup cost: $300,000—with additional outlays to come. To buttress his ambitious plan, Hale included a photograph of the Orion Nebula taken by George Ritchey with the Yerkes twenty-four-inch reflector. If a telescope two feet in aperture, on the lowlands outside Chicago, can generate such an exquisite image, imagine what a five-foot instrument might do, elevated above the hazy underlayment of the atmosphere.

Hale's frontal assault on the Carnegie Institution's treasure was firmly repulsed by its executive board, who weighed proposals from eighteen fields of study running the gamut from geophysics to nutrition. Not only was he up against luminaries in these other branches of science, but also traditional-astronomy strongmen like Simon Newcomb and Lewis Boss. Even his nominal ally, Edward Pickering, found the plan extravagant. The Carnegie Institution questioned the wisdom of granting such a massive infusion to a single project, especially in a field—astrophysics—that its veteran advisors characterized as yet embryonic. Instead, it commissioned Hale, Campbell, and Hussey to seek favorable sites in southern California for a new solar observatory.

Standing on Mount Wilson in June 1903, Hale knew that he had found the home for his astrophysical research center. His technical report to the Carnegie Institution executives painted as clear a picture of its crystal skies as one could without resorting to poetry. But he was not about to wait for their judgment or their money. "I am a born adventurer," he once confessed, "with a roving disposition that constantly urges me toward new long chances."

Six months later, Hale moved with his family into a rented bungalow in Pasadena, eight miles from the base of Mount Wilson. The nominal aim of his so-called Yerkes Expedition was a more thorough assessment of the summit's solar-observing potential. In truth, Hale pooled his family savings with loans from his brother and his uncle and began construction of his mountaintop observatory.

Among the first to hitch his star to Hale's was former Yerkes graduate student, now staff astronomer, Walter S. Adams. Brilliant and convivial, Adams had completed his doctoral studies at Munich to great acclaim. "The prospects of a bohemian year on Mt. Wilson with you appeal to me very strongly indeed," Adams confessed, "and I should be only too glad to join you for what would cover 'Omar's loaf of bread and jug of wine.'" Hale paid Adams out of his own pocket. With a small grant from the Carnegie Institution, he also supported Ferdinand Ellerman, his longtime photographic assistant, and George Ritchey, the gifted optical designer and guardian of the sixty-inch mirror. Ellerman embraced the rough life, donning a ten-gallon hat, high boots, and a cartridge belt with a loaded revolver and a hunting knife. The sight of his heavily armed friend startled Adams, who envisioned "a struggle for existence on the wild mountain top." Visiting astronomers were welcomed: A trainee from Padua recalled hiking the trail to the summit while Hale recited verses from the *Divine Comedy* in fluent Italian.

Walter S. Adams.

Adams's recollections of the observatory's formative days convey the spirit of adventure felt by Hale and his staff. They were pioneers of astrophysical science, drawn to far-off geography to gain best vantage on the luminous rustlings of the cosmos. At the time, Pasadena was barely two decades old, a rail-nurtured city of twenty-five thousand, with grand hotels, architect-designed mansions, and East Coast aspirations. The citified core dwindled rapidly toward its outskirts, giving way to dirt roads, bungalows, and farm fields.

Hale made frequent trips up the mountain, bicycling to its base and overnighting at a ramshackle cabin near its summit, where he watched the diurnal roll of the night sky through a hole in the roof. With local hands and muled-in supplies, he restored the cabin, installing a granite-block fireplace in anticipation of winter. A minilibrary of astrophysics and poetry books capped off the renovation. On the steep-sided promontory Hale had charted a year earlier, workers completed construction of his astronomers' lodge: the Monastery.

In mid-1904, after initial testing with a solar telescope imported from Yerkes, the Carnegie Institution granted Hale $40,000 to build a larger version of the instrument. Again, no money was included for completion of the sixty-inch reflector. Hale pushed ahead with the solar research in the hope that the results would stimulate a larger infusion of funds. Once asked about his uncanny ability to procure private and institutional support for his work, Hale responded, "The gods bring threads to a web begun."

On December 20, 1904, a year to the day after his arrival in Pasadena,

The Monastery building on Mount Wilson, photographed in 1905.

the threads arrived. Hale was ascending Mount Wilson when he was called to a trailside way station. Over a crackly telephone line, the mountain's only direct link to the outside world, an operator read out a telegram from Washington, DC: the Carnegie Institution had just allocated the full request of $310,000 to establish and operate the Mount Wilson Solar Observatory, with explicit authorization to mount the sixty-inch reflector. Hale resigned from Yerkes two weeks later to take up his new directorial post. (Hale's final visit to Yerkes occurred in 1932. On his way out, he stepped into the dome of the great refractor, doffed his hat, and uttered, "Noble instrument.")

The following summer, the state-of-the-art Snow Solar Telescope, formerly in operation at Yerkes, was transferred to Mount Wilson. Its thirty-inch, motorized, flat mirror directed sunlight horizontally to a second flat mirror, which in turn illuminated an image-forming concave reflector of sixty-feet focal length. The six-inch-wide solar image could be viewed directly or photographed; alternatively, the light could be deflected into a stationary spectrograph, either for line analysis or to photograph the Sun's disk at a chosen wavelength. Secured to a concrete footing, the spectrograph used by Hale was far larger than any designed to hang from the end of a movable telescope.

In 1908, Hale built the world's first solar tower telescope—essentially a vertical version of the Snow instrument, sixty feet tall—to mitigate the image-distorting effects of heat rising from the ground. Four years later came a 150-foot tower telescope, capable of projecting a sixteen-inch-wide image of the Sun. Its spectrograph was fully twenty times as precise as the unit Hale had used at his original Kenwood Observatory. With the various solar instruments, Hale determined that sunspots are centers of intense magnetism, the first sign of magnetic activity on another world and arguably Hale's greatest discovery. His observations also provided a springboard toward an understanding of the temperature-based variance of stellar spectral lines.

Meanwhile, the sixty-inch reflector was readied in George Ritchey's workshop in Pasadena, then hauled to the summit by mule train, along with 150 tons of construction and mounting materials. On December 20, 1908, four years after being funded, the sixty-inch took its first photograph. Like its groundbreaking antecedents by Henry Draper, Andrew Common, and Isaac Roberts, this latest exposure of the Orion Nebula immediately raised the measure of excellence in deep-space imaging. The instrument's unprecedented light-gathering capacity permitted photographic and spectroscopic studies of the brighter spiral nebulae; the presence of novae—eruptive stars—in these systems plus the Andromeda Nebula's Sun-like spectrum convinced Hale that spirals are remote stellar aggregations like the Milky Way. And as utilized by a young Harlow Shapley, Harvard Observatory's future director, to assess the spatial distribution of globular star clusters, the sixty-inch helped ascertain the size of our galaxy and Earth's location within it.

Still in operation, the sixty-inch reflector is a masterpiece of engineering, featuring innovations that came to define research telescopes in its wake. For superior stability, the one-ton mirror nestles close to the instrument's stubby, cast-iron support arms. The accumulated weight of glass and metal—altogether some twenty-two tons—floats in a trough of mercury. The driving gear alone spans ten feet, its periphery notched with 1,080 teeth. The telescope's most radical design element at the time was not its gaping aperture, but its multifaceted utility: one instrument to serve a variety of research needs, from low-magnification, wide-field imaging to high-magnification, high-dispersion spectroscopy. The various optical paths are enabled through a set of exchangeable secondary mirrors, each configuration optimized for the particular sort of photographic or spectroscopic work being conducted. Gathered photons can be

Mount Wilson Observatory's sixty-inch reflector fitted with a spectrograph, circa 1910.

deflected out the side of the tube near the top, into lightweight cameras or spectrographs; or they can emerge near the bottom, which is better suited to bulkier apparatus. For ultrahigh-dispersion spectral work, photons can even be directed through the mounting's hollow polar axis into an auxiliary, temperature-controlled chamber, housing a large, fixed spectrograph.

With the sixty-inch, no longer was it necessary for heavy, astrophysical equipment to jut from the tail of a movable telescope tube. And no longer did astronomers have to balance atop a ladder to access the high-up Newtonian focus. The telescope's compactness reduces the size, hence, the cost of the observatory structure. Yet its multireflection optical path confers effective focal lengths up to 150 feet.

With the various telescopes up and running, and his family ensconced below, George Hale spent much of his time in the rough, granite splendor of the Monastery. Life atop Mount Wilson alternated between the exhilarating and the mundane. Hale worked his willing staff year-round in unheated observatories; during the coldest nights, fingers and toes grew numb and lashes froze to the eyepiece. Midnight lunch consisted of hardtack and hot chocolate. Women, liquor, and coffee were forbidden.

Mount Wilson Observatory staff pictured in front of the Monastery in 1906. George Ellery Hale, photographic assistant Ferdinand Ellerman, Walter S. Adams, and Edward Emerson Barnard are seated, respectively, third, fifth, sixth, and seventh from the left.

Mountain lions roamed the peak. Yet to Hale and his single-minded constituency, Mount Wilson was heaven.

When supplies ran low, Hale hiked down the old trail, bicycled into town, then hiked back up. (When off-mountain, he sped around on a three-wheeled Indian motorcycle, trading it in for a car after a collision with a trolley.) Hale viewed his resident colleagues as a scientific elite. Dinner in the Monastery was jacket-and-tie formal. Personalized napkin rings were laid around the table in rank order, with that evening's large-telescope observer always seated at the table's head. After dinner, Hale led discussions about astronomy or recited poetry. There was, according to staff astronomer Harold Babcock, "a sense of great events in the making."

The formative years at Mount Wilson were not without emotional challenge to George Hale and his family. In supporting her husband's scientific efforts, Evelina Hale bore the brunt of the household responsibilities and the raising of their children. In 1906, entirely spent, she checked herself into a sanitarium for a rest cure. Hale was similarly seized by a complex array of maladies his physician summed up as "brain exhaustion." Insomnia, headaches, and agitation were his constant companions; an innate melancholy turned into bouts of depression. Despite his active social and professional engagement, he feared he would sink into the reclusive habits of his mother and wind up a room-confined invalid. The doctor's proposed cure was as distressing as the illness: cut back on work, adopt a balanced life. Discontented with his much-lauded accomplishments and his part-time fatherhood, Hale proved unable to rein in the full-gallop pace forced by his own outsized aspirations. His symptoms only worsened when he embarked—for a *third* time—on a quest to build the largest telescope in the world.

During the summer of 1906, with Evelina confined to the sanitarium and the sixty-inch reflector still two years shy of completion, Hale spent a weekend at the estate of Los Angeles businessman John D. Hooker. After dinner, talk turned to astronomy. Hale could not help but notice Hooker's keen interest in the subject. Sensing an opportunity, he brought up the construction of the record-setting sixty-inch telescope, then launched into a recitation about what might be accomplished with an even larger reflector of, say, eighty-four inches. Before the week was out, Hooker offered $45,000 to cast the new glass disk and shape it into an astronomical mirror. But he insisted on a diameter of *one hundred* inches, worried that anything smaller might be surpassed within a few years. Hooker further stressed that the additional cost of designing, mounting, and housing the instrument was Hale's problem. (That sum, $600,000, was eventually to be granted by the Carnegie Institution.)

By the autumn of 1906, George Hale found himself with a half-finished sixty-inch reflector and a four-and-a-half-ton, one-hundred-inch glass lozenge on order from France; a pair of mammoth, untested telescope mounts, the latter a design concept utterly dependent on the anticipated success of the former; an institutional budget, vast by contemporary standards, yet too paltry to realize his grand schemes; a world-weary wife and a boy and a girl grown accustomed to his all-too-frequent good-byes; and, in the face of all these obstacles, an irrepressible will to push farther and fainter into the cosmos and bring the denizens of those unexplored realms under the purview of the human mind.

The one-hundred-inch disk completed its overseas and overland journey from the Saint-Gobain foundry in France to Pasadena on December 7, 1908, the same day the sixty-inch mirror was lowered into its mountaintop cradle. Peeling back the protective coverings, Hale and Ritchey were horrified to find that the gargantuan glass was flawed. Three melts had been required to fill the capacious mold, and this had left a bubble-strewn layer with each successive pour. How such a tripartite structure might react to changes in temperature or spatial orientation was anyone's guess. At Hale's insistence, a second disk was poured at Saint-Gobain, but this one cracked during cooling. In yet a third attempt, the bubbles appeared again.

By 1910, against Ritchey's stiff opposition, Hale decided to proceed with the disk at hand. As best he could determine, the worrisome bubbles would not pock the mirror's surface after it was scoured to its proper one-and-a-quarter-inch depth. As to the disk's structural integrity—with no viable alternative, it was a risk Hale was willing to take. Five years of troublesome grinding, polishing, and testing ensued. Ritchey grew combative, insisting to everyone who would listen—including John D. Hooker, the project's funder—that the massive effort would end in disaster. Hale was forced to ban Ritchey from the workspace and hire a replacement. During this period, Hale's mental state deteriorated. "Congestion of my head," he wrote a colleague, "is caused chiefly by worry, excitement, responsibility, discussion of any scientific subject, attendance at scientific meetings, lecturing . . . and continued mental work."

In January 1911, Andrew Carnegie injected another ten million dollars into his institution, expressing among his wishes that the one-hundred-inch telescope be completed. To Hale, the news was at once affirming and anxiety provoking. In this most vulnerable period of his life, he had staked both his reputation and his self-worth on a high-risk, technological gamble. Evelina grew despondent over her husband's behavior, confiding to a sympathetic acquaintance: "He immediately began to make plans for years ahead and so worked himself up to almost the breaking point. I wished Carnegie could keep his millions to himself." Barely six months after Carnegie's endowment, Hale admitted himself to a sanitarium, the first of four

Mount Wilson Observatory's one-hundred-inch reflector telescope.

eventual breakdowns. In the years-long aftermath, as his telescope crept toward completion, Hale strayed in and out of his self-described "neurasthenic quagmire."

The final bolt was tightened on the one-hundred-inch reflector in the autumn of 1917. The assembled behemoth weighed a hundred tons and sat on a house-sized pier of reinforced concrete. Within the pier were a mirror-maintenance shop and photographic darkroom, as well as the instrument's star-tracking clockwork, which impelled a seventeen-foot drive wheel. The open-sided tube pivots within a rectangular steel cradle whose ends rotate almost friction-free upon mercury bearings. Despite the mirror's extensive—and favorable—laboratory testing, Ritchey's dire predictions of disaster hung in the air. Ultimate judgment rested on a direct view of the cosmos. Hale had to see with his own eyes whether his years of effort had paid off.

On November 2, 1917, George Hale, Walter Adams, George Ritchey, and visiting English poet Alfred Noyes gathered for the Hooker telescope's "first light." Also present was a phalanx of laborers—machinists, carpenters, electricians—who had come to witness the birth of their monumental creation. After nightfall, an array of electric motors slewed the instrument eastward toward Jupiter. As the planet's light beamed out of the eyepiece, Hale's face must have betrayed his anguish. In place of the majestic Jovian globe appeared a kaleidoscopic mish-mash of six overlapping images that swelled across the field of view. The cause was unmistakable: the mirror was warped into a half-dozen facets, each generating its own off-kilter likeness of the planet.

Hale suggested hopefully that the mirror might have expanded from the Sun's heat when workmen opened the dome earlier that day; more cooling-off time might bring it back to shape. The alternative was almost too awful to contemplate: the mirror might be warping under its own weight. There was scant improvement over the next several hours, when the men adjourned to the Monastery. Around 3:00 a.m., Hale and Adams reconvened under the dome. This time, the telescope was directed at a bright star. Hale peered into the eyepiece, then shouted with joy at the prospect: a single, razor-sharp image. The mirror had equilibrated with the cool night air, its delicate arc restored. Glimpses of the Moon and Saturn clinched the case. Hale's present nightmare was over, his dream at last ready to unfold.

Mount Wilson's new reflector had nearly three times the light-collecting area of the sixty-inch, and from its mile-high vantage point, was capable of probing depths of space beyond the reach of any other instrument in the world. No surprise that the attention of astronomers was turned to Mount

Wilson as its great glass eye probed the heavens. Hale built the instrument, not for his own use, but for deep-sky specialists of the current and coming generations. At his insistence, the design pressed the technological limits of the era, and expressed his confidence that optical and engineering obstacles would be surmounted as they arose. The one-hundred-inch reflector was conceived as a telescope of the future as much as of the present.

So elastic was the new instrument's capabilities that Nobel-Prize-winning physicist Albert Michelson and astronomer Francis Pease realized that they could use it to measure the angular diameter of stars outside the solar system. To secure their results, the scientists bolted a twenty-foot-long optical device called an *interferometer* across the upper end of the telescope's tube. In 1920, Michelson and Pease measured the angular width of Betelgeuse, in Orion, long suspected of being a giant star. Given an estimate of the star's distance, they transformed the angular diameter into a physical span, confirming that Betelgeuse is indeed gargantuan compared to the Sun. The measurement created a sensation in the scientific community: an explicit example of the power of astrophysical observation. (Michelson relied on Pease for the astronomical component of the work. Once asked to identify a particular bright star in the sky—none other than Betelgeuse itself!—Michelson replied, "How the devil should I know?")

Finding himself unable to sustain his multitude of responsibilities, Hale resigned as director of Mount Wilson Observatory in 1922. He spent much of the subsequent time traveling, reading, and writing for the general public. Now and then, his interior eye ventured into the blackness beyond the reach of the Hooker reflector and conjured an instrument to overleap that frontier. He and his colleagues speculated about telescopes as large as three hundred inches. "Starlight is falling on every square mile of the earth's surface," Hale lamented in *Harper's Magazine* in April 1928, "and the best we can do at present is to gather up and concentrate the rays that strike an area one hundred inches in diameter." Readers learned that development of a much bigger reflector was feasible—if only sufficient money could be found.

On a whim, Hale sent a copy of his article to the Education Board at the Rockefeller Foundation. Their intense interest in the idea took him by surprise. He wrote to Lick Observatory's Robert G. Aitken, "An article of mine on large telescopes, shot like an arrow into the blue, seems to have hit a 200" reflector." By year's end, Hale had negotiated a joint appropriation of six-million dollars from the Rockefeller Foundation, Carnegie Institution, and California Institute of Technology for construction of the giant instrument. Completed in 1947, nine years after Hale's death, and named in his

George Ellery Hale at his desk in the Monastery at Mount Wilson Observatory around 1905.

honor, the two-hundred-inch reflector on Palomar Mountain would mark the fourth time George Hale had fostered the largest telescope in the world. In a draft introduction to his book, *Ten Years' Work of a Mountain Observatory*, Hale characterizes a scientist's path as "steep and beset with difficulties, but it leads to heights which continually unfold new prospects of ever increasing charm."

For the better part of a century, astrophysical pioneers had promoted a vision of cosmic study distinctly different from that of their traditional counterparts. The astronomical enterprise they espoused exchanged census-oriented methodologies for techniques founded on the collection and analysis of light. Thus, stars became more than mere markers of position and motion, but active, evolving engines of physics—each a full-blown sun, only diminished to a pinpoint by the vastness of space. And nebulae, those mysterious smudges of luminance in the eyepiece, acquired their true identity as free-floating reservoirs of cosmic chemicals. Further clues to the physical properties of these far-flung populations were sought in celestial photons that drizzle over Earth. Forward-thinking observers like George Ellery Hale pushed for a new generation of instruments to intercept the faintest emissions, even though the key to decode them did not yet exist.

Mount Wilson's one-hundred-inch reflector was the culminating element in a triad of astrophysical innovation that evolved over the latter half of the nineteenth century. With its completion, the finely honed tools of celestial photography and celestial spectroscopy could be applied to a host of heavenly objects, primary among them the spiral nebulae. Nearly eight decades had passed since Lord Rosse's Leviathan telescope revealed this pinwheel species of space cloud, and two decades since James Keeler's deep-sky pictures exposed their incredible ubiquity. The convergence of photographic, spectroscopic, and telescopic technologies during the early twentieth century promised to resolve the heated debate about the nature of spiral nebulae: Are they aggregations of gas and infant stars within our Milky Way, or full-fledged galactic islands that populate the entire universe?

With the one-hundred-inch reflector nearing completion, Hale looked to expand the Mount Wilson staff. Among the candidates was a recent doctoral recipient from Yerkes whose photographic study of variations in a faint nebula had been published in the *Astrophysical Journal*. In the spring of 1917, Hale offered the young man a position, with full access to the big telescope. The reply arrived shortly afterward in the form of a telegram: "Regret cannot accept your invitation. Am off to the war." Signed, Edwin Hubble.

Chapter 26

Size Matters

> Once one has exhausted all possibilities for error, one is finally forced to abandon a prejudice, and redefine what one means by 'correct.' So painful is this experience that one does not forget it. The subsequent replacement of an old prejudice by a new one is what constitutes a gain in real knowledge. And that is what we, as scientists, continually pursue.
>
> —Douglas Gough, "Impact of Observations on Prejudice and Input Physics," 1993

Onto the stage set by George Ellery Hale strode another player attuned to historic possibilities: Edwin Powell Hubble. There was a dashing, almost cinematic, quality to Hubble when he arrived at Mount Wilson in September 1919. Six-foot-two and leading-man handsome, Hubble wore his mantle of authority comfortably, whether engaging with colleagues, hosting dignitaries in the observatory, or posing for the cover of a magazine. Colleague Milton Humason recalled his first vision of Hubble under the dome of the sixty-inch telescope: "His tall, vigorous figure, pipe in mouth, was clearly outlined against the sky. A brisk wind whipped his military trench coat around his body and occasionally blew sparks from his pipe into the darkness of the dome. . . . He was sure of himself—of what he wanted to do, and of how to do it."

Edwin Powell Hubble.

If Edwin Hubble had thus far led a charmed life, it had been at his own insistence. His ascension among his peers had been a calculated campaign, starting with his Rhodes Scholar education at Oxford, through his graduate astronomical training at Yerkes and meteoric rise to major in the army during World War I, and presently to his appointment at Mount Wilson. In conversation, Hubble was apt to inject accounts of his exploits which, over time, accreted into a breathless exaltation that obscured the divide between man and myth. An amateur pugilist during his college days, he once dispatched a knife-wielding thug by knocking him senseless—after being stabbed in the back. He rescued a professor's wife from drowning by hoisting her onto his shoulders and striding along the sandy bottom until he emerged Neptune-like from the waves. When the country slid into depression, the charismatic Hubble diverted attention toward America's noble quest to comprehend the universe. Edwin Hubble became astronomy's John Wayne, a living Rushmore of American mettle, who seemed to overstride obstacles through sheer resolve.

At Oxford, and later on Mount Wilson, Hubble adopted the dress and manner of a highborn Englishman, complete with Norfolk jacket, plus-fours, high-top boots, and faux accent. It was a presentation designed to impress, if not intimidate. Colleagues tolerated his pretensions, while the press found this martial icon of science irresistible. Despite an abiding interest in astronomy—he worked at Yerkes Observatory while an undergraduate at the University of Chicago—Hubble acceded to his father's wishes and studied jurisprudence at Oxford. From the start, it was an ill fit, the calcified subject matter stifling to one who sought wider horizons. While Hubble kept his father apprised of his studies and his many athletic prizes, he poured out his disillusionment to his mother: "I sometimes feel that there is within me, to do what the average man would not do, if only I find some principle, for whose sake I could leave everything else and devote my life."

After Oxford, Hubble idled for a year as a high school physics teacher before his outsize ambitions erupted into a full-bore drive into astronomy at the University of Chicago. In 1914, when Hubble began his graduate training, the Yerkes Observatory was a ghost of its former self. George Ellery Hale was a decade gone, living out his professional dream at Mount Wilson. Gone with him, the cream of the Yerkes staff: Walter Adams, George Ritchey, Ferdinand Ellerman, Francis Pease. Hale's successor, the avuncular Edwin Frost, conducted stellar spectroscopy with the forty-inch refractor, but was going blind. Edward Barnard, the noted celestial photographer, lacked the academic training to direct a doctoral candidate, especially one with Hubble's aspirations. Rounding out the Yerkes staff were several junior astronomers of no particular distinction.

Rather than vie with his superiors for observing time on the great refractor, Hubble opted for George Ritchey's sidelined twenty-four-inch reflector. Although diminutive, the instrument's fast optics proved key to Hubble's photographic foray into the realm of the nebulae. Among his first targets was a striking, fan-shaped nebulosity in Monoceros numbered 2261 in J. L. E. Dreyer's 1888 *New General Catalogue of Nebulae and Clusters of Stars*. NGC 2261 beckons the viewer with its triangular form, intense star-like concentration at the southern apex, and streamers trending off to the north. Although motionless to the eye, there is an impression of activity, as though one had frozen the flutter of a candle flame.

Hubble compared his images of NGC 2261 from 1915 and 1916 to those taken during the previous decade, including one by photographic pioneer Isaac Roberts in 1900. He was amazed to find that the nebula had indeed changed, both in its outline and in the form and placement of its internal features. Variable nebulae had been observed before, but never had such dramatic alterations been captured over so brief a time span. In an era when cosmic distance measurement was problematic at best, Hubble asserted that NGC 2261 must lie relatively close to the solar system for its transformation to be so manifest. Supporting his contention was the demonstrable movement of faint stars associated with the nebula, shifts otherwise imperceptible if the cloud were situated far away. A pair of well-turned articles about Hubble's Variable Nebula, as it would come to be known, brought Yerkes's rising star into professional view.

In October 1916, with the one-hundred-inch mirror soon to be mounted, George Hale met with Edwin Hubble in Chicago and offered him a staff position at Mount Wilson, pending completion of his doctoral degree. Hubble accepted, no doubt keen to wield a telescope four times the aperture of his current one. Yet it would be three years before he reached California. On April 6, 1917, the United States declared war on Germany, and Hubble was itching to get into the fight.

Hubble rushed an update on NGC 2261 to the *Astrophysical Journal*, featuring his latest exposure of the nebula. Enlarged twenty-four times from its half-millimeter extent on the plate, the bright, broad triangle of a year earlier had shriveled into a cometary wisp. Hubble could not say whether the changes in appearance stemmed from movement of gas and dust within the nebula or from variable lighting of the material by its starry nucleus. (NGC 2261 is indeed a cosmic shadow play: dust clouds circling the illuminating star sweep zones of darkness throughout the nebula.)

Hubble typed up his dissertation—a minimalist seventeen pages—and submitted it to Edwin Frost at the beginning of May. He ignored Frost's plea to fatten the skimpy production by incorporating his paper on NGC 2261. "[I]t does not add appreciably to the sum of human knowledge," he avowed, then added prophetically, "Someday I hope to study the nature of these nebelflecken to some purpose."

Hubble's dissertation outlines the scientific conundrums presented by the thousands of faint nebulae that crowd photographic plates, especially the spirals, and the need for a meaningful classification scheme for the various nebular forms. Significantly, he notes that the distribution of faint nebulae (excluding those which clearly belong to our own Milky Way galaxy) is nonuniform; instead, they show a tendency to cluster, in seeming analogy to gravitationally bound star clusters within the Milky Way. More than a summary of results, Hubble's paper is a road map for the future of extragalactic studies. In an oblique attempt, perhaps, to justify the glaring insufficiency of his own work, Hubble writes, "These questions await their answers for instruments more powerful than those we now possess." Hubble passed his hastily scheduled oral examination with high honors, then telegraphed George Hale, who assured him that his staff position at Mount Wilson would be held until after the war.

Just three months into his military training, twenty-eight-year-old Private Hubble became Captain Hubble, and shortly thereafter, Major Hubble, in charge of six hundred officers and infantrymen. His unit did not arrive in France until September 1918 and, in the six weeks preceding the armistice, saw no direct combat. "I barely got under fire," Hubble complained to Frost, "and altogether I am disappointed in the matter of war." A frustrated Edwin Hubble, still in officer's uniform, made an extended stopover in Cambridge, where he bathed in the attentions of England's astronomical elite. A stern reminder from George Hale about his prior commitment dislodged Hubble from his adoring throng. In August 1919, he boarded a ship for America, seeking new realms to conquer.

When Edwin Hubble arrived at Mount Wilson, one of the most incendiary problems in astronomy remained the nature of spiral nebulae: Are they assemblages of stars and gas situated within the borders of the Milky Way, or are they galaxies in their own right, external to our own? In other words, is the universe one in which the Milky Way is the central, dominating entity, or one in which it is a mere member of the horde? The arguments on both sides were compelling, often fundamentally incompatible. And while intuition drew an increasing number of astronomers toward the multigalaxy, "island-universe" scenario, resolution of the matter awaited some breakthrough piece of observational evidence.

The stumbling block in studying the faint spirals that appeared on plates in such number is how to fix their distances. To ascertain, from the vantage point of your window, whether a tree is situated within your backyard or your neighbor's, the tree's location must be determined relative to the property line. Likewise, to ascertain whether a nebula is situated within our galactic system or in extragalactic space, the nebula's location must be determined relative to the borders of the Milky Way. Thus, the nature of the spiral nebulae hinges on two parameters: the nebula's distance and the extent of our Galaxy, in effect, the galactic property line. That the latter is a challenge to fix stems from the fact that Earth's placement deprives astronomers of an external perspective on the Milky Way's form and span. They are constrained to observe the Galaxy from within, yielding a depthless compression of celestial bodies onto the night sky.

In the late eighteenth century, William Herschel attempted to locate the limits of our galactic system by an observational scheme he called "star-gaging." Having assumed that the distribution of stars is, on average, homogeneous throughout the Milky Way, Herschel used his workhorse nineteen-inch reflector (the "twenty-feet telescope," after its focal length) in a visual count of stars in twenty-four hundred directions in the sky. A geometry-based analysis of the count revealed the relative distances at which stars give over to empty space, that is, the location of the Galaxy's edge in that particular direction. The resultant stellar map is roughly elliptical in cross-section, with a series of clefts along the galactic plane and Earth prominently situated near the center.

Herschel came to realize that the presence of star clusters and the clear variation in star density along different sight-lines were at odds with his supposition of uniformity. Furthermore, each gain in telescope aperture brought to light a multitude of previously unseen stars. "By these observations," he concluded, "it appears that the utmost stretch of the space-penetrating power of the 20 feet telescope could not fathom the Profundity of the milky way." Even Herschel's gaping forty-eight-inch reflector showed a background haze, which he took to be the gathered glow of yet more remote stars.

In the early twentieth century, Dutch astronomer Jacobus C. Kapteyn, at the University of Groningen, carried out the photographic successor of Herschel's star census using plates obtained from other observatories. Kapteyn found that the galaxy is disk-shaped, no more than thirty thousand light-years across and five thousand light-years thick, with Earth centrally situated. (A light-year is the distance light travels through space in a year, about six trillion miles.) Although Kapteyn's Milky Way would ultimately prove to be a pipsqueak version of the real thing, a span of thirty thousand light-years seemed suitably immense for a galaxy at the time.

Unknown to Herschel, Kapteyn, and other galactic surveyors, interstellar space is peppered with silicate and carbon dust that absorbs and scatters starlight in transit, often to the point of extinction. The dust renders invisible the more remote sectors of our galactic disk, as stellar photons are plucked from their paths before they reach our telescopes. Harvard astrophysicist Cecilia Payne-Gaposchkin likened the problem of surveying the galaxy to "drawing a map of New York City on the basis of observations made from the intersection of 125th Street and Park Avenue. Although it would be clear to an observer that the city is a big one, any statement as to its extent and layout would clearly be impossible. London would offer an even better analogy, for the neighborhood is not only congested but foggy." Given the aggregate evidence of light-absorbing material in outer space, some astronomers feared that the census-based Kapteyn model understated the galaxy's girth. What they needed was an alternative method to size up the galactic system—one that would circumvent the hypothetical obscuration of starlight by interstellar dust.

Among the night sky's grandest spectacles are the globular star clusters: magnificent, spherical swarms of tens of thousands to hundreds of thousands of stars, each held fast by the unseen hand of gravity. The nature of these systems was disputed well into the twentieth century. Cambridge astrophysicist James Jeans speculated that globulars are the condensed, stellar end-products of spiral nebulae. Fellow Cantabrigian Arthur Eddington mused, in the style of the times, about their spatial relationship to the Milky Way: "The question of whether they are to be regarded as coequal empires, or as dependent but nearly self-governing colonies, must await more precise evidence."

In the early 1800s, William Herschel's son John noticed that globular clusters are distributed differently than the majority of isolated stars. While one hemisphere of the sky contains scores of clusters, the opposite contains almost none. Fully one-third of known globulars occupy a region in Sagittarius comprising only 2 percent of the celestial sphere. Many are also found

The globular star cluster Messier 15, in a two-hour exposure by Isaac Roberts from November 4, 1890.

far from the luminous band of the Milky Way, where the bulk of galactic stars lie. The upshot of this overtly skewed distribution is that, in the compact Kapteyn model, the system of globular clusters appears to hover in spatial exile at one end of the galaxy, an asymmetry offensive to reason.

In 1914, five years before Edwin Hubble's arrival at Mount Wilson, George Hale recruited an ambitious Princeton graduate, Harlow Shapley, to study variable stars in globular clusters with the sixty-inch reflector. Globular clusters are rich in Cepheids, named for the category prototype Delta Cephei, whose variability was discovered in 1784 by English astronomer John Goodricke. These highly luminous stars cycle in brightness over periods ranging from about a day to seventy days. The character of the light variation is easily identifiable: a rapid rise to peak brightness, followed by a gradual dimming. Citing recent evidence that Cepheids are stars of immense diameter, Shapley proved that the observed brightness changes cannot stem from binary-star eclipses, as had been postulated: the requisite

Harlow Shapley.

orbits would be smaller than the stars themselves. Instead, the variation of a Cepheid's light must arise from radial pulsations of a single star, acting as a classical heat engine. (The specifics of the energy-generating mechanism were not elucidated for decades.)

Cepheids are sufficiently luminous to be visible at great distances within the Milky Way, the Southern Hemisphere's famed Magellanic Clouds, and globular clusters. In a study of variable stars in the Small Magellanic Cloud, Harvard astronomer Henrietta Leavitt found that the period of a Cepheid's brightening-dimming cycle correlates with its average light output: Cepheids of similar brightness have similar periods; and more significantly, the longer the period, the more luminous the star. This latter dependence, called the period-luminosity law, is typically rendered in graphical form: the Cepheids' periods arrayed along the x-axis and their corresponding luminosities along the y-axis. Once calibrated, the period-luminosity law provides astronomers with a stepwise method of computing a Cepheid's distance—and, by extension, that of any host system in which the Cepheid happens to reside.

First, the star's period is determined through repeated visual or photographic observations. The period-luminosity law, in turn, reveals the star's absolute light output. Because the intensity of a light source diminishes with the square of its distance from the viewer, a quantitative comparison of the Cepheid's absolute brightness versus its perceived brightness yields its distance. Thus, Harlow Shapley had a means to gauge the remoteness of any globular star cluster that harbors even a single Cepheid.

To dispel the problematic, off-kilter distribution of globular clusters in space, Shapley asserted that these starry aggregations constitute a spheroidal halo symmetric about the galactic core; it is our solar system that lies in the Galaxy's outskirts. Situated as many of them are, above or below the dusty galactic plane, globular clusters can be seen to much greater distances than stars within the disk. "To the measurer of the sidereal universe," Shapley posited, "star clusters are beacon lights. They point the way to the center of the Galaxy and to its edges . . . The globular clusters are a sort of framework—a vague skeleton of the whole Galaxy—the first and still the best indicators of its extent and orientation." By estimating the distances of globular clusters via their Cepheids, Shapley computed the locus of their distribution in space. And with that central point established, the Galaxy's overall size and Earth's placement within it followed.

In his 1919 paper, "Remarks on the Arrangement of the Sidereal Universe," Shapley proposed a breathtaking, tenfold increase in the Milky Way's diameter, to three hundred thousand light-years. The solar system, he maintained, lies some sixty thousand light-years from the nucleus of this lenticular array of matter. Shapley acknowledged the precariousness of his results: "It is probable that the further accumulation of observations will modify to some extent the views outlined . . . The present data may in some cases be susceptible of alternative interpretation, or possibly the conclusions may be questioned in the belief that the material is insufficient. But the greater part of the hypothesis proposed is merely the most direct and simple reading of the observations." While time would indeed scale down Shapley's numbers, the bijou Kapteyn galaxy was now arrayed against a formidable competitor.

Shapley believed that his Big Galaxy subsumed both the globular star clusters and the spiral nebulae—in particular, that spirals are subordinate in scale to the Milky Way: "From the new point of view, our galactic universe appears as a single, enormous, all-comprehending unit. . . . The adoption of such an arrangement of sidereal objects leaves us with no evidence of a plurality of stellar 'universes.'" Shapley's model was substantially fortified by the photographic work of his Mount Wilson associate, Adriaan van Maanen, who claimed to have detected rotation in comparing plates of several brighter spirals. As Shapley realized, if spirals are Big Galaxy analogs seen at great distance, their purported angular spin would translate into absurdly large stellar velocities toward their periphery, approaching or even exceeding the speed of light. Basking in the logical certainty that spiral nebulae must therefore lie within the Milky Way, Shapley dashed off a congratulatory note to van Maanen: "Between us we have put a crimp in the island universes, it seems,—you by bringing the spirals in and I by pushing the Galaxy out. We are indeed clever, we are." Further affirmation of the Big Galaxy concept followed from a host of arguments based on what were subsequently proven to be wrongly categorized stars or misinterpreted observations.

Evidence against Shapley's megagalaxy-as-universe model was readily found. His edifice rested on a rickety foundation: that stars in globular clusters are identical to similar-looking stars in the disk of the Milky Way. In particular, Shapley asserted that Cepheid variables in clusters follow the same period-luminosity relation as those elsewhere in the Galaxy. If untrue, then the huge distances he attributed to them are incorrect. (Indeed, it would later be proven that the stellar population of globular clusters—including the Cepheids—is older than that found in the Milky Way's disk, and have different observational properties.)

Furthermore, eruptive stars, or novae, had been photographed in almost a dozen spiral nebulae, and were invariably much fainter than novae within the Milky Way. This marked diminishment in apparent brightness—on average, one ten-thousandth the radiance of a galactic nova—signaled the remoteness of the spirals. The Andromeda Nebula, considered one of the nearer spirals, was pegged at a distance of a million light-years on this basis. At the very least, the occurrence of novae within otherwise diffuse spiral nebulae infers the presence of stars, rendered irresolvable by distance.

Also seemingly counter to Shapley's single-galaxy universe were the hyperkinetic movements of spiral nebulae. Vesto Melvin Slipher, at the Lowell Observatory in Flagstaff, Arizona, had measured galactic radial velocities with a twenty-four-inch Clark refractor and Brashear spectrograph. His presentation to the American Astronomical Society in August 1914 drew a standing ovation, and ultimately, medals from the Paris Academy, the Royal Astronomical Society, and the Astronomical Society of the Pacific. (First-semester graduate-student Edwin Hubble was in the audience.) The acclaim stemmed in part from Slipher's astonishing news—spirals are streaking through space at up to six hundred miles per second, much faster than any galactic star or globular cluster—but also from collegial admiration of the technological feat underlying those measurements.

The Lowell telescope and spectrograph had been optimized for planetary observations, where light was plentiful and centrally condensed. Faint, diffuse objects, such as spiral nebulae, required exceedingly long exposures to register on the photographic plate; recording their dispersed spectra was far more challenging. Through arduous experimentation, Slipher settled on a single-prism spectrograph, sacrificing dispersion for a gain in transmitted light. He added a super-fast camera lens and chemically boosted the photographic emulsions to reduce exposure times. The result was a 450-pound analytical engine capable of revealing nebular spectral lines, as well as their Doppler shifts.

Vesto Melvin Slipher, pictured in 1932.

Even with Slipher's upgrades,

seven hours might be required to capture a measurable image of the spectrum of the Andromeda Nebula. Fainter spirals entailed exposures of twenty to forty hours spanning several nights. As Slipher's list of three- and then four-figure radial velocities grew from its initial fifteen to twenty-five by 1917, astronomers pondered whether such fleet objects could be confined by the gravitational field of the Milky Way, even one as generously proportioned as Shapley's. Shapley countered by suggesting that spirals are discrete, gaseous outflows from our galaxy, sailing through space on the gentle pressure of starlight.

Spiral nebulae also display a peculiar avoidance of the Milky Way's otherwise highly populated disk; instead, they occupy two broad cones of space perpendicular to the galactic plane. In an address to the American Association for the Advancement of Science in December 1816, Lick Observatory Director W. W. Campbell points out that, were spirals internal to our galaxy, as Shapley's model places them, their observed distribution is perverse: "spirals live close to the right of us and close to the left of us, but . . . they avoid getting between us and the Milky Way structure." Much more likely, Campbell surmises with prophetic insight, is that obscuring material in the galaxy's disk extinguishes the light of remote nebulae along those lines of sight. As to the impasse over the nature of spiral nebulae, he adds, "*We are not certain how far away they are; we are not certain what they are*. However, the hypothesis that they are enormously distant bodies, that they are independent systems in different degrees of development, is one which seems to be in best harmony with known facts."

On April 20, 1920, astronomers' divergent impressions on the scale of the cosmos received a formal airing at the annual meeting of the National Academy of Sciences in Washington, DC. The event, billed as a debate but more a pair of sequential lectures, pitted Mount Wilson's rising star, Harlow Shapley, against Lick's island-universe advocate, Heber D. Curtis. In the running for the directorship at Harvard Observatory (whose representatives were in attendance), Shapley plugged his own contributions and aspirations in the research arena. He spoke at length about the distribution of globular clusters and the size of the Milky Way, and gave short shrift to extragalactic matters. Taking the rostrum, Curtis laid out a highly technical, point-by-point rebuttal of Shapley's assumptions and cited a string of evidence in support of the multigalaxy universe.

Little notice was taken of the event at the time. No minds were changed, no breakthrough discoveries announced. Only after the publication in 1921 of conjoined articles by Shapley and Curtis did the scientific community come to recognize the symbolic import of the so-called Great Debate. Implicit in the pages of observational data and esoteric arguments was the weighty issue at hand: humanity's conception of the universe. As in their long-ago wrangle over the geocentric and heliocentric cosmologies, astronomers stood on either side of a scientific divide. Each camp believed in the facts as they saw them, each was aware that their respective models were mutually exclusive: only one of the cosmic constructions cohered with reality. Yet as fervent as their arguments, all participants recognized that the ultimate arbiter of the verity of their ideas was better observational evidence—nature's testimony, more clearly rendered.

Although the Great Debate failed to resolve the core cosmological issues, it did cast a spotlight on the blistering pace of technological achievement over the previous decades. Astrophysical methods only recently embryonic now probed once inaccessible realms of space. Where photographic pioneers had been hard-pressed to capture the amalgamated glow of a globular cluster, their successors were recording subtle light variations of its constituent stars. Where spectroscopists had once struggled to tease out Doppler shifts in the spectra of bright stars, they were now measuring line displacements in the dispersed light of spiral nebulae. Where astronomers had previously inspected the heavens with telescopes the size of a rolled-up carpet, now they deployed sixty-inch and one-hundred-inch mountaintop reflectors on a nightly basis. By 1920, photography, spectroscopy, and modern megatelescopes had merged into a unitary instrument of enormous analytical power. And where technology leaps ahead, discovery is apt to follow.

Chapter 27

A Night to Remember

What are galaxies? No one knew before 1900. Very few people knew in 1920. All astronomers knew after 1924.

—Allan Sandage, introduction to the *Hubble Atlas of Galaxies*, 1961

An astronomer's observing notebook is a window onto a scientist's working habits. Every pertinent detail is recorded: date and time; the celestial target's designation, sky coordinates, angle from the meridian; if imaged, the photographic plate identifier (nowadays, a computer file name) plus duration of exposure; description or enumeration of the "seeing," the cutely nontechnical term for the clarity of the sky. Errors are crossed out, never erased or obliterated, corrections scrawled in nearby.

Seemingly little can be gleaned about the astronomer's state of mind from such a telegraphic spreadsheet. Family stresses, concerns over promotion, physical or mental fatigue intrude nowhere in these coldly rational, alphanumeric rows and columns. Yet there is a personal dimension implicit in this ritualistic recording of nights' passings under the stars: the supreme devotion to exploration, the relentless, sleep-deprived, frostbite-be-damned yearning to commune with the dimly lit wonders of deep space, each an astral Juliet beckoning the observer from an impossibly high balcony.

Every line of the astronomer's observing notebook includes space for remarks, queries, and out-of-the-ordinary circumstances: whether the camera has been modified, whether the target nebula resembles a cat's eye, whether a star appears brighter than it did the previous night. On rare occasions, the terse prose of the remarks sections gives way to an entry whose wording is more detailed than usual, more attentive to the vagaries of interpretation—no longer a note to self, but a declaration to posterity. Here, in a more deliberately rendered hand, is where the heightened pulse of the observer is almost palpable. Here is the moment of discovery.

On the evening of October 5, 1923, Edwin Hubble opened the dome of Mount Wilson's one-hundred-inch reflector, unaware that tonight's observing run would have momentous consequences, much less spark national headlines. It was his tenth allocation of time on the big telescope this year, the intervening nights spent on the sixty-inch and smaller instruments. With the shutters parted, the view of the sky through the vertical breach was unpromising: on a "seeing" scale of 1 to 5, Hubble jotted "1" in his observing notebook—barely worth opening the dome at all.

The previous four years had been rife with developments in Hubble's career, as well as those of his Mount Wilson compatriots. Hubble was a veteran observer by now, fully integrated into the research juggernaut fostered by George Ellery Hale. Having suffered his fourth mental breakdown, Hale had relinquished his directorship to his capable second, Walter S. Adams, and left with his family for an extended tour of Europe and the Middle East. Big Galaxy crusader Harlow Shapley had absconded for Harvard, taking with him a barely concealed antipathy toward Hubble's Oxfordian manner, conservative politics, and dawning celebrity. (Both men were native Missourians.)

Hubble kept his eye on the sky conditions as he prepared the camera for the evening's initial target, NGC 6729, a fan-shaped emission nebula in Corona Australis. He could reflect, with considerable satisfaction, on his accomplishments of the past two years. With near-fanatical thoroughness, he had amassed a photographic database of nebulae: a visual menagerie as seemingly diverse as any in the biological realm. His 1922 paper, "A General Study of Diffuse Galactic Nebulae," summarized the nature, form, and distribution of these deep-space clouds, and noted any apparent physical relation to proximate stars. Hubble's nebula classification scheme was widely adopted in the astronomical community.

His "galactic" types are clearly associated with stars, and include both planetaries—compact, often rounded forms whose moniker is a Herschel-era relic—and diffuse nebulae like the famed luminescent billow in Orion. The "nongalactic" category comprises clouds of spiral, spindle, ovate, globular, or irregular form. With the exception of the random flare-up of a nova, these nebulae lack discrete, fully resolved images of stars; to the eye, they appear misty throughout. (Hubble was noncommittal on the vexing issue of the spirals' remoteness, specifying that his "nongalactic" does not imply "extragalactic.")

Several months after this landmark paper, Hubble released a tour de force analysis of the physical engine that sustains a galactic nebula's glow. His conclusion: gaseous nebulae fluoresce under the withering glare of their embedded stars; the energy breaching the surface of these stellar powerhouses is absorbed by the surrounding gas, then reemitted in equal amount into the void. In Hubble's model, planetary nebulae arise from the prior expulsion of a star's atmosphere, creating a gassy cocoon energized by its parent star's light. Sprawling Orion-like complexes are made visible by the radiation of multiple hot stars, both within and adjacent to the bodies of gas. In the absence of such stars, a diffuse nebula might yet reveal itself in sharp silhouette against more distant banks of luminous matter.

As the October evening progressed, Hubble discerned an improvement in the sky conditions: he entered "2" in the *Seeing* column of his observing notebook. Slewing the telescope to the starry, ragged-bordered nebula NGC 6822, he took a sixty-minute exposure. By now, the night sky over Mount Wilson verged on its storied crystal clarity: the seeing had risen to "3+". The next plate was reserved for the Andromeda Nebula. After midnight, now October 6, Hubble acquired a forty-five-minute portrait while the great spiral cloud closed in on the meridian.

Not surprisingly, the quality of the developed plates reflected the improvement in sky conditions, Andromeda's being the best. Scanning its magnified image, Hubble's eye was drawn, in turn, to three faint stars, barely discernible amid the nebula's light-speckled background. Still, they hailed their presence to Hubble, for whom Andromeda's stellar field was virtually etched into memory. These were stars he had not seen before, each one a possible nova.

Novae had been discovered in a number of spiral nebulae, their relative dimness compared to those in the Milky Way suggestive of great distance. But the physical process underlying these stellar eruptions was unknown. Nor had any nova occurred close enough to Earth that its distance, hence its absolute light output, might be ascertained. Brilliant as these starry beacons are, in Hubble's day, they could not be adopted as "standard candles"—calibrated energy emitters whose comparative apparent and absolute brightnesses allow computation of cosmic distance. Nevertheless, Hubble dutifully took a pen to the Andromeda plate, marked each nova's location with a dash, and wrote the letter *N* alongside.

Over the succeeding months, Hubble continued to monitor Andromeda with both the one-hundred-inch and the sixty-inch reflectors. He collected additional images of the nebula dating back as far as 1909 from Mount

Wilson's photographic plate archive. Two of the three suspected novae exhibited the characteristic behavior of this stellar class: a transient outburst that gradually fades from view. However, the third proved remarkably durable: it was visible in every picture, old or new. Yet its prominence, relative to the stars around it, was different from plate to plate. This was no one-shot nova, Hubble realized, but some type of cyclical variable star.

On a sheet of graph paper, Hubble plotted the star's light-curve, a series of points tracing out the star's brightness with the passage of time. A distinctive, undulating pattern emerged: rapid rise to maximum brilliance, followed by gradual decline to minimum, the cycle repeating with clocklike regularity every 31.415 days. This inconspicuous mote of light was a Cepheid variable star, the first of its kind ever confirmed in a spiral nebula. Hubble had turned up an astronomical standard candle flickering in the arms of the Andromeda Nebula. He retrieved his observing notebook, paged to the line for October 5, 1923, and appended to the *Remarks* entry, "On this plate (H335H), three stars were found, 2 of which were novae, and 1 proved to be a variable, later identified as a Cepheid—the 1st to be recognized in M31 [the Andromeda Nebula]." Then, like a spotlight heralding discovery, Hubble drew a brawny arrow pointing at his five cramped lines of text. On the historic plate H335H (the first H standing for the Hooker telescope, the second for Hubble), he crossed out the label "N" that stood alongside the newly identified Cepheid and inked in "VAR!"

Using Harlow Shapley's published calibration of the period-luminosity law for Cepheids in the Milky Way, Hubble computed the distance to Andromeda's lone stellar milepost: one million light-years. He weighed the various factors that might alter or even nullify this gross inflation

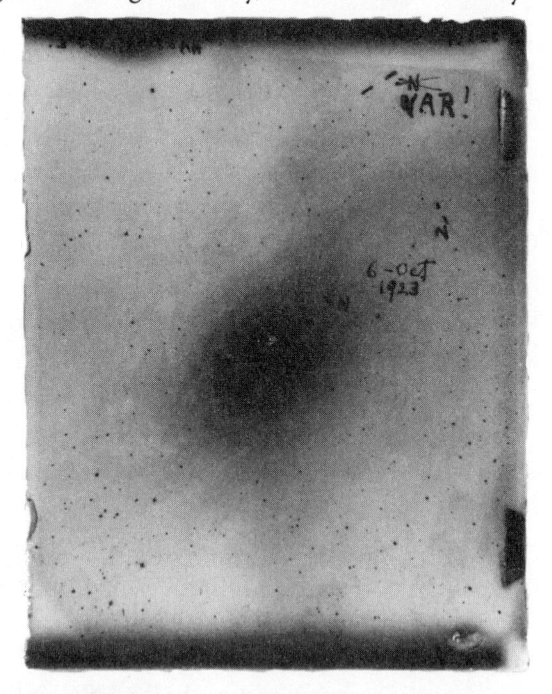

Edwin Hubble's photographic negative H335 of the Andromeda nebula, dated October 6, 1923. The label "VAR!" marks his discovery of a Cepheid variable star.

of the cosmic landscape. There was scant evidence of significant light extinction along the line of sight, which would make the star appear dimmer, hence, farther, than it really is. Furthermore, regions of sky adjacent to the Andromeda Nebula were devoid of Cepheids; thus, the newfound variable most likely resided within Andromeda and was not a stellar interloper projected against it. (Subsequent observations confirmed this assertion.)

At base, Hubble's distance estimate rested on the belief that Cepheids in Andromeda are equivalent to their Milky Way compeers—that Shapley's period-luminosity calibration holds wherever Cepheids are found. Like numerous others before him, Hubble postulated a uniformity of nature, which, he explains, "seems to rule undisturbed in this remote region of space. This principle is the fundamental assumption in all extrapolations beyond the limits of known and observable data, and speculations which follow its guide are legitimate until they become self-contradictory."

Hubble's distance estimate for Andromeda was a thunderous blow to proponents of the single-galaxy cosmos. The night sky's most prominent spiral nebula, he asserted, lies well beyond the borders of the Milky Way, even Shapley's hyper-extended Big Galaxy version. Thus, Andromeda is a galaxy unto itself, subsumed by no other, its dimensions and stellar population comparable to, if not larger than, our own. Can the thousands of fainter spirals be anything other than independent Milky Ways and Andromedas strewn throughout the vastness of space?

In late February 1924, Hubble alerted Shapley to his discovery of a Cepheid in the Andromeda Nebula. Included in the envelope was a copy of the star's telltale light-curve. Sitting in his office with his graduate student, Shapley lamented, "Here is the letter that has destroyed my Universe." Nevertheless, he replied with a critique of Hubble's assumptions and measurements. The purported Cepheid, he insisted, must be anomalous, if it exists at all; spurious variable stars can appear whenever plates of different exposure are compared.

Hubble dismissed Shapley's allegations of error, yet he understood that his controversial distance estimate could not rest on the measurement of a single star. He continued to mine Andromeda for variables, and simultaneously delved more intensively into the loosely wound spiral Messier 33 in Triangulum and the irregular nebula NGC 6822 in Sagittarius. By year's end, having hardly tapped Andromeda's rich vein of variables, he had raised the Cepheid count to a dozen; Messier 33 eventually yielded up thirty-five Cepheids, and NGC 6822 another eleven. Without exception, the stars confirmed the extragalactic nature of their host systems. The corresponding luminosities he derived for other key stellar components, such as novae and prominent blue stars, cohered with those found in the Milky Way and the Magellanic Clouds. This mutual consistency girded the reasonableness of the island-universe model.

While Hubble doggedly searched for extragalactic Cepheids, word of his transformative measurements flooded the professional grapevine. Even before publication, Princeton's Henry Norris Russell, dean of American stellar astrophysics and a lapsed Shapley ally, pronounced Hubble's work "undoubtedly among the most notable scientific advances of the year." The *New York Times* alerted its readers on November 23, 1924, that "'Island Universes' Similar to Our Own" had been discovered by Mount Wilson's "Dr. Hubbell." Yet astronomers who scoured the journals wondered alike: When was Edwin Hubble going to publish his findings?

In a congratulatory note, Russell encouraged Hubble to present his results at the end-of-the-year meeting of the American Astronomical Society in Washington, DC. He added that the paper would likely garner the annual $1,000 research prize of the Association for the Advancement of Science. Yet as the meeting neared its conclusion, there was no correspondence from Hubble, who remained at Mount Wilson. "Well, he is an ass," Russell muttered to an associate. "With a perfectly good thousand dollars available he refuses to take it." On December 31, 1924, as Russell prepared a last-ditch missive to his recalcitrant colleague, he spied a large envelope bearing his name behind the hotel's front desk. The much-anticipated paper had arrived. The following day, Russell stood before the assembled scientists—including Harlow Shapley—and read aloud Hubble's momentous report, "Cepheids in Spiral Nebulae." (The paper shared the AAAS research prize with a study of protozoans in the digestive tracts of termites.)

Hubble's hesitancy to publish, he soon confessed, stemmed from a reluctance to confront his senior Mount Wilson colleague Adriaan van Maanen, whose observations of spiral-nebula rotation were irreconcilable with the Cepheid-based distances. The rotation data had garnered weighty support among astronomers, as it derived from straightforward plate measurements and bore a lesser burden of assumptions than Cepheid-derived

Adriaan van Maanen.

results. Hubble knew that publication of his work would be a scientific slap of the gauntlet: an open charge that van Maanen's measured rotations were spurious. True, Lowell Observatory's Vesto Slipher and others had observed the spectroscopic signature of rotation in a few inclined spirals: a telltale tilt of spectral lines along each nebula's long axis. And spirals' swept-back arms might reflect a whirling action. But given the enormous distances Hubble had assigned the spiral nebulae, such roundabout displacements are all but immeasurable on photographic plates, even over a span of centuries. Before going public with his incendiary results, Hubble had wanted to nail down the source of what he already believed were illusory signs of rotation.

Later in 1925, Hubble bolstered the island-universe model with his seven hundred thousand light-year distance to the Magellanic-Cloud analog NGC 6822. The following year, in a paper knowingly titled "A Spiral Nebula as a Stellar System," he situated Messier 33 approximately 850,000 light-years from Earth. Hubble applied the full resolving power of the one-hundred-inch telescope to the system's profuse, star-like condensations, which some astronomers classed as stellar-nebular hybrids, perhaps gassy protostars. The vivid exposures showed beyond doubt that these were ordinary stars, their photographic images as small and round as any previously recorded. Hubble's summary report on the Andromeda Nebula appeared in 1929, with data from forty Cepheids sustaining its status as a far-flung star system.

On the other hand, the inferred diameters of both Andromeda and Messier 33 were a mere one-tenth that of Shapley's Milky Way. In consequence, Shapley offered a multitiered arrangement, with a swollen, central Milky Way accompanied by subordinate systems like globular clusters, the Magellanic Clouds, and spiral nebulae. The counterpunch to this "metagalaxy" concept would land in incremental blows, as Hubble identified ever-more-remote spiral nebulae.

Hubble had earlier urged Harlow Shapley to abandon his opposition to the island-universe model, suggesting that "the straws are all pointing in one direction and it will do no harm to begin considering the various possibilities involved." Now through his latest results, he appeared to be informing both Shapley and van Maanen that the battle of the universes was over. In fact, the controversy spilled into the next decade, with increasingly strained explanations of the systematic movements that nobody, it seemed, but van Maanen could detect. The issue was settled only after Hubble published his own thumbs-down reassessment of spiral rotation, using some of van Maanen's own plates. Left unsaid, but clearly implied: van Maanen had seen what his own preconceptions had led him to see.

In a companion paper, a chastened van Maanen conceded that his prior results should be viewed "with reserve."

In his autobiography, Harlow Shapley attributes his lingering intransigence to his friendship with van Maanen and his misplaced confidence in the Big Galaxy model. Had he not left Mount Wilson for a telescope-disadvantaged Harvard in 1921, he might have expanded his variable-star search from globular clusters to spiral nebulae—and he might have basked in the acclaim accorded Hubble for the discovery of Andromeda's Cepheids. Shapley long harbored regret over his lost opportunity, declaring in 1969, "The work that Hubble did on galaxies was very largely using my methods. . . . He never acknowledged my priority, but there are people like that."

From the 1930s onward, astronomers fashioned a Procrustean bed of research-derived constraints that shrank Shapley's Big Galaxy and stretched Hubble's cosmic distances. The dimming of starlight by dust in the galaxy's disk was found to be more pronounced than had been assumed, requiring various degrees of correction to stellar distances. Of equal import, the Cepheid species was found to be bipartite, based on the circumstances of their origin. Older Cepheids, which inhabit globular clusters and the bulging nucleus of spiral galaxies, formed from a different brew of chemical elements than their younger counterparts in the galactic disk. (This spatial segregation also holds for the general stellar population in spiral galaxies; evidently the old stars in the halo and nucleus reflect the initial sphericity of a protogalaxy before its gas condenses into a flattened disk.) In other words, the Cepheids observed by Hubble in the disks of extragalactic systems are a significant variant of those observed by Shapley in globular clusters: For a given period of variation, Hubble's Cepheids are fully four times more luminous than Shapley's. Thus, there are two period-luminosity laws, each of which was substantially refined and recalibrated over the years.

The twin revelations about galactic light extinction and the Cepheid dichotomy became the agents of change that brought conflicting evidence into concordance. The diameter of Shapley's Milky Way saw a threefold decrease to a more moderate one hundred thousand light-years. Meanwhile, Hubble's distances to extragalactic star systems doubled, with a commensurate increase in their size. The Andromeda Galaxy (formerly, Nebula) is more than two million light-years away, its diameter now in full measure worthy of the galactic appellation.

Hubble continued to range over the extragalactic spacescape during the mid-1920s, developing a morphological classification system for galaxies that is still in use today. (In his own writings, Hubble eschewed the term "galaxy," preferring the more traditional "extra-galactic nebula.") Honors were bestowed upon him, audiences crowded his public lectures, newspapers featured his eminently heroic accomplishments. In 1928, Hubble and his wife Grace toured Europe for five months, visiting tourist sites and scientific institutions. Hubble arrived home, energized by exchanges with colleagues at an astronomical conference in Holland.

One discussion harked back to the memorable meeting Hubble had attended in 1914 when Vesto Slipher debuted his measures of the radial velocities of spiral nebulae—motions that far outpace even the fastest stars in the Milky Way. As startling as their breakneck speeds was the curious fact that thirteen of the fifteen spirals in Slipher's sample display redshifted spectral lines: the radial component of their motion is directed away from Earth, as though in collective flight from our planet. The skewed velocity spread was sustained in Slipher's follow-up report of 1917, with twenty-one of twenty-five spirals in recession. The data also suggested a possible correlation between a spiral's faintness and its redshift, as though more distant spirals are systemically receding at greater velocity. The inference could not be proven at the time; there was no reliable way to gauge extragalactic distances prior to Hubble's discovery of Cepheids in spiral nebulae.

In this hint of a coordinated flight of galaxies, Hubble sensed opportunity—a timely convergence of his own interests, experience, and resources. By now, Slipher had painstakingly accumulated the radial velocities of more than forty spirals, having extended the reach of Lowell Observatory's spectrograph as far into space as it could go. But the twenty-four-inch refractor perceived only the collective glow of these systems, and was too small to resolve their constituent stars. To detect Cepheids and other all-important standard candles required a larger telescope. With Mount Wilson's one-hundred-inch reflector, Hubble had the means to complete the motional characterization of extragalactic systems: to wed each of Slipher's radial velocities to its corresponding distance. In late 1928, following Vesto Slipher's leading thrust, Major Hubble marshaled his resources and launched an all-out assault on the realm of the galaxies.

Chapter 28

Oculis Subjecta Fidelibus

It is . . . an enormous advantage to be in the right place at the right time and to know that one is in that enviable position.

—Malcolm Longair, "History of Astronomical Discoveries," 2009

Humanity inhabits a spinning globe, coursing round a central star, variously tugged by a nearby moon and neighboring planets, all moving through space among a multitude of stars that themselves bob and weave as they circle a disk-shaped galaxy. The notion of cosmic place is a complex construct, amenable only to sophisticated measurement and subject to an agreed-upon definition of a reference frame. Immobile as human senses make this earthly station appear, only by the determination of positions and movements of celestial objects do we infer our planet's relative location, course, and speed.

For centuries, the goal was to situate Earth within the solar system and ascertain its orbital parameters, along with those of its companion planets. During the 1800s, astronomers extended their reach into space, assembling over many decades a crude, three-dimensional map of the solar neighborhood. Photographers tracked the movement of stars in the plane of the sky, while spectroscopists used the Doppler shift to obtain the corresponding radial velocities. Astronomers were thus able to animate their once-fixed stellar map.

With estimable sagacity, Harlow Shapley used globular star clusters to lay out the form and dimensions of our galaxy, and situate the solar system within it. Astronomers discovered a telltale streaming of stars in our vicinity, confirming that objects in the disk of the Milky Way revolve according to the same Keplerian principles as do the Sun's family of planets. They believed that galaxies might form an external reference frame against which to refine the solar system's movement through space: if galaxies are free-floating motes in the void, their helter-skelter motions would average to zero, establishing a three-dimensional coordinate grid as practical as an array of fixed points. In essence, the Earth-centric viewpoint was abandoned for an imaginary aerie that overlooks the realm of the galaxies. Astronomers learned to their surprise that, while each galaxy exhibits a degree of random motion, galaxies collectively move in coordinated fashion: their radial velocity increases with distance. Thus, while assessing Earth's location and heading from an extragalactic perspective, observers stumbled upon a cosmic mystery with far-reaching implications.

Empowered by theoretical developments in physics and mathematics during the early twentieth century, scientists came to realize that time and space are inseparable: a cosmic fourth dimension exists, accessible only in a notional sense, yet real. The galaxies in this vision became luminous surveyor's stones that traced out the gravitational contours of deep space and led, ultimately, to a purely mathematical representation of the extragalactic landscape. With newfound conceptual tools, astronomers could probe beyond the limits of their instruments—indeed, beyond the limits of any conceivable instrument—and ponder the universe in its entirety. In more than the physical sense, gravity pulled it all together.

Isaac Newton's mathematical theory of gravitation, derived in the late seventeenth century, proved extremely successful in accounting for the motions of comets, moons, planets, and stars. In Newton's conception, celestial bodies inhabit a three-dimensional universe governed by the prosaic rules of Euclidean geometry. A unique and uniform flow of time reigns over the entirety of space. That gravity's origin and mechanism of transmission remained a mystery was of no consequence: Newton's celestial mechanics generated results that cohered with the observed movements of bodies in the cosmos. However, while the model worked well for planetary and stellar systems (for example, the orbits of binary stars), it prompted a troubling conundrum when applied on the more expansive cosmological scale. Newton found that an infinite space uniformly sown with matter is unstable against the self-gravity of its material contents. It will invariably draw itself together. Our universe—if not infinite, then sufficiently large that Newton's analysis applies—cannot remain static; absent any countervailing force, it must be in a state of contraction or expansion. Thus, in Newtonian physics, there was no explanation as to how the observed and presumably infinite universe remained stable.

Albert Einstein's General Theory of Relativity, whose initial development culminated in 1917, provides a pathway to a solution of the cosmological quandaries posed by Newtonian mechanics. Einstein abandoned the Euclidean constraints under which Newton had labored and instead applied more modern, non-Euclidean geometries to the notions of space and time—a radical inquiry that would have profound scientific and philosophical ramifications.

In Einstein's world, parallel lines might intersect or diverge, and the angles in a triangle might sum up to more or less than the standard 180 degrees. Time forms the basis of a fourth dimension that stands on an equal footing with the three familiar dimensions of space; hence, the introduction of the catch-all term space-time. To Einstein, gravity is not a force emanating from a massive body, but the result of a warp in the invisible continuum of space–time surrounding the body. A moving object that strays into such a region follows a predetermined path dictated by the local contour of space–time. The impression of a gravitational pull is an artifact of this distorted geometry, which is the true agent that compels the object to deviate from its intended path. Denser bodies deform their space–time environment to a greater degree, imparting the sensation of a stronger gravitational force. The field equations of Einstein's General Theory of Relativity provide the mathematical means to compute the shape of space–time around a given configuration of matter. Conversely, the observed path of, say, a comet or a satellite reveals the local distribution of matter that governs its movement.

Einstein pointed out that his relativistic concepts could just as well be applied to the universe as to a star or a planet. To avoid the same cosmological instability that afflicted Isaac Newton's formulation of gravity, Einstein amended his field equations, inserting an ad hoc term to keep his model universe in stasis. Although legitimate in the abstract, Einstein's mathematical counterpoise—his so-called cosmological constant—had no obvious manifestation in the physical world. Yet its inclusion was compelled by astronomers' general belief that the universe must be static.

Einstein solved the field equations for a cosmos encompassing a uniform distribution of matter at rest. (Whether the material is fluidlike, as Einstein had posited, or bunched in discrete units, such as stars and nebulae, is of no consequence, as long as it is evenly sown.) The resultant curvature of this model is spherical: the space–time continuum closes in on itself, forming a finite universe whose size depends on the amount of matter it contains. The transmission of light among widely separated objects is governed by the spherical geometry: light beams must travel along the surface of the sphere and cannot take the shortcut through the sphere's interior. Indeed, a light beam might eventually circumnavigate the bounded space and arrive back at its place of origin. The shortest path between two points is no longer a straight line, à la Euclid, but an arc; hence, cosmic distances are construed differently than in a "flat" Euclidean world.

Perhaps the most significant philosophical and scientific ramification is the conceptual shift away from the infinite Newtonian universe and its attendant complexities. Einstein even roughed out the diameter of his cosmological model—one hundred million light-years—as well as its mean density of matter.

In 1917, Dutch astronomer Willem de Sitter offered an alternative cosmological solution to Einstein's field equations, one that maintained the imperative of a static cosmos. However, the conceptual price for this mandatory stasis was high: de Sitter's model was pure space, devoid of matter. Equally strange, if a hypothetical star or nebula were plunked into this vacuum, its light would appear redshifted to a distant (and likewise hypothetical) observer. Some astronomers noted the curious parallel of de Sitter's theorized redshifts and Vesto Slipher's growing list of redshifted spiral galaxies. Furthermore, de Sitter found that his predicted redshifts grew as the distance from the observer increased, a linkage not yet evident in Slipher's sparse data set. In response to the obvious criticism that our universe is far from empty—witness the planets, stars, and galaxies—de Sitter pointed to the sheer abundance of space; given a sufficiently large volume, the overall density of matter might be low enough to approximate his mass-starved model.

In 1922, Russian physicist Alexander Friedmann introduced yet other solutions to Einstein's field equations. While preserving Einstein's notion of a closed, hyperspherical world, one of Friedmann's models allowed an evolution of its overall curvature: the radius of the universe changes over time. Starting from a geometric point eons ago—"the time that has passed since Creation," Friedmann termed it—the universe might expand for a while, then reverse and gradually draw back in on itself. Such a universe might be reborn, cycling through its life sequence without end. In a different scenario, the universal fabric has an underlying hyperboloidal form, like de Sitter's model, only seeded with matter: a one-shot cosmos destined to expand forever.

Theoretical physicists offered only *possible* cosmologies; which, if any, of their proposals might apply to the real universe would have to be determined by astronomical observations. But astronomers were slow on the uptake. Though intrigued by the various cosmological implications of general relativity, most were confounded by its mathematical complexity. Mount Wilson's director, George Ellery Hale, confessed to a colleague that "the complications of the theory of relativity are altogether too much for my comprehension. If I were a good mathematician I might have some hope of forming a feeble conception of the principle, but as it is I fear it will always remain beyond my grasp." (Many modern-day researchers have the same reaction to cosmic string theory.)

The few astronomers who delved into the esoteric world of general relativity during the 1920s focused their attention on the static-universe models of Einstein and de Sitter. Alexander Friedmann's published papers from 1922 and 1924 were ignored. So were the nonstatic cosmologies of the Belgian cleric-mathematician George Lemaître, who made an explicit argument for a linear increase of galaxian velocity with distance. The key advantage of the Friedmann–Lemaître cosmologies is that they permit the universe to contain matter (unlike de Sitter's empty model), while providing a theoretical basis for the observed redshifts of galaxies. Yet their ideas languished, the scholarly current behind the static models too strong to deflect into alternative channels of cosmological thought.

A devout observationalist, Edwin Hubble was agnostic about the various cosmological theories. Whether Einstein's, de Sitter's, or some other world model prevailed was of marginal concern to him. All that mattered were their observational ramifications to his research program. That the majority of galaxies exhibit large redshifts had been established to Hubble's satisfaction. And de Sitter's model, he knew, predicted a correlation between redshift and distance in an ultra-low-density universe. Indeed, several observers had asserted a linear correspondence, using galactic diameter or faintness as a makeshift substitute for distance. But in Hubble's mind, there could be no substitute for a calibrated measurement of a galaxy's distance.

With the one-hundred-inch reflector, he could fix the distances of galaxies, derived from the brightness of their constituent stars. Then, swapping the camera for a spectrograph, he could obtain redshifts of extragalactic systems beyond the reach of Vesto Slipher's twenty-four-inch refractor. Given the urgency to publish—much attention had converged of late on the issue of cosmological redshifts—Hubble knew there was insufficient time to do both: he decided to couple his own distance estimates with Slipher's velocities. (Fully forty-one of the forty-six radial velocities of nearby galaxies had been acquired by Slipher.) As his own observations progressed, Hubble needed a proven partner to photograph the spectra of remote galaxies; even a single, additional redshift might prove decisive in swaying the collective opinion of astronomers. To the strategic enterprise, veteran Mount Wilson observer Milton Humason brought a grammar-school education, a fondness for gambling, a desk-drawer flask of "Panther Pacifier"—and a supreme gift for deep-space photography.

Milton Humason, circa 1930.

Coaxing his pack mule along the treacherous path up Mount Wilson in 1910, Milton Lasalle Humason could hardly have suspected that one day he would plumb the cosmic depths with the world's largest telescope, much less coauthor a landmark paper in the *Astrophysical Journal*. Humason had worked in the Pasadena area for many years, and its rugged hillsides felt like home. After stints as a bellboy, handyman, and citrus farmer, Humason returned to the mountain in 1917 as a janitor. The following summer, he learned the rudiments of celestial photography from an undergraduate intern, and became so adept at it that Harlow Shapley recommended him for a staff position. By the time Edwin Hubble came calling in 1928, Humason was Mount Wilson's foremost imaging expert and self-appointed guardian of the facility, instructing night assistants—and the occasional astronomer—on the proper use and care of the equipment.

Hubble detailed his proposed project, spelling out the division of labor: he would focus on galaxies' distances and Humason on their velocities. Hubble envisioned a cosmological distance ladder extending into extragalactic space, each rung defined by a standard candle whose calibration rests upon that of the prior rung. Thus, Cepheids disclose the distances of the nearest galaxies, which collectively establish the average luminosity of a galaxy's most radiant stars. These stellar beacons, in turn, become the standard candle to assess the distances of more remote galaxies, whose Cepheids are imperceptible to the camera. The final rung in Hubble's stepwise scheme is set by a galaxy's overall glow. At this extreme range, stars are irresolvable, and merge into an amorphous whorl of light. In relatively short order, Hubble told Humason, he could gauge the distances of a sufficient number of Slipher's galaxies to justify publication.

Although narrower in scope, Humason's role was vital to Hubble's plan. Humason would photograph spectra of several nearby galaxies, and validate his own measured radial velocities against Slipher's. Assuming agreement between their respective data sets, Humason was to venture at once into the depths beyond the range of the Lowell refractor. He would target the galaxy NGC 7619, in Pegasus, a system whose dimness was indicative of great distance. This solitary data point, although beyond what Hubble termed the "realm of positive knowledge," might strengthen the case for a linear correspondence between galactic distance and velocity.

Within a year, Hubble had extended his cosmological distance ladder more than six million light-years into space, encompassing twenty-four galaxies. (Subsequent recalibrations have septupled his initial range estimate.) He presented his results to the National Academy of Sciences on January 17, 1929, in a paper titled, "A Relation Between Distance and Radial Velocity Among Extra-Galactic Nebulae." The measurements of galactic distance were his own, yet fully twenty of the two dozen radial velocities were Slipher's (whose name appears nowhere in the paper). Compared to prior works, Hubble's made a compelling case for the predicted linear escalation of a galaxy's radial velocity with distance, an astronomical precept that would ultimately bear his name. In graphical form, the data points congregate loosely about a rising straight line drawn through their midst. Knowing his audience, Hubble intuited that his diagram called out for more data to establish the reality of the line.

As instructed, Humason had affirmed that his spectroscopic readings were in accord with Slipher's. With that, he took a multinight, thirty-three-hour

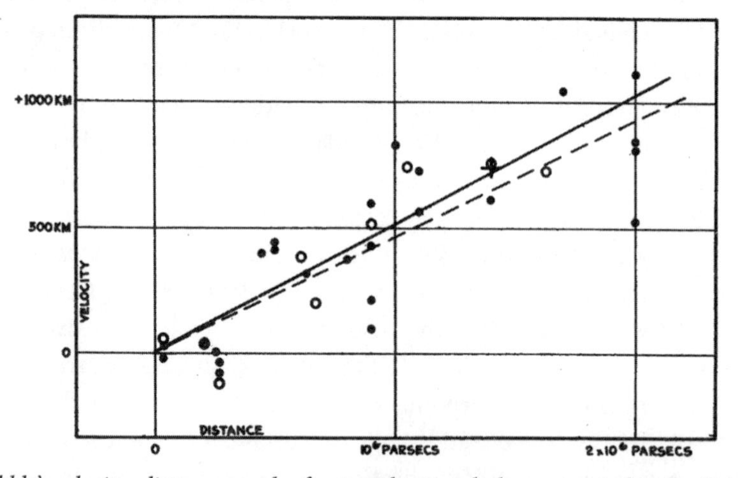

Hubble's velocity–distance graph of extragalactic nebulae, presented to the National Academy of Sciences on January 17, 1929. The solid and dashed lines represent alternative straight-line interpolations to the data points.

time-exposure of the spectrum of the remote galaxy NGC 7619, followed by one of forty-five-hour duration. The result: a recession velocity of twenty-four hundred miles per second, more than twice that of any previously recorded. Paired with Hubble's own distance estimate of twenty-four million light-years, the data point for NGC 7619 indeed became the graphical anchor that secured the linear linkage between galactic velocity and distance. So critical was this measurement that Hubble featured Humason's announcement as the lead-in article to his own. Two years later, in 1931, a promised follow-up paper appeared, in which Hubble and Humason add forty new radial velocities to their previous list and extend their survey of galaxies sixteenfold in distance. (Hubble makes amends for his prior snub by citing Vesto Slipher's "great pioneer work.")

With the release of the 1931 paper, there was near unanimity among astronomers that the extragalactic velocity-distance relationship was real. Even longtime critic Harlow Shapley joined the ranks of the believers, although he claimed (rightly) that Hubble's distances were skewed by mistaking compact, gaseous nebulae for a galaxy's brightest stars. Albert Einstein conferred with Hubble during a visit to Mount Wilson in 1931 and declared his once-ascendant static-universe model null and void.

Ninety years earlier, upon Friedrich Bessel's presentation of the first credible measurement of a star's distance, astronomer John Herschel challenged the Royal Astronomical Society to overcome their doubts and accept the result. "*Oculis subjecta fidelibus*," Herschel offered, echoing the lines of the Roman poet Horace about the trustworthy eye. "If all this does not carry conviction along with it, it seems difficult to say what ought to do so." In the latter-day case of systematic cosmic motion, the astronomical community likewise decided that the requisite threshold of observational evidence had been reached. Yet Hubble himself remained noncommittal with regard to the physical interpretation of galactic redshifts. Indeed, he used the term "apparent radial velocity" to indicate that the observed spectral-line shifts might stem from a cause other than a headlong recession of galaxies; the confounding theories of general relativity and quantum mechanics had alerted him to the manifold ways in which nature upends logical expectations.

Hubble's stark, pen-and-ink diagram, with its canted, empirically derived line, endowed James Keeler's turn-of-the-century extragalactic spacescapes with profound meaning. The line and its associated mathematical formula, V = HR, later dubbed Hubble's law, became twin avatars of what astronomers took to be an expanding universe. The universe is dynamic, they asserted, space itself billowing to vaster proportions, sweeping apart its luminous

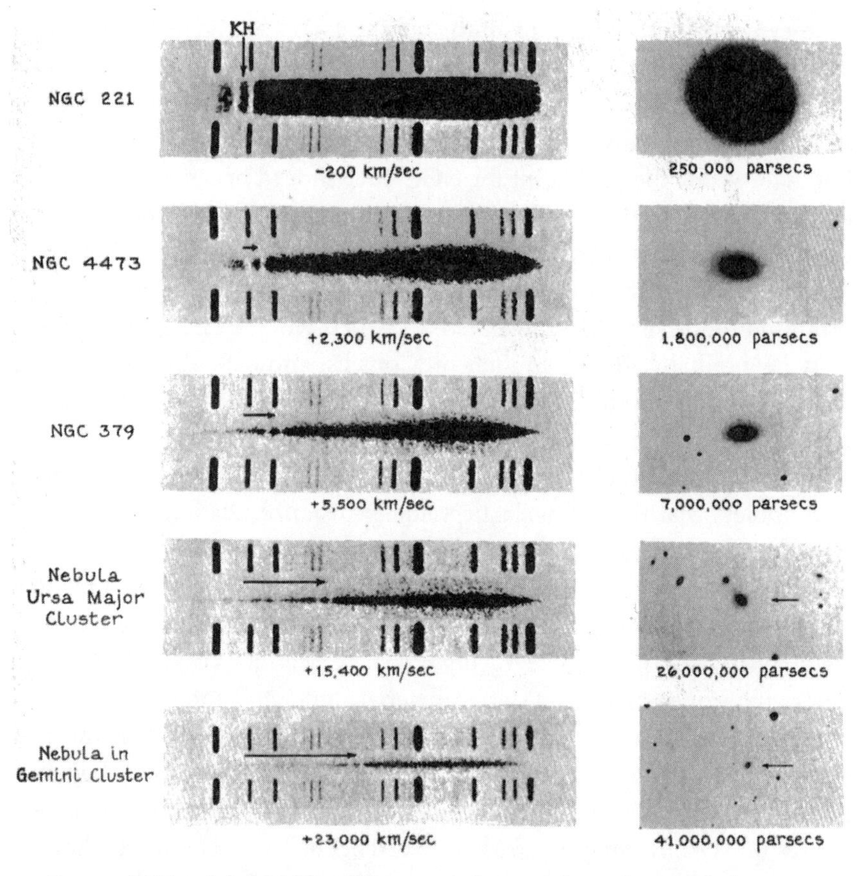

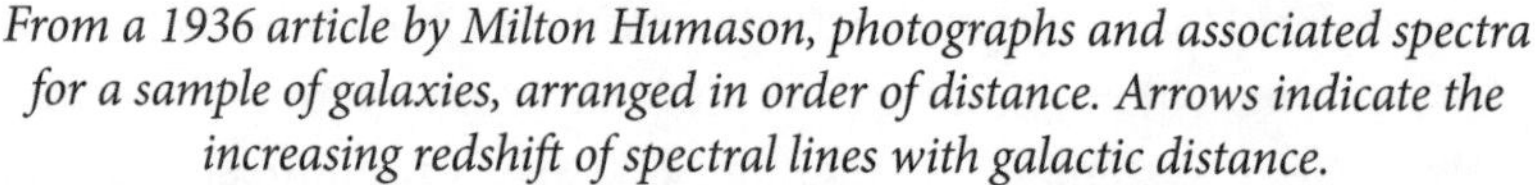

From a 1936 article by Milton Humason, photographs and associated spectra for a sample of galaxies, arranged in order of distance. Arrows indicate the increasing redshift of spectral lines with galactic distance.

points of reference—the galaxies. In the time machine of one's imagination, the cosmic clock can be driven backwards, the observed dispersal of galaxies reversed until atoms and photons meld into an infinitesimal, primeval amalgam. Thus, universal expansion compels a beginning: an ab initio, hyperdense fireball from which all cosmic energy and matter emerged—in modern-day parlance, the Big Bang.

Edwin Hubble did not discover the expanding universe so much as lend the accumulated weight of his professional authority to the observed velocity–distance relation from which this phenomenological inference followed. Once Einstein and de Sitter signed on—and credited Edwin Hubble as its observational midwife—the expanding universe became the reigning paradigm among cosmological cognoscenti. That Hubble's work paralleled, and in some cases duplicated, the efforts of others does not diminish the galvanizing effect his publications had upon the course of extragalactic astronomy. In the end, whether Hubble's law or Slipher's law or Lemaître's law or some permutation of these, the iconic formula V = HR represents nature's declaration of the way things are, heedless of any label affixed by inhabitants of one or another of its surfeit of planets.

The constant H, likewise named for Hubble, expresses the current rate of cosmic expansion and is computed from the array of observed galactic radial velocities and distances. A large value of H implies rapid expansion, thus, a relatively youthful universe, whereas a small value of H indicates a slower growing, Methuselan cosmos. Hubble's initial estimates of his eponymous constant tended toward the former: for each million-parsec increment in distance—equivalent to 3.26 million light-years—a galaxy's speed ratcheted up a staggering five hundred kilometers per second (eight hundred miles per second). Had the universe maintained this frenzied rate of growth from the start, it would have come into existence a mere two billion years ago. To scientists' chagrin, contemporary estimates of Earth's geological time line considerably exceeded the derived age of the universe: illogically, our own planet appeared to predate the very space it occupies. Subsequent revisions have cut the Hubble constant from its original five hundred to about seventy kilometers per second per million parsecs, raising the estimated age of the universe to some fourteen billion years, in full accord, not only with Earth's age, but that of the galaxy's oldest stars.

In recent decades, astronomers have detected the cosmic microwave background, the remnant radiation from an early, high-temperature phase of the universe. They have posed a compelling theory that the infant universe underwent a burst of hyperinflation, doubling its dimensions more than one hundred times over a tiny fraction of a second. Observers have confirmed the gravitational signature of dark matter—distributed material whose overall bulk far exceeds all the visible stars and gas clouds in galaxies. And in a remarkable development, astronomers have learned that the universal expansion is accelerating, as though space were being propelled to ever-larger scales by a mysterious antigravity. This so-called dark energy now stands at the forefront of cosmological research.

In a series of lectures at Yale University in 1935, published the following year as *The Realm of the Nebulae*, Hubble declared that the "history of astronomy is a history of receding horizons." On the one hand, these horizons might be spatial: a record-setting number of light-years into the void. However, with the advent of sensitive electronic cameras, advanced telescopes, and computer-based image processing, an optical horizon also

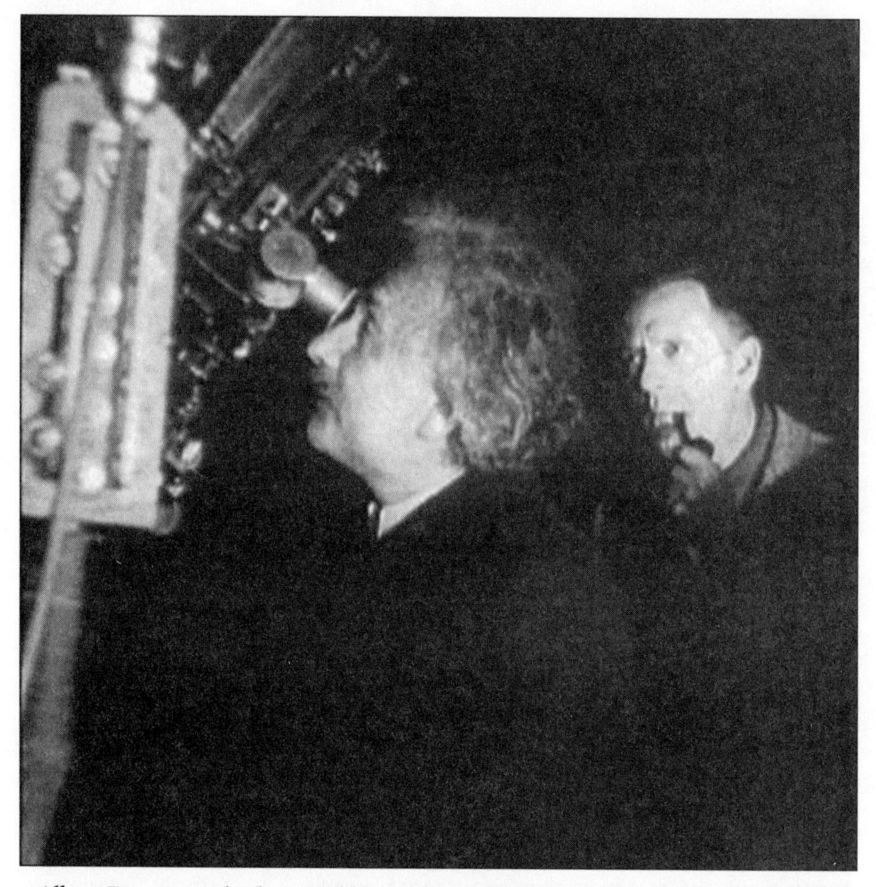

Albert Einstein and Edwin Hubble at Mount Wilson's one-hundred-inch telescope on the evening of January 29, 1931.

beckons: the ultrafaint, resolution-limited threshold of detectability. One need not target only the farthest reaches of the observable universe, when exoplanets await discovery around nearby stars and black holes gorge themselves on matter in our own galaxy. By their nature, Hubble explained to the audience, frontier explorations of space lead to a state of uncertainty: "We are, by definition, in the very center of the observable region. We know our immediate neighborhood rather intimately. With increasing distance, our knowledge fades, and fades rapidly. Eventually," Hubble concluded, with the assurance of one who has sailed over the horizon and returned, "we reach the dim boundary—the utmost limits of our telescopes. There, we measure shadows, and we search among ghostly errors of measurement for landmarks that are scarcely more substantial."

Epilogue

Curiosity is the spark of inspiration, inspiration the fuel of passion, and passion—even more than brilliance—the engine of scientific progress. Over the span of eight decades, into the early twentieth century, a succession of determined innovators guided astronomy through a metamorphosis that redefined its methodologies, its social structure, and arguably its very reason for being. Once handmaiden to the practical imperatives of navigation and calendar-keeping, the explication of the cosmos became an end in itself. The classical astronomer's question, "Where is a star?" evolved into the astrophysicist's more profound inquiry, "What is a star?"

Amateur astronomers and inventors, immune to the strictures of professional advancement, drove the early stages of this transformation. Like all scientists, they sought the joy that comes with concentrated effort to solve a problem. Many also strove for self-validation through acceptance into the ranks of scholarly associations—at least until the complexity of astrophysical research outpaced their aspirations and abilities. In the private observatories of Warren De La Rue, Henry Draper, Lewis Rutherfurd, William Huggins, Andrew Common, and Isaac Roberts, photons from outer space no longer illuminated the astronomer's retina, but were funneled into opto-mechanical devices borne of human ingenuity and scientific necessity. But rather than being shunted aside by the camera and the spectroscope, the observer's eye was unshackled from its evolutionary bonds and endowed with unprecedented acuity. One by one, institutional astronomers were swept up by the tide of technological innovation, moved to action by the successes of their amateur counterparts and by the widening adoption of the new methods by their peers.

Yet as George Ellery Hale realized at the close of the nineteenth century, celestial photography and spectroscopy formed a two-legged stool: until the telescope grew dramatically in aperture and precision, these valuable tools would fail to reach their full potential. Over the course of three decades, Hale transformed a rugged California mountaintop into one of the world's premier research institutions. By 1920, Mount Wilson Observatory occupied the apex of a pyramid of technological achievement, an edifice piled with stones of innovation and experiment, bound in place by the cement of physics, chemistry, and inventor's sweat. From this remarkable convergence of the photographic, spectroscopic, and telescopic crafts, the once-fledgling field of observational astrophysics would mature into the scientific powerhouse that it is today.

During this astrophysical awakening, the Sun's blinding surface yielded to photochemical dissection, its underlying structure and processes opened up to theoretical scrutiny. The light of stars was unwound, revealing the atmospheric composition and radial velocities of luminaries light-years distant. Nebulae disclosed their manifold forms to the camera and their elemental makeup and physical properties to the spectrograph. Of the last, the spiral nebulae would come to hold special historical significance. From Lord Rosse's first crude hand-renderings of spiral nebulae, to James Keeler's revelatory photographs proving their vast numbers, to Vesto Slipher's spectroscopic measurement of galactic redshifts, and finally to Edwin Hubble's synthesis of evidence regarding their nature and cosmological significance, the rise of astrophysics transformed observational cosmology into a full-fledged, quantitative science. As the twentieth century unfolded, breakthroughs in theoretical physics interfaced with telescopic investigations, stimulating their mutual advancement.

Today, astronomers explore the universe with mountaintop reflectors whose dimensions dwarf those of previous generations. Robotic instruments tirelessly ply the night sky for asteroids and exploding stars. Space telescopes circle Earth, yielding cosmic images unblurred by our planet's atmosphere. Photosensitive electronic chips have replaced chemical plates in the camera and the spectrograph. Supercomputers churn through masses of observational data generated nightly in observatories around the world. With all these means, modern astronomers have determined the precise age of the universe, its physical state a moment after birth, and its probable time line into the future; created numerical models that depict the energy-generating mechanisms in stars; tracked the telltale movements of objects under the gravitational thrall of a black hole; tallied an ever-expanding roster of extrasolar planets; and photographed youthful galaxies billions of light-years away. These accomplishments, as well as the cosmic conundrums that yet remain, hark back to an earlier era of astrophysical exploration.

Hubble's determination of the Andromeda Galaxy's distance in 1923 epitomizes both the enormous gulf that separates us from the rest of the universe and the astronomer's ever-widening reach into space—the "receding horizons" Hubble referred to in the 1930s. Our knowledge of the physical world is likewise circumscribed by a horizon. This horizon, too, recedes with time, as human inquisitiveness exerts continual pressure against the margins of the unknown. With each increment of acquired knowledge, our time-tested assumptions about nature are jostled, and should they fall, are replaced by unbidden enigmas. In fact, history has proven that scientists' overall state of comprehension goes hand-in-hand with their overall state of ignorance; the greater our understanding of the universe and its constituent objects, the more unanswered questions that present themselves.

Like a cognitive cornucopia, science pours forth new questions as rapidly as it lays aside the old. This endless stream of conundrums has sustained inquiring minds for thousands of years. The amateurs and professionals who together strove to modernize astronomy exemplify the scientist's ineffable need to confront and vanquish the unknown. So it has been, so it will remain. As English poet and classicist A. E. Housman declared in his introductory lecture at University College, London, in 1892, "Other desires perish in the gratification, but the desire of knowledge never: the eye is not satisfied with seeing nor the ear filled with hearing. . . . The sum of things to be known is inexhaustible, and however long we read we shall never come to the end of our storybook."

References

Part I: Picturing the Heavens

p. 19 "By applying . . ."—Seaton (1899), p. 154.

Chapter 1. True Eye and Faithful Hand

p. 21 "There is no one . . ."—Clerke (1902), p. 5.
p. 22 "for many succeeding ages."—Newell (1806), p. 12.
p. 22 "We seemed to be . . ." —"An Account of the Total Eclipse . . ." (1806), p. 245.
p. 22 "angels had been . . ."—*Columbian Centinel* (June 21, 1806), p. 2.
p. 23 "the neatness, patience, and accuracy . . ."—George Phillips Bond, in "Sketch of William Cranch Bond" (1895), p. 401.
p. 23 "in despair . . ."—George Phillips Bond, in Holden (1897), p. 10.
p. 23 "Then and there . . ."—E. Bond (1938).
p. 24 "by which Earth-bound observers . . ."—Becker (2011), p. 13.
p. 24 "watching the appearances . . ."—Morus (2005), p. 198.
p. 25 "depends entirely . . ."—Loomis (1856), p. 52.
p. 26 "once in the workshop . . ."—Bessel, F. W., and Schumacher, H. C., ed. P*opuläre Vorlesungen über wissenschaftliche Gegenstände*. Hamburg: Perthes-Besser & Mauke (1848), p. 17.
p. 27 "a tube with an eye . . ."—Milham (1937), p. 531.
p. 27 "It is with no feeling of pride . . ."—Loomis (1856), p. 26.
p. 28 "nothing in this act . . ."—Loomis (1856), p. 26.

Chapter 2. The Ingenious Mechanic of Dorchester

p. 29 "No living man . . ."—Mitchell (1851), p. 278.
p. 30 "[H]is optic nerve . . ."—E. Bond (1938).
p. 30 "ingenious mechanic . . ."—Jones and Boyd (1971), p. 28.
p. 31 "must be such as to enable . . ."—W. C. Bond (1856), p. iii.
p. 32 "a piece of ordnance . . ."—Oliver Wendell Holmes, *The Poet at the Breakfast-Table*, Houghton, Mifflin and Company, 1887, p. 257.
p. 32 "The time had not yet arrived . . ."— W. C. Bond (1856), p. v.
p. 33 "His antipathy . . ."—Holden (1897), p. 15.
p. 34 "To watch the motions . . ."—Holden (1897), p. 13.
p. 34 "There is something to my mind . . ."—Baker (1890), p. 13.
p. 35 "his habits were not adapted . . ."—Holden (1897), p. 17.
p. 35 "Mr. Bond was well established . . ."—Holden (1897), pp. 17–18.
p. 35 "there is no disparagement . . ."—Holden (1897), p. 18.
p. 36 "In regard to the principle features . . ."—Jones and Boyd (1971), p. 53.
p. 36 "was approached by terraces . . ."—E. Bond (1938).

p. 37 "An astronomical observer . . ."—Jones and Boyd (1971), p. 56.
p. 37 "An excavation was first made . . ."—Mitchell (1851), p. 5.
p. 37 "the first appearance . . ."—Jones (1968), p. 56.
p. 37 "It is delightful to see . . ."—Jones (1968), p. 59.
p. 39 "One observer, with a sharp pencil . . ."—Jones & Boyd (1971), p. 54.
p. 39 "They came by the Hundreds . . ."—Jones (1968), p. 66.
p. 39 "perfect Babel"—Jones (1968), p. 66.

Chapter 3. Writing with Light

p. 40 "[Photography] seemed to epitomize . . ."—Rudisill (1971), pp. 73, 76.
p. 43 "As night fell . . ."—Newhall (1961), p. 16.
p. 44 "What then was there so wonderful . . ."—Daguerre (1971), p. 16.
p. 45 "one of these Parisians . . ."—Barger and White (1991), p. 20.
p. 45 "there should be found . . ."—Newhall (1982), p. 17.
p. 46 "without any notion . . ."—Daguerre (1838).
p. 47 "You can now say . . ."—Daguerre (1971), Introduction, p. 18.
p. 48 "that compared to these masterpieces . . ."—Newhall (1982), p. 23.
p. 48 "It is hardly saying . . ."—Barger and White (1991), pp. 26–27.
p. 48 "must have been very difficult . . ."—Report to the Chamber of Peers of the French Parliament, July 13, 1839; in Daguerre (1971), p. 33.
p. 49 "opticians' shops were crowded . . ."—Newhall (1982), p. 23.
p. 50 "We have seen the views . . ."—Lewis Gaylord Clark, editor, *The Knickerbocker*, New York, 1839, in Taft (1964), p. 3.
p. 50 "[W]e distinguish the smallest details . . ."— "Chemical and Optical Discovery." *Boston Courier* (February 28, 1839).
p. 50 "It is not merely the likeness . . ."—Manuscript letter by Elizabeth Barrett (Browning) to Mary Russell Mitford, December 7, 1843. Wellesley College Library, Special Collections. [www.daguerreotypearchive.org].
p. 51 "daguerreotype likenesses . . ."—*National Police Gazette* (New York) 1 (June 27, 1846): 355.
p. 53 "curious and ingenious specimen of art"—Humphrey (1850), p. 14.

Chapter 4. Summits of Silver

p. 54 "It clear'd off afternoon . . ."—Jones and Boyd (1971), p. 71.
p. 54 "doing things with nothing to do with . . ."—Grant (1851), p. 94.
p. 55 "thrown upon the scene as large as life."—Pierce (1987), p. 18.
p. 56 "By this most simple means . . ."—Snelling (1888), p. 562.
p. 56 "a fixed fact for the naturalist . . ."—Grant (1851), p 95.
p. 56 "Mr. Whipple was satisfied . . ."—Jones and Boyd (1971), p. 71.
p. 57 "we do not despair . . ."—Jones and Boyd (1971), p. 75.
p. 57 "Nothing could be more interesting . . ."—Whipple (1853), p. 66.
p. 59 "had a tendency to move . . ."—Whipple (1853), p. 66.
p. 61 "the sea breeze, the hot and cold air . . ."—Whipple (1853), p. 66.
p. 62 "It is our purpose to pursue the subject . . ."—Hoffleit (1950), p. 27.
p. 62 "much troubled by clouds . . ."—Hoffleit (1950), p. 27.
p. 62 "The effect was at once apparent . . ."—Hoffleit (1950), p. 27.
p. 62 "We have rarely seen anything . . ."—David A. Wells, in *Annual of Scientific Discovery*, Boston: Gould and Lincoln, 1852, p. 135.

p. 62 "Fringes of darkness casting themselves off . . ."—"Photography in the United States." Photographic Art-Journal (1853), p. 338.
p. 64 "The change which takes place . . ."—Jones and Boyd (1971), pp. 102–103.
p. 64 "taken as old Sol lights . . ."—Whipple (1853), p. 66.

Chapter 5. The Man with the Oil-Can

p. 65 "In bringing before the Association . . ."—De La Rue (1860), p. 130.
p. 65 "Anyone who is not satisfied . . ."—Holden (1897), pp. 87–88.
p. 66 "I am the man with the oil-can."—Huggins (1889), p. 249.
p. 67 "If I were asked what course of practice . . ."—Nasmyth (1883), p. 327.
p. 68 "charmed"—De La Rue (1859–60), p. 180.
p. 72 "We will not attempt to state . . ."—Taft (1938), p. 211.
p. 72 "[T]o be an amateur in wet plate days . . ."—Taft (1938), p. 374.
p. 73 "The inventor of Collodion died . . ."—June 13, 1857, in Hughes (2010).
p. 75 "[I]t was not easy to find . . ."—De La Rue (1857), p. 16.
p. 75 "numberless impediments sufficient to damp . . ."—De La Rue (1859–60), p. 181.

Chapter 6. The Evangelists

p. 76 "The wonderful exactness of the photographic record . . ."—Turner (1905), p. 78.
p. 77 "Most patiently he taught us the names . . ."—Holden (1897), p. 50.
p. 77 "one of the most remarkable applications . . ."—http://www.gap-system.org/~history/Biographies/Maxwell.html.
p. 77 "On a fine night . . ."— G. P. Bond (1890), pp. 301–302.
p. 78 "George is, and has been for months . . ."—Jones & Boyd (1971), p. 82.
p. 80 "surface of the globe must be explored . . ."—G. P. Bond (1859), p. 78.
p. 80 "There is nothing, then, so extravagant . . ."—G. P. Bond (1890), pp. 301–302.
p. 82 "[N]o one need hope for even moderate success . . ."—De La Rue (1860), pp. 133–134.
p. 82 "I have never any failure . . ."—De La Rue (1860), p. 135.
p. 83 "as if a giant with eyes . . ."—LeConte (2011), p. 20.
p. 84 "The accounts [of the corona and prominences] . . ."—Rothermel (1993), p. 148.
p. 85 "the retina which never forgets . . ."—Lockyer (July 30, 1874), p. 255.

Chapter 7. The Aristocrat and the Artisan

p. 86 "I take great pleasure in bringing . . ."—Rutherfurd (1848), p. 437.
p. 86 "almost shrinking modesty . . ."—Gould (1892), p. 32.
p. 87 "from spare parts . . ."—*Morris-Rutherfurd family papers, 1717–1889.*
p. 87 "189 feet N.W. from Second Avenue . . ."—Rutherfurd (1865), p. 304.
p. 88 "Men have been known to go . . ."—Chapman (1998), p. 175.
p. 88 "I was soon on the balcony . . ."—Lankford (1984), p. 215.
p. 89 "My largest is an achromatic telescope . . ."—Rutherfurd (1848), p. 437.
p. 90 "The making of the best negative . . ."—"American Photographical Society . . ." (1862), p. 379.
p. 90 "If we look with a reflector at a bright star . . ."— Common (1884), p. 39.
p. 91 "a labor . . ."—Rutherfurd (1865), p. 307.
p. 93 "[W]e are filled with mingled wonder . . ."—"Rutherfurd's Photograph . . ." (1866), p. 37.
p. 93 "I have made many thousand photographs . . ."—Reingold (1964), p. 254.
p. 94 "[I]nstead of being restricted . . ."—Gould (1878), p. 15.
p. 95 "rush into print."—Rees (1892), p. 694.

p. 95 “by far the most distinguished . . .”—“Lewis Morris Rutherfurd.” (1889), p. 375.

Chapter 8. Passion Is Good, Obsession Is Better

p. 96 “When Henry and Anna Draper rode home . . .”—Schucking (1982), p. 306.
p. 97 “the rest sat silent . . .”—Plotkin (1982), p. 322.
p. 97 “had for a companion, friend, and teacher . . .”—Youmans (1882), p. 756.
p. 97 “On one side was the sincerest filial devotion . . .”—Young (1883), p. 30.
p. 98 “No astronomical drawing . . .”—Smyth (1846), p. 75.
p. 100 “His lectures are so interesting . . .”—Young (1883), p. 31.
p. 100 “Lamentations being useless . . .”—Gould (1878), p. 17.
p. 101 “It becomes a pleasant . . .”—H. Draper (1864a), p. 24.
p. 101 “A current of cold air . . .”—H. Draper (1864a), p. 9.
p. 102 “An uninterrupted horizon . . .”—Martin (1992), p. 6.
p. 103 “This is the first observatory . . .”—H. Draper (1861), p. 64.
p. 104 “[I]nstead of injuring the photograph . . .”—H. Draper (1864a), p. 34.
p. 105 “[I] can see no reason . . .”—H. Draper (1864a), p. 1.
p. 105 “My experience in the matter . . .”—H. Draper (1864a), p. 55.
p. 105 “I had become acquainted . . .”—Brashear (1988), p. 55.
p. 106 “our wedding trip”—Anna Palmer Draper to Edward C. Pickering, April 10, 1887, in Boyd (1969), p. 97.
p. 107 “From what I see here . . .”—John Draper to Henry Draper, December 19, 1870; Plotkin (1982), p. 324.
p. 108 “By a little pushing you might . . .”—John Draper to Henry Draper, February 9, 1871, in Plotkin (1982), p. 325.
p. 108 “was as good as any in existence . . .”—Barker (1888), p. 101.
p. 109 “So great was [Anna Draper’s] interest . . .”—Cannon (1915), p. 381.
p. 109 “probably the most difficult and costly . . .”—Youmans (1882), p. 756.
p. 109 “Observations made by its means . . .”—Clerke (1902), p. 234.
p. 110 “God forbid that astronomy . . .”—Meadows (1984), p. 61.

Chapter 9. From Closet to Cosmos

p. 112 “Collodion—slow old fogey . . .”—Newhall (1982), p. 124.
p. 114 “twenty-two thousand miles on steamer . . .”—Newhall (1982), p. 124.
p. 116 “[H]ow cold my feet were . . .”—Ashbrook (1984), pp. 384–385.
p. 116 “The exposure of the Orion nebula . . .”—Gingerich (1980), p. 365.
p. 118 “the best representation . . .”—Holden (1882), p. 227.
p. 118 “Dr. Draper’s negative was made . . .”—Holden (1882), p. 227.
p. 118 “The photographic plate . . .”—Pickering (1886), p. 181.
p. 119 “It is only a short time since . . .”—H. Draper (1882b), p. 341.
p. 119 “I think we are by no means . . .”—Whitney (1972), p. 180.
p. 119 “It is hard to avoid the appearance . . .”—Young (1882), p. 333.

Chapter 10. Leaves of Glass

p. 120 “The camera is an encroaching instrument . . .”—Clerke (1888), p. 28.
p. 120 “was always at the telescope”—Turner (1904), p. 313.
p. 121 “full of enterprise . . .”—Turner (1903), p. 306.
p. 121 “half submerged by the fogs . . .”—Clerke (1888), p. 46.
p. 122 “the world was naturally . . .”—Turner (1903), p. 307.
p. 122 “The work of correcting was tedious . . .”—Calver (1879), p. 19.
p. 122 “I see no obstacles . . .”—Calver (1879), p. 20.

p. 123 "The stars were seen as lines . . ."—Stone (1884), pp. 221–222.
p. 123 "delusion"—Keeler (1908), p. 10.
p. 125 "epoch-making"—Ball (1904), p. 361.
p. 125 "assumed the office of historiographer . . ."—Clerke (1893), p. 490.
p. 126 "'Streamers and fleecy masses' . . ."—Clerke (1886), p. 42.
p. 126 "cloud-like, curdling masses."—Roberts (1887), p. 89.
p. 127 "[W]e ought, with all gratitude . . ."—Roberts (1889b), p. 297.
p. 128 "a candle shining through . . ."—King (1955), p. 46.
p. 128 "two luminous cones or pyramids . . ."—de Vaucouleurs (1987), p. 595.
p. 128 "a mystery never in . . ."—de Vaucouleurs (1987), p. 596.
p. 128 "one of the most valuable photographs . . ."—Macdonald (2010), p. 243.
p. 129 "lines of small stars shown . . ."—Ranyard (1889), p. 76.
p. 129 "united luster of millions . . ."—de Vaucouleurs (1987), p. 596.
p. 129 "Here one might see a new solar system . . ."—Smith (2008), p. 96.
p. 130 "library, not of books . . ."—Common (1884–85), p. 39.
p. 131 "[T]he facility of reproduction . . ."—Barker (1888), p. 83.

Chapter 11. The Grandest Failure

p. 132 "If we could first know . . ." —Lincoln, Abraham. "House Divided Speech," 1858, [www.abrahamlincolnonline.org].
p. 132 "rigorously orthodox kind"— Clerke (1888), p. 32.
p. 132 "of two points marking a frontage . . ."— Gill (1891), p. 604.
p. 133 "[H]owever perfect an instrument may be . . ."—Tenn (1990a), p. 85.
p. 133 "no dreamy contemplation . . ."—Gill (1891), p. 603.
p. 133 "betake themselves to bed . . ."— Gill (1891), p. 603.
p. 133 "only a merging into a rich yellow . . ."—Gill (1882a), p. 20.
p. 135 "Should they fade and vanish . . ."—Clerke (1888), p. 33.
p. 136 "If we can get over the distortion . . ."—Darius (1983), p. 48.
p. 136 "so united that often at the Observatory . . ."—Callandreau (1903), p. 558.
p. 138 "We should hardly be willing . . ."—Lankford (1984), p. 32.
p. 138 "rival scheme"—Turner and Common (1889), p. 310.
p. 138 " . . . thus neither treated the resolutions . . ."—Turner and Common (1889), pp. 309–311.
p. 139 "It is the first time I have ever felt obliged . . ."—Jones and Boyd (1971), p. 276.
p. 139 "[W]hatever can be done to promote the work . . ."—Jones and Boyd (1971), p. 277
p. 139 "It is difficult to believe . . ."—Pickering (1889), p. 375.
p. 139 "I am disgusted . . ."—Jones and Boyd (1971), p. 276.
p. 139 "[S]uch protests . . ."—Gill (1888), p. 321.
p. 140 "had the consciousness of a power . . ."—Whittingdale (1943), p. 75.
p. 141 "advancement of the physical side of astronomy . . ."—Plotkin (1990), p. 47.
p. 142 "inquiries of this [physical] sort . . ."—Young (1887), p. 354.
p. 142 "greatest triumph . . ."—Evans (1984), p. 156.
p. 143 "vermin of the skies"—Ashbrook (1984). p. 297.

Chapter 12. An Uncivil War

p. 144 "It is a remarkable and highly significant fact . . ."—Keeler (1899a), p. 128.
p. 145 "the Dictator . . ."—Osterbrock (1984), p. 81.
p. 146 "a pile of junk."—Osterbrock (1984), p. 158.
p. 146 "as antiquated . . . as Noah's ark"—"Mountain Homes Shut . . ." (1897), p. 23.

p. 146 "a monstrosity. Despite . . ."—Shane (1964), p. 83.
p. 146 "peace commission is already . . ."—"War Once More . . ." (1897), p. 14.
p. 146 "rather wish the accident . . ."—"Mountain Homes Shut . . ." (1897), p. 23.
p. 146 "[T]here is civil war . . ."—Turner and Hollis (1897), p. 296.
p. 146 "Why should a Derby-winner end . . ."—Turner and Hollis (1897), p. 300.
p. 147 "sparsely wooded mountain land . . ."—Hansen (1947), pp. 188–189.
p. 147 "Keeler doesn't claim . . ."—Campbell (1900), p. 240.
p. 148 "The difficulties here referred to . . ."—Keeler (1900), p. 326.
p. 148 "On one of the fine nights . . ."—Keeler (1899c), p. 200.
p. 151 "does not tire, as the eye does . . ."— Holden (1886), p. 468.

Part II. Seeing the Light

p. 153 "The physicist and the chemist . . ."—De La Rue (1861), p. 130.

Chapter 13. The Odd Couple

p. 155 "The most important discovery . . ."—Roscoe (1900), p. 530.
p. 155 "produces instantaneous tingling . . ."—Roscoe (1900), pp. 517–518.
p. 155 "had a very salamanderlike power . . ."—Roscoe (1900), pp. 545–546.
p. 156 "with the true perseverance . . ."—Roscoe (1900), p. 513.
p. 156 "Why Heidelberg . . ."—Turgenev, Ivan. *Fathers and Sons*. Oxford University Press, 1998, p. 68.
p. 156 "Beneath the stone floor . . ."—Roscoe (1900), p. 545.
p. 156 "I thought I had dropped . . ."—Roscoe (1900), p. 544.
p. 156 "It was quite in keeping . . ."—"Professor R. W. Bunsen" (1900), p. 101.
p. 157 "Heaven forbid, when I should . . ."— Oesper (1927), p. 435.
p. 157 "The only value . . ."— Oesper (1927), p. 434.
p. 157 "the man who came nearest . . ."—Oesper (1927), p. 439.
p. 157 "Ah, so many sit . . ."—Roscoe (1906), p. 54.
p. 158 "Alas, no, my untimely death prevented . . ."—Roscoe (1900), p. 542.
p. 158 "A chemist who is not a physicist . . ."—Oesper (1927), p. 438.
p. 159 "I am currently quite annoyed . . ."—Oldham (2008), p. 51.
p. 160 "It will do me some good . . ."—Oldham (2008), p. 164.
p. 160 "My stay in Breslau has recently become . . ."—Warburg (1929), p. 207.
p. 161 "first notables of science"—Jungnickel and McCormmach (1990), p. 288.
p. 161 "two scientists who by working together . . ."—Jungnickel and McCormmach (1990), p. 288.

Chapter 14. What's My Line?

p. 163 "The world is moved along . . ." —Keller, Helen. *Optimism: An Essay*. 1903, [www.gutenberg.org]. Adapted from a letter by English historian John Richard Green, December 21, 1870.
p. 163 "Light it self is a . . ."—Newton (1671), p. 3079.
p. 165 "By candle-light . . ."—Wollaston (1802), p. 380.
p. 167 "In such a way . . ."—von Rohr (1926), p. 286.
p. 168 It would be of great importance . . ."—Fraunhofer (1823), pp. 290–291.
p. 169 "I will prepare the apparatus . . ."—Jackson (1996), p. 14.
p. 169 "On looking as I was directed . . ."—Jackson (1996), p. 14.
p. 170 "I have seen with certainty in the spectrum . . ."—Hearnshaw (1987), p. 323.
p. 171 "A cotton wick is soaked . . ."—Talbot (1826), p. 78.
p. 171 "a glance at the prismatic spectrum . . ."—James (1981), p. 34.

p. 172 "[Salt] floats in the air . . ."—Clerke (1902), p. 132.
p. 172 "though doubtless very accurate . . ."—James (1983), p. 32.

Chapter 15. Laboratories of Light

p. 173 "[I]n order to examine the composition . . ."—Kirchhoff and Bunsen (1860), p. 107.
p. 173 "At present, Kirchhoff and I . . ."—Gingerich (1992), p. 171.
p. 176 "for if bodies should exist . . ."—Kirchhoff and Bunsen (1860), p. 107.
p. 177 "I made some observations . . ."—Kirchhoff (1860), p. 195.
p. 178 "that the dark lines . . ."—Kirchhoff (1860), p. 195.
p. 179 "The sun possesses an incandescent . . ."—Kirchhoff (1861a), pp. 185–186.
p. 180 "if anyone asked me . . ."—Meadows (1970), pp. 135–136.
p. 180 "we approach the question of the habitability . . ."—*More Worlds Than One*, London: John Murray (1854), p. 97.
p. 180 "at once destroyed, at a blow, the idea . . ."—Lockyer (1881), p. 269.
p. 180 "victory over space"—Schuster (1881), p. 468.
p. 181 "In the lower half of the field . . ."—Roscoe (1862), p. 328.
p. 182 "if we were to go to the sun . . ."—De La Rue (1861), p. 130.
p. 182 "our readers will feel an interest . . ."—Crookes (1861), p. 184.
p. 183 "In these expressions . . ."—Kirchhoff (1863), p. 252.
p. 183 "I have laid Prof. Miller's diagrams . . ."—Kirchhoff (1863), p. 255.
p. 183 "read Miller's words . . ."—Kirchhoff (1863), p. 261.
p. 183 "It is seen that the proposition . . ."—Kirchhoff (1863), p. 258.
p. 183 "had clearly propounded this question . . ."—Kirchhoff (1863), p. 256.
p. 184 "I have recently read, with very great interest . . ."—De La Rue (1861), p. 131.
p. 185 "The real importance of . . ."—James (1985), p. 13.
p. 185 "What do I care for gold . . ."—Helmholtz (1888), p. 537.
p. 185 "Look here . . ."—Helmholtz (1888), p. 537.

Chapter 16. Deconstructing the Sun

p. 186 "It is not an uncommon thing . . ."—Sutton (1986), p. 429.
p. 187 "I certainly have never seen any thing . . ."—Rutherfurd (1878), p. 43.
p. 188 "On reaching New York I called . . ."—Meadows (1984a), p. 62.
p. 188 "picture is absolutely unretouched . . ."—Draper (1873), p. 417.
p. 188 "I am glad that you have stated so clearly . . ."—John Browning to Henry Draper, January 4, 1874, in Plotkin (1982), p. 40.
p. 189 "superiority is so great there . . ."—Hentschel (1999), p. 208.
p. 190 "There is argon in the gas . . ."—Kragh (2009), p. 174.
p. 191 "The slit of my spectroscope was placed . . ."—Frost (1910), p. 96.
p. 191 "The phenomenon was so sudden . . ."—Meadows (1970), p. 221.

Chapter 17. A Strange Cryptography

p. 194 "[W]hen a molecule of hydrogen vibrates . . ."—Morus (2005), p. 214.
p. 195 "Talent and zeal, untiring devotion . . ."—"Report of the Council to the Thirty-Eighth General Meeting of the Society." *Monthly Notices of the Royal Astronomical Society* (1858), p. 110.
p. 196 "This news was to me like the coming upon . . ."—Becker (2011), p. 331.
p. 198 "enable the observer to determine . . ."—Huggins and Miller (1864), p. 414.
p. 201 "the numerous and closely approximated fine lines . . ."—Huggins and Miller (1864), pp. 413–414.

p. 201 "[T]he positions which he ascribes . . ."—Huggins and Miller (1864), p. 414.
p. 203 "strange cryptography of unravelled starlight"— From the original, "Within this unravelled starlight exists a strange cryptography." Huggins (1897), p. 909.
p. 203 "that the stars, while differing . . ."—Huggins and Miller (1864), p. 434.
p. 203 "surrounded by planets . . ."—Huggins and Miller (1864), p. 434.
p. 204 "if it could be successfully applied . . ."—Huggins (1864c), p. 437.
p. 205 "armed with the spectrum apparatus"—Huggins (1864c), p. 438.
p. 205 "At first I suspected some derangement . . ."—Huggins (1864c), p. 438.
p. 205 "It is obvious . . ."—Huggins (1864c), p. 442.
p. 206 "star clusters grown misty through . . ."—Clerke (1893), p. 484.
p. 206 "science will be more advanced by the slow . . ."—Huggins (1865a), p. 449.
p. 207 "[H]e was frequently called upon to speak . . ."—Dreyer and Turner (1923), p. 153.
p. 207 "Forward"—Becker (2011), p. 77.

Chapter 18. Trumpets and Telescopes

p. 208 "This time was, indeed, one of strained expectation . . ."—Huggins (1897), p. 913.
p. 212 "[I]n considering the importance of his principle . . ."—Hearnshaw (1992), p. 162.
p. 213 "if the colours were really tinged . . ."—Becker (2011), p. 131.
p. 215 "It has something to do with change . . ."—Becker (2011), p. 121.
p. 215 "a new method of research . . ."—Huggins (1897), pp. 921–922

Chapter 19. Burn This Note

p. 217 "I want to tell Huggins how much you have done . . ."—Becker (2011), p. 180.
p. 217 "scientific housemaid"—Tenn (1986), p. 10.
p. 218 "I had to teach myself what to do . . ."—Tenn (1986), pp. 9–10.
p. 218 "in which case I should help in arranging instruments . . ."—Tenn (1986), p. 10
p. 218 "One is interesting with a lump of engineer's waste . . ."—Tenn (1986), p. 10.
p. 218 "I am very glad to learn . . ."—Becker (2011), p. 179.
p. 219 "was occupied on all favorable days . . ."—Becker (2011), p. 185.
p. 220 "Huggins is very pleasant & everything . . ."—Edward S. Holden to Henry Draper, August 2, 1876. Becker (2011), p. 180.
p. 220 "The research is difficult . . ."—H. Draper (1877), p. 95.
p. 220 "You can either make star spectra . . ."—J. W. Draper to Henry Draper, August 15, 1876. Plotkin (1972), p. 46.
p. 221 "He was greatly surprised at my spectra . . ."—William Huggins to Charles A. Young, January 31, 1883. Becker (2011), p. 224.
p. 221 "You cannot imagine the pain . . ."—Becker (2011), p. 224.
p. 222 "grand rhythmical group"—Huggins (1897), p. 927.
p. 222 "whether these lines are not intimately connected . . ."—Huggins (1880), p. 678.
p. 223 "experiments and the preparations for them . . ."—Draper (1879), p. 420.
p. 223 "it may be possible for you after a time . . ."— Becker (2011), p. 225.
p. 224 "physical side of astronomy"—Plotkin (1990), p. 48.
p. 224 "This star [Epsilon Lyrae] appears double . . ."—Plotkin (1980), p. 285.
p. 224 "I urged upon him the importance of an early publication . . ."—Edward Pickering to Anna Palmer Draper, January 13, 1883. Jones and Boyd (1971), p. 220.
p. 224 "yet I feel so very incompetent for the task . . ."—Anna Palmer Draper to Edward Pickering, January 17, 1883. Jones and Boyd (1971), pp. 220–221.

p. 225 "It is not necessary that the paper should . . ."—Boyd (1969), p. 81.
p. 225 "*very wild indeed* . . ."—William Huggins to Edward Pickering, March 12, 1884. Jones and Boyd (1971), p. 224.
p. 225 "Dr. Huggins' arguments, that results . . ."—Edward Pickering to Anna Palmer Draper, March 31, 1884. Boyd (1969), p. 85.
p. 225 "Your publication [of 1880] does not enable . . ."—Edward Pickering to William Huggins, March 31, 1884. Jones and Boyd (1971), p. 225.
p. 226 "I cannot tell whether you have been led astray . . ."—William Huggins to Edward Pickering, April 1884. Jones and Boyd (1971), p. 225.
p. 226 "I felt very sorry that you should have been subjected . . ."—Anna Palmer Draper to Edward Pickering, April 30, 1884. Jones and Boyd (1971), pp. 225–226.
p. 227 "I wonder what Mr. Huggins will say . . ."—Anna Palmer Draper to Edward Pickering, January 23, 1887. Jones and Boyd (1971), p. 230.
p. 227 "induce other astronomers to undertake . . ."—Jones and Boyd (1971), p. 229.
p. 227 "I quite agree with you in feeling . . ."—Jones and Boyd (1971), pp. 230–231.
p. 227 "I have just received a paper . . ."—Becker (2011), p. 225.
p. 229 "I do think we have in photography . . ."—Lockyer (July 30, 1874), p. 255.

Chapter 20. A Spectacle of Suns

p. 230 "Astronomy paints its picture in the brighter colors . . ."—Keeler (1897), p. 275.
p. 234 "When one component is approaching . . ."—Pickering (1890), pp. 46–47.
p. 234 "[I]s it not a sufficient argument . . ."—Edward Pickering to Anna Palmer Draper, December 8, 1889. Jones and Boyd (1971), p. 244.
p. 236 "Translating the mathematical formulae . . ."—Keeler (1891), p. 48.

Chapter 21. The Cloud That Wasn't There

p. 237 "The born astronomer . . ."—Newcomb (1897), pp. 305–306.
p. 239 "pictorial and qualitative rather than metrical . . ."—Frost (1899), p. 367.
p. 239 "One could always tell how the night had been . . ."—Struve and Zebergs (1962), p. 47.
p. 240 "They are intended to be looked at . . ."—Barnard (1895), p. 63.
p. 241 "it will be seen . . ."— Barnard (1895), pp. 64–65.
p. 241 "These nebulosities . . . have been amply verified . . ."—Barnard (1899), p. 155.
p. 241 "a little unreasonable to suppose that Herschel . . ."— Barnard (1903), p. 78.
p. 243 "Most of the great curved nebula . . ."—Barnard (1903), p. 80.
p. 243 "with the same instruments described in his present . . ."—Barnard (1903), p. 80.
p. 243 "showed, besides the great Orion Nebula . . ."—Roberts (1903d), p. 158.
p. 243 "Difficulty seems to have a peculiar . . ."—Hills (1914), p. 389.
p. 245 "I shall not be sorry to have him go . . ."—McDonald (2010), p. 256.
p. 245 "The new Society is designed to be popular . . ."—Bracher (1989), p. 7.
p. 246 "Many ladies are interested in astronomy and own telescopes . . ."—Pickering (1882), p. 4.

Chapter 22. The Union of Two Astronomies

p. 247 "The domains of the physical sciences . . ."—Keeler (1897), p. 271.
p. 247 "there may be some who view . . ."—Keeler (1897), p. 277.
p. 248 "The majestic elder astronomy . . ."—Clerke (1888), pp. 29–30.
p. 248 "I regret having to give . . ."—W. W. Campbell to James Keeler, April 4, 1894. Lankford (1997), p. 179.
p. 248 "the superior attractiveness of astrophysical . . ."—Lankford (1997), p. 179.

p. 249 "Astronomers must have been a group . . ."—Stebbins (1947), p. 412.
p. 249 "who were at sword's points before . . ."— Stebbins (1947), p. 412.
p. 250 "a good-sized check on a bank . . ."— Stebbins (1947), p. 405.
p. 250 "The new astronomy . . ."—Clerke (1888), p. 30.
p. 250 "Although the work is the most acute and absolutely . . ."—Plotkin (1990), p. 49.
p. 251 "Life is work, and work is life"—Becker (2011), p. 300.
p. 252 "Of the sciences in America . . ."—Lankford (1997a), pp. 375–376.
p. 253 "kind of wealthy pauperism"—Plotkin (1984), p. 124.
p. 253 "If any millionaire be interested . . ."—Carnegie (1889), p. 687.
p. 254 "The great irony of the project . . ."— Lankford (1997a), pp. 397–398, 399.
p. 255 "virtually killing the ambitions . . ."— Struve (1943), pp. 474–475.
p. 256 "Light is all-important . . ."—Keeler (1897), p. 277.

Part III. Money, Mirrors, and Madness

p. 257 "The doors of the observatory are never closed . . ."—Chant (1907), p. 263.

Chapter 23. Mr. Hale of Chicago

p. 259 "[George Ellery Hale was] slight in figure . . ."—Wright (1966), p. 132.
p. 259 "A queer man lives nights in that cheese box"—Wright (1966), p. 36.
p. 259 "seemed bent on going where . . ."—Adams (1947), p. 197.
p. 260 "I was born an experimentalist . . ."—Adams (1938), p. 372.
p. 260 "George always wanted things yesterday"—Wright (1966), p. 30.
p. 260 "nervously organized"—Wright (1966), p. 33.
p. 260 "Our delights were enhanced by frightening . . ."—Wright (1966), p. 31.
p. 261 "I cannot think without excitement . . ."—Wright (1966), p. 42.
p. 261 "The odor of the disulphide . . ."—Adams (1938), p. 372.
p. 262 "Mr. Hale of Chicago"—Wright (1966), p. 43.
p. 262 "any grating of mine should be good . . ."—Adams (1938), p. 371.
p. 262 "To say that I have been busy . . ."—George Ellery Hale to Harry Goodwin, October 8, 1891. Sheehan and Osterbrock (2000), p. 96.
p. 263 "The red portion of the spectrum . . ."—Young (1890), p. 197.
p. 263 "Setting the spectroscope upon this latter..."—Young (1890), p. 199.
p. 264 "Not only is a large amount . . ."—Hale (1890b), p. 314.
p. 266 "I am treated like a *Grand-Duke*"—George Hale to Harry Goodwin, August 21, 1891. Wright (1966), p. 82.
p. 266 "The enclosed looks like business"—Wright (1966), p. 75.
p. 267 "I would not consider . . ."—George E. Hale to Harry Goodwin, July 17, 1892. Wright (1966), p. 92.
p. 268 "Boodler"—Wright (1966), p. 100.
p. 268 "lick the Lick"—Wright (1966), p. 98.
p. 270 "telescope is not only the largest . . ."—George Hale to Charles Yerkes, May 31, 1897. Wright (1966), p. 129.
p. 270 "logical formulation was shaped . . ."—Seares (1939), p. 266.
p. 270 "a fairly accurate analysis . . ."—Adams (1947), p. 199.
p. 271 "This is a very peaceful region . . ."—Wright (1966), p. 137.
p. 271 "two sets of underwear and one pair of pants . . ."—Wright (1966), p. 139.
p. 271 "moving cylinder of fur coats . . ."—Wright (1966), p. 126.
p. 273 "None of the beams they collect . . ."—Clerke (1893), p. 516.

Chapter 24. The Universe in the Mirror

p. 274 "It has often been asserted . . ."—Ritchey (1901), p. 228–229.
p. 277 "It was a mighty bewilderment . . ."—Holmes, Oliver Wendell. *The Poet at the Breakfast-Table.* Boston: James R. Osgood and Co., 1872, p. 257.
p. 278 "I was never more struck with the conviction . . ."—King (1955), p. 224.
p. 278 "The nights are as remarkable . . ."—Lassell (1852), p. 14.
p. 279 "I was not without a pang or two . . ."—Lassell (1877), p. 179.
p. 281 "to have been one of the greatest calamities . . ."—Osterbrock (1985), p. 88.
p. 282 "For the English, mine does not exist"—Gascoigne (1996), p. 108.
p. 282 "After some five years' constant experience . . ."—Browning (1870), p. 31.
p. 283 "reflector is so seriously influenced . . ."—Turner (1912), p. 25.
p. 283 "No combination of lenses . . ."—Hale (1897), p. 123.
p. 284 "It would be interesting to think . . ."—Ritchey (1901), p. 233.

Chapter 25. Threads to a Web

p. 285 "The stars looked like jewels on black velvet . . ."—Hale (1912), p. 198.
p. 286 "discover the exceptional man . . ."—Wright (1966), p. 159.
p. 287 "I am a born adventurer . . ."—Christianson (1995), p. 170.
p. 287 "The prospects of a bohemian year . . ."—Walter S. Adams to George E. Hale, February 12, 1904. Wright (1966), p. 182.
p. 288 "a struggle for existence . . ."—Wright (1966), p. 183.
p. 288 "The gods bring threads . . ."—Seares (1939), p. 266.
p. 289 "Noble instrument"—Wright (1966), p. 201.
p. 292 "a sense of great events . . ."—Wright (1966), p. 205.
p. 294 "Congestion of my head . . ."—Christianson (1995, p. 169.
p. 294 "He immediately began to make plans . . ."— Wright (1966), p. 265.
p. 295 "neurasthenic quagmire"—Wright (1966), p. 275.
p. 296 "How the devil should I know"—Adams (1947), p. 302.
p. 296 "Starlight is falling on every square mile . . ."—Hale (1928a), p. 640.
p. 296 "An article of mine on large telescopes . . ."—George E. Hale to Robert G. Aitken, June 26, 1928. Wright (1966), p. 390.
p. 297 "steep and beset with difficulties . . ."—Wright (1966), p. 284.
p. 298 "Regret cannot accept your invitation . . ."—Mayall (1970), p. 180.

Chapter 26. Size Matters

p. 299 "Once one has exhausted all possibilities . . ."—Gough, D.O. "Impact of Observations on Prejudice and Input Physics." Weiss, W. W., and Baglin, A., eds. *Inside the Stars. Astronomical Society of the Pacific Conference Series* 40 (1993), p. 775.
p. 299 "His tall, vigorous figure, pipe in mouth . . ."—Humason (1954), p. 291.
p. 300 "I sometimes feel that there . . ."—Christianson (1995), p. 67.
p. 302 "[I]t does not add appreciably . . ."—Edwin Hubble to Edwin Frost, May 1, 1917. Christianson (1995), p. 101.
p. 302 "These questions await their answers . . ."—Hubble (1920), p. 69.
p. 302 "I barely got under fire . . ."—Edwin Hubble to Edwin Frost, August 14, 1919. Christianson (1995), p. 109.
p. 303 "By these observations . . ."—Herschel (1817), p. 326.
p. 304 "drawing a map of New York City . . ."—Struve and Zebergs (1962), p. 408.
p. 304 "The question of whether they are to be regarded . . ."—Fernie (1970), p. 1214.

p. 306 "To the measurer of the sidereal . . ."—Shapley (1930), p. 155.
p. 307 "It is probable that the further accumulation . . ."—Shapley (1919), p. 312.
p. 307 "From the new point of view . . ."—Shapley (1919), p. 311.
p. 307 "Between us we have put a crimp . . ."—Harlow Shapley to Adriaan van Maanen, June 8, 1921. Bartusiak (2009), p. 164.
p. 309 "spirals live close to the right . . ."—Campbell (1917), p. 532.

Chapter 27. A Night to Remember

p. 311 "What are galaxies? . . ." —Sandage (1961), p.1.
p. 314 "On this plate (H335H), three stars were found . . ."—Berendzen, *et al* (1984), p. 135.
p. 315 "seems to rule undisturbed . . ."—Hubble (1925b), p. 432.
p. 315 "Here is the letter that . . ."—Christianson (1995), p. 159.
p. 316 "undoubtedly among the most notable . . ."—Berendzen and Hoskin (1971), p. 5.
p. 316 "undoubtedly among the most notable scientific . . ."—"Finds Spiral Nebulae . . ." (1924), p. 6.
p. 316 "Well, he is an ass . . ."—Gingerich (1987), p. 126.
p. 317 "the straws are all pointing in one direction . . ."—Edwin Hubble to Harlow Shapley, August 25, 1924. Christianson (1995), p. 159.
p. 318 "with reserve"—van Maanen (1935), p. 337.
p. 318 "The work that Hubble did on galaxies . . ."—Shapley (1969), p. 57.

Chapter 28. Oculis Subjecta Fidelibus

p. 320 "It is . . . an enormous advantage . . ."—Longair (2009), p. 242.
p. 323 "the time that has passed since Creation"—Belenkiy (2012), p. 40.
p. 324 "the complications of the theory . . ."—Smith (1979), p. 140.
p. 326 "realm of positive knowledge"—Hubble (1935), p. 4.
p. 327 "great pioneer work"—Hubble and Humason (1931), pp. 57–58.
p. 327 "If all this does not carry conviction . . ."—Herschel, John F. W. "Address Delivered at the Annual General Meeting of the Royal Astronomical Society, February 12, 1842 on Presenting the Honorary Medal to M. Bessel." *Memoirs of the Royal Astronomical Society*, 12 (1842), pp. 442–454. *Segnius irritant animos demissa per aurem, / Quam quæ sunt oculis subjecta fidelibus:* "What we learn merely through the ear makes less impression upon our minds than what is presented to the trustworthy eye."—Horace, *Ars poetica*.
p. 329 "history of astronomy is . . ."— Hubble (1958), p. 21.
p. 330 "We are, by definition, in the very center . . ."—Hubble (1958), pp. 201–202.

Epilogue

p. 333 "Other desires perish in the gratification . . ."—Housman, A. E. *Introductory Lecture, Delivered Before the Faculties of Arts and Laws and of Science in University College, London, October 3, 1892.* London: University Press, 1933.

Time Line

1839 François Arago announces the daguerreotype photographic process. William Henry Fox Talbot announces his calotype photographic process.

1840 John W. Draper takes the first daguerreotype of the Moon that shows surface details.

1843 John W. Draper takes the first daguerreotype of the Sun's spectrum.

1845 A-H-L Fizeau and Léon Foucault take the first daguerreotype of the Sun that shows sunspots. The Third Earl of Rosse completes the Leviathan reflector telescope in Ireland. William Lassell builds a twenty-four-inch, equatorially mounted reflector telescope outside Liverpool, which he reassembles on the island of Malta in 1852.

1847 William and George Bond and John Whipple commence their experiments in celestial photography at Harvard.

1850 First daguerreotype of a star (Vega) is obtained at Harvard.

1851 Harvard's daguerreotype of the Moon is displayed to great acclaim at the Crystal Palace exhibition in London. First daguerreotype of a planet (Jupiter) is obtained at Harvard. First daguerreotype of a total solar eclipse showing the Sun's corona is taken by Berkowski at Königsberg. Frederick Scott Archer announces his wet-collodion photographic process.

1852 Warren De La Rue produces the first wet-collodion photographs of the Moon.

1856 Justus von Liebig develops a chemical method to deposit a coating of silver onto glass. Léon Foucault produces the first silvered-glass reflector telescopes.

1857 George Bond obtains a wet-collodion photograph of the double star Mizar and Alcor.

1858 William Usherwood photographs Comet Donati using a tripod-mounted camera. Warren De La Rue produces stereoscopic pictures of the Moon; constructs a photoheliograph to record daily images of solar surface activity.

1859 Robert Bunsen and Gustav Kirchhoff announce their findings on the origin and significance of the solar spectral lines, opening up the Sun and stars to chemical analysis. William Lassell builds a forty-eight-inch, equatorially mounted reflector telescope, which he reassembles on the island of Malta in 1861.

1863 William Huggins and William Allen Miller publish their initial visual study of stellar spectra.

1864 Henry Draper publishes his manual describing the construction of a silvered-glass reflector telescope and its use in celestial photography. Based on a visual spectroscopic study, William Huggins concludes that some nebulae consist primarily of distributed, luminous gas, not discrete stars.

1865 Lewis Rutherfurd commences his long-term project to obtain images of star clusters with a refractor telescope optimized for photography.

1869 The Great Melbourne Telescope, the last of the large speculum-metal reflectors, is put into service.

1871 Richard Leach Maddox announces the dry-plate photographic process.

1872 Henry Draper obtains the first photographic spectrum of a star (Vega) that depicts spectral lines.

1880 Henry Draper photographs the Orion Nebula.

1882 David Gill's photograph of the Great Comet of 1882 captures an unexpected backdrop of stars. Henry Draper photographs the spectrum of the Orion Nebula.

1883 Andrew Common's photograph of the Orion Nebula depicts features not seen visually through a telescope.

1887 The first-ever international conference of astronomers convenes in Paris to discuss the *Carte du Ciel*, a comprehensive photographic map of the night sky.

1888 Isaac Roberts displays his revelatory three-hour exposure of the Andromeda Nebula before the Royal Astronomical Society. Henry Rowland publishes a map of the Sun's spectrum depicting some twenty thousand absorption lines.

1889 Edward C. Pickering and Hermann Vogel independently discover spectroscopic binary stars, based solely on photographs of the periodic Doppler shift of their spectral lines. Edward E. Barnard commences his wide-field photographic survey of the Milky Way. George E. Hale invents the spectroheliograph.

1890 First volume of Harvard's Henry Draper catalog of stellar spectra is published.

1891 Maximilian Wolf makes the first photographic discovery of an asteroid.

1892 Edward E. Barnard makes the first photographic discovery of a comet.

1895 The *Astrophysical Journal* begins publication.

1897 Completion of the forty-inch refractor at Yerkes Observatory.

1899 James E. Keeler obtains photographs with Lick Observatory's Crossley reflector that reveal the ubiquity of spiral nebulae. First meeting of the Astronomical and Astrophysical Society of America.

1904 Mount Wilson Observatory founded by George E. Hale.

1908 Mount Wilson Observatory's sixty-inch reflector telescope is put into service.

1914 Vesto M. Slipher presents his spectroscopic study revealing that the radial velocities of galaxies are unexpectedly large.

1917 Mount Wilson Observatory's one-hundred-inch reflector telescope is put into service. Albert Einstein publishes his General Theory of Relativity.

1919 Harlow Shapley proposes a tenfold increase in the diameter of our galaxy based on observations of its globular star clusters.

1920 The Great Debate about the dimensions of the galaxy and the scale of the universe is held in Washington, DC.

1923 From observations of its Cepheid variable stars, Edwin Hubble concludes that the Andromeda Nebula lies well outside the Milky Way Galaxy.

1929 Edwin Hubble and Milton Humason announce their preliminary finding that the radial velocities of galaxies increase linearly with distance, a quantitative relationship that would become known as Hubble's law.

1931 Hubble and Humason present additional observations that confirm the extragalactic velocity–distance relationship, which many astronomers accept as evidence that the universe is expanding.

Glossary of Names

Adams, Walter S. (1876–1956) — Staff astronomer and, later, director of the Mount Wilson Observatory in California. George Ellery Hale's right-hand man in the establishment and operation of the facility.

Airy, George Biddell (1801–1892) — England's seventh Astronomer Royal, Airy was a prime mover in the improvement of positional astronomy and an influential figure in other branches of astronomy.

Arago, François (1786–1853) — Director of the Paris Observatory and dean of the French scientific establishment in the mid-1800s. Publicly announced the daguerreotype photographic process in 1839.

Archer, Frederick Scott (1813–1857) — Announced his wet-collodion photographic process in 1851.

Barnard, Edward Emerson (1857–1923) — Pioneering wide-field astrophotographer at Lick and Yerkes Observatories. Noted for his panoramic images of the Milky Way.

Bond, George Phillips (1825–1865) — Son of William Cranch Bond and second director of the Harvard College Observatory. Early advocate for the inclusion of photography in astronomical research.

Bond, William Cranch (1789–1859) — First director of the Harvard College Observatory. An expert in practical astronomy related to navigation, Bond participated in early attempts to photograph the Moon through a telescope.

Brashear, John (1840–1920) — Prominent American maker of astronomical instruments during the late-nineteenth and early twentieth centuries.

Bunsen, Robert Wilhelm (1811–1899) — Noted professor of chemistry at the University of Heidelberg. With physicist Gustav Kirchhoff, conducted fundamental research regarding the use of spectral lines to establish the chemical composition of the Sun's atmosphere.

Campbell, William Wallace (1862–1938) — Third director of the Lick Observatory and organizer of its survey of stellar radial velocities.

Clark, Alvan (1804–1887) — Together with his sons, Alvan Graham Clark (1832–1897) and George Bassett Clark (1827–1891), foremost makers of refractor telescopes in the United States during the latter half of the nineteenth century.

Clerke, Agnes (1842–1907) — Noted late-nineteenth-century historian of astronomy.

Common, Andrew Ainslie (1841–1903) — English amateur astronomer and proponent of celestial photography with large-aperture reflector telescopes.

Daguerre, Louis (1787–1851) — One of the inventors of photography. His daguerreotype process was announced in 1839.

De La Rue, Warren (1815–1889) — English amateur astronomer who during the 1850s demonstrated the value of the reflecting telescope in lunar photography and subsequently developed the photoheliograph to obtain daily images of the Sun's surface.

Doppler, Christian (1803–1853) — Austrian mathematician who theorized in 1842 that the perceived frequency of a wave is altered by the relative motion between the source and the observer.

Draper, Henry (1837–1882) — American amateur astronomer who captured the first photograph of a nebula and the first spectrum of a star other than the Sun. Assisted by his wife, Anna Palmer Draper.

Draper, John William (1811–1882) — New York chemist who took the first successful photograph of the Moon and of the Sun's spectrum. Father of Henry Draper.

Fitz, Henry (1808–1863) — Mid-nineteenth-century New York telescope maker and close colleague of astrophotographer Lewis Morris Rutherfurd.

Fizeau, Armand-Hippolyte-Louis (1819–1896) — French pioneer of celestial photography. With Léon Foucault, took the first successful daguerreotype of the Sun.

Fleming, Williamina (1857–1911) — Late-nineteenth-century Harvard astronomer. With E. C. Pickering, developed an early spectral classification system for stars, subsequently revised and greatly expanded by Annie J. Cannon.

Foucault, Léon (1819–1868) — French pioneer of celestial photography and silvered-glass reflector telescopes. With his countryman Fizeau, took the first successful daguerreotype of the Sun.

Fraunhofer, Joseph (1787–1826) — Early nineteenth-century German telescope maker and spectroscopist who cataloged the dark lines in the Sun's spectrum.

Gill, David (1843–1914) — In 1882, as director of the Royal Cape Observatory in South Africa, recorded images of background stars on a photograph of a comet. Leading figure in the international *Carte du Ciel* sky-mapping project.

Hale, George Ellery (1868–1938) — American solar astronomer and key figure in the establishment of Mount Wilson and Palomar Observatories.

Herschel, John (1792–1871) — Son of astronomer William Herschel, continued his father's deep-sky surveys in the Southern Hemisphere and became a leading figure in the advancement of science in England during the first half of the nineteenth century.

Herschel, William (1738–1822) — Discoverer of the planet Uranus and the most famous large-telescope builder and celestial observer of the eighteenth century.

Holden, Edward S. (1846–1914) — First director of the Lick Observatory, Mount Hamilton, California.

Hubble, Edwin (1889–1953) — Noted extragalactic astronomer at Mount Wilson Observatory, and later at Palomar, whose observations during the 1920s and 1930s led to the conclusions that spiral nebulae are galaxies and that the universe is expanding.

Huggins, William (1824–1910) — Pioneer in astronomical spectroscopy, who proved that the light of many nebulae arises from diffuse, glowing gas, not discrete stars. In later research, collaborated with his wife Margaret Lindsay Huggins.

Humason, Milton (1891–1972) — Staff astronomer at Mount Wilson Observatory, and later at Palomar. Assisted Edwin Hubble in obtaining spectra of faint galaxies that led to the distance–velocity relationship known as Hubble's law.

Janssen, Pierre Jules (1824–1907) — French solar observer who, coincident with English astronomer J. Norman Lockyer, developed the means to observe outbursts around the Sun's limb without an eclipse.

Kapteyn, Jacobus C. (1851–1922) — Dutch astronomer who used star counts in selected areas of the sky to develop his eponymous small-galaxy model of the Milky Way.

Keeler, James E. (1857–1900) — Second director of California's Lick Observatory, who used the thirty-six-inch Crossley reflector to obtain photographs that depicted the ubiquity of spiral nebulae.

Kirchhoff, Gustav (1824–1887) — Noted professor of physics at Heidelberg and later Berlin. With chemist Robert Bunsen, conducted fundamental research regarding the use of spectral lines to establish the chemical composition of the Sun's atmosphere.

Lassell, William (1799–1880) — English builder of large, equatorially mounted reflector telescopes during the mid-1800s, including one of aperture forty-eight inches on the island of Malta.

Leavitt, Henrietta (1868–1921) — Harvard astronomer who discovered the period-luminosity relation for Cepheid variable stars, subsequently refined by Harlow Shapley and used to gauge the size of the galaxy.

Lockyer, J. Norman (1836–1920) — English astronomer who discovered helium in the solar spectrum and who, coincident with French scientist Jules Janssen, developed the means to observe outbursts around the Sun's limb without an eclipse.

Maddox, Richard Leach (1816–1902) — Developed the gelatin dry-plate photographic process in 1871.

Niépce, Joseph Nicéphore (1765–1833) — One of the inventors of photography, took the oldest extant photograph in 1825.

Pickering, Edward C. (1846–1919) — Harvard Observatory director who established large-scale research programs in stellar photometry, photography, and spectroscopy. Discovered spectroscopic binary stars.

Ritchey, George W. (1864–1945) — Innovative designer of large-aperture reflector telescopes, first at Yerkes Observatory, later at Mount Wilson.

Roberts, Isaac (1829–1904) — British amateur astronomer whose exquisite 1890s photographs of nebulae and star clusters received wide acclaim.

Rosse, Third Earl of (William Parsons; 1800–1867) — Built the world's largest telescope, a six-foot speculum-metal reflector, in 1845, with which he discovered spiral nebulae.

Russell, Henry Norris (1877–1957) — Prominent twentieth-century stellar astrophysicist at Princeton University.

Rutherfurd, Lewis Morris (1816–1892) — New York amateur astronomer who conducted groundbreaking experiments in the use of refractor telescopes for celestial photography and spectroscopy.

Secchi, Angelo (1818–1878) — Italian astronomer who created an early classification scheme for stellar spectra.

Shapley, Harlow (1885–1972) — American astronomer at Mount Wilson, and later at Harvard, who developed observational techniques to deduce the size of the galaxy and the solar system's position within it. In the Great Debate of 1920, advocated his large-galaxy model of the Milky Way.

Slipher, Vesto M. (1875–1969) — Astronomer at Lowell Observatory in Flagstaff, Arizona, whose spectroscopic measurements revealed the large radial velocities of galaxies.

Talbot, William Henry Fox (1800–1877) — In 1839, announced his calotype photochemical technique, forerunner of many subsequent photographic processes.

van Maanen, Adriaan (1884–1946) — Dutch-American astronomer who claimed (mistakenly) that spiral nebulae exhibit measurable rotation, hence, lie within or near our galaxy.

Vogel, Hermann Carl (1841–1907) — German astronomer who developed spectroscopic techniques to measure stellar radial velocities to high precision.

Whipple, John Adams (1822–1891) — Boston commercial photographer who initiated early experiments in celestial photography at the Harvard College Observatory.

Wolf, Maximilian (1863–1932) — German astronomer who conducted wide-field photography of the Milky Way, including the first photographic discovery of an asteroid.

Bibliography

Abbreviations:

ApJ	*Astrophysical Journal*
BMNAS	*Biographical Memoirs, National Academy of Sciences*
JHA	*Journal for the History of Astronomy*
MNRAS	*Monthly Notices of the Royal Astronomical Society*
PA	*Popular Astronomy*
PASP	*Publications of the Astronomical Society of the Pacific*
PRS	*Proceedings of the Royal Society of London*
PTRS	*Philosophical Transactions of the Royal Society*
S&T	*Sky and Telescope*

Abetti, Giogio. "Father Angelo Secchi: A Noble Pioneer in Astrophysics." *Astronomical Society of the Pacific Leaflets* 8, No. 368 (1960): 135–142.

———. "Recollections of George Ellery Hale." *Astronomical Society of the Pacific Leaflets* 8, No. 387 (1961): 287–294.

Abrahams, Peter. "Henry Fitz, American Telescope Maker." *Journal of the Antique Telescope Society* 6 (1994, revised 1995, 2000). [www.europa.com/~telscope/fitz.txt]

Adams, Walter S. "George Ellery Hale, 1868–1938." *ApJ* 87 (1938): 369–388.

———. "Memoir of George Ellery Hale, 1868–1938." *BMNAS* 21 (1939): 181–241.

———. "Early Days at Mount Wilson." *PASP* 59 (1947): 213–231; 285–304.

———. "Some Reminiscences of the Yerkes Observatory, 1898–1904." *Science, New Series* 106 (1947): 196–200.

———. "Obituary: Dr. Edwin P. Hubble." *Observatory* 74 (1954): 32–35.

Airy, G.B. "Report on the Progress of Astronomy During the Present Century." *Report of the First and Second Meetings of the British Association for the Advancement of Science.* London: John Murray, 1833.

Albers, Henry, ed. *Maria Mitchell: A Life in Journals and Letters.* Clinton Corners, NY: College Avenue Press, 2001.

"The American Photographical Society, Thirty-Fourth Meeting." *American Journal of Photography* 4 (1862): 378–381.

"The American Photographical Society, Sixty-First Meeting." *American Journal of Photography* 7 (1865): 539–541.

"An Account of the Total Eclipse of the Sun, June 16, 1806." *The Balance, and Columbian Repository* 5 (August 5, 1806): 244–245.

Arago, Francois. “Fixation des images qui se forment au foyer, d’une chambre obscure.” *Comptes rendus de l’Académie des sciences* 8 (1839**a**): 4–7.

———. “Le Daguerréotype.” *Comptes rendus de l’Académie des sciences* 9 (1839**b**): 250–267.

Archer, Frederick Scott. “On the Use of Collodion in Photography. *The Chemist, New Series* 2 (March 1851): 257–258.

———. *The Collodion Process on Glass*. London: Printed for the Author, 1854.

Ashbrook, Joseph. “The Great September Comet of 1882.” *S&T* 22 (December 1961): 331–332.

———. “The Star Photographs of Lewis M. Rutherfurd.” *S&T* 57 (February 1979): 141–156.

———. *The Astronomical Scrapbook: Skywatchers, Pioneers, and Seekers in Astronomy*. Cambridge, MA: Sky Publishing Corp., 1984.

Bailey, Solon I. *The History and Work of Harvard Observatory, 1839 to 1927*. New York: McGraw-Hill, 1931.

Baker, Daniel W. *History of the Harvard College Observatory during the period 1840–1890*. Cambridge, MA, 1890.

Ball, Robert S. “Isaac Roberts (1829–1904).” *PRS* 75 (1904): 356–363.

Banta, Melissa. *A Curious and Ingenious Art: Reflections on Daguerreotypes at Harvard*. Iowa City, IA: University of Iowa Press, 2000.

Barger, M. Susan, and White, William B. *The Daguerreotype: Nineteenth-Century Technology and Modern Science*. Washington, DC: Smithsonian Institution Press, 1991.

Barker, George F. “Henry Draper.” *American Journal of Science* 25 (1883): 89–96.

———. “Memoir of John William Draper, 1811–1882.” *BMNAS* 2 (1886): 349–388.

———. “Memoir of Henry Draper, 1837–1882.” *BMNAS* 3 (1888): 81–139.

Barnard, Edward E. “On the Extended Nebulosity about 15 Monocerotis.” *MNRAS* 56 (1895): 63–66.

———. “Development of Photography in Astronomy.” *PA* 6 (1898): 425–455.

———. “Note on the Exterior Nebulosities of the Pleiades.” *MNRAS* 59 (1899): 155.

———. “Diffused Nebulosities in the Heavens.” *ApJ* 17 (1903): 77–80.

Bates, Ralph S. “Henry Fitz – Early American Telescope Maker.” *S&T* 1 (November 1941): 18.

Becker, Barbara. “Celestial Spectroscopy: Making Reality Fit the Myth.” *Science, New Series* 301 (September 5, 2003): 1332–1333.

———. “From Dilettante to Serious Amateur: William Huggins’ Move into the Inner Circle.” *Journal of Astronomical History and Heritage* 13 (2010): 112–119.

———. *Unravelling Starlight: William and Margaret Huggins and the Rise of the New Astronomy*. Cambridge: Cambridge University Press, 2011.

Belenkiy, Ari. “Alexander Friedmann and the Origins of Modern Cosmology.” *Physics Today* 65 (2012): 38–43.

Bell, Trudy. "Lick Observatory." *The General History of Astronomy*, Vol. 4: *Astrophysics and Twentieth-Century Astronomy to 1950*, Part A. Gingerich, Owen, ed. Cambridge: Cambridge University Press, 1984.

———. "The Great Telescope Race." *S&T* 121 (June 2011): 28–33.

Bennett, J. A. "The Spectroscope's First Decade." *Scientific Instrument Society Bulletin* No. 4 (1984): 3–6.

———. "The Era of Newton, Herschel and Lord Rosse." *Experimental Astronomy* 25 (2009): 33–42.

Bennett, J. A., and Hoskin, Michael. "The Rosse Papers and Instruments." *JHA* 12 (1981): 216–229.

Berendzen, Richard. "Origins of the American Astronomical Society." *Physics Today* 27 (December 1974): 32–39.

Berendzen, Richard, Hart, Richard, and Seeley, Daniel. *Man Discovers the Galaxies*. New York: Columbia University Press, 1984.

Berendzen, Richard, and Hoskin, Michael. "Hubble's Announcement of Cepheids in Spiral Nebulae." *Astronomical Society of the Pacific Leaflets* 10 (1971): 425–440.

Bessel, Friedrich W., and Schumacher, Heinrich C. "Über den gegenwärtigen Standpunkt der Astronomie." *Populäre Vorlesungen über wissenschaftliche Gegenstände*. Hamburg: Perthes-Besser & Mauke, 1848, 1–33.

Black, Robert L. "The Cincinnati Telescope." *PA* 52 (1944): 70–79.

Boltzmann, Ludwig. **"**Gustav Robert Kirchhoff.**"** *Keynote Lecture at the Celebration of the 301st Anniversary of the Founding of the Karl-Franzens-University of Graz, November 16, 1887*. Leipzig: Barth, 1888. [www.ub.uni-heidelberg.de/helios/fachinfo/www/math/htmg/Boltzmann/kirchhoff.pdf]

Bond, Elizabeth. "Sketches of the Bond Family." *Papers of the Bond Family, 1845–1872*. Cambridge, MA: Harvard University Archives, c. 1938.

Bond, George P. "Letter [on Stellar Photography] from Mr. Bond, Director of the Observatory, Cambridge, U. S., to the Secretary." *MNRAS* 17 (1857**a**): 230–232.

———. "[On Stellar Photography.] May 12th, 1857, Monthly Meeting." *Proceedings of the American Academy of Arts and Sciences* 3 (1857**b**): 386–389.

———. "Stellar Photography." *Astronomische Nachrichten* 47 (1857**c**): 1–6; 48 (1858**a**): 1–14; 49 (1858**b**): 81–100.

———. "Stellar Photography." *American Almanac* (1858**c**): 81–82.

———. "Celestial Photography." *American Almanac* (1859): 77–84.

———. "On the Results of Photometric Experiments Upon the Light of the Moon and of the Planet Jupiter Made at the Observatory of Harvard College." *Memoirs of the American Academy of Arts and Sciences, New Series* 8 (1861): 221–286.

———. "The Future of Stellar Photography (From a letter written in 1857 to William Mitchell)." *PASP* 2 (1890): 300–302.

Bond, William C. *History and Description of the Astronomical Observatory of Harvard College*. Cambridge, MA: Metcalf & Co., 1856.

"Boston Daguerreotypists." *The Daguerreian Journal* 2 (1851): 114–115.

Boyd, Lyle G. "Mrs. Henry Draper and the Harvard College Observatory." *Harvard Library Bulletin* 17 (1969): 70–97.

Bracher, Katherine. *The Stars for All: A Centennial History of the Astronomical Society of the Pacific*. San Francisco: Astronomical Society of the Pacific, 1989. [https://astrosociety.org/about/history.html]

Brashear, John A. *A Man Who Loved the Stars: The Autobiography of John A. Brashear*. Pittsburgh, PA: University of Pittsburgh Press, 1988.

Brewster, David. "Memoir of the Life of M. Le Chevalier Fraunhofer, the Celebrated Improver of the Achromatic Telescope, and Member of the Academy of Sciences at Munich." *Edinburgh Journal of Science* 7 (1827): 1–11.

Browning, John. *A Plea for Reflectors, 4th ed.* London, 1870.

Bruce, Robert V. *The Launching of Modern American Science 1846–1876*. New York: Alfred A. Knopf, 1987.

Brück, H. A. "P. Angelo Secchi, S. J. 1818–1878." *Spectral Classification of the Future*, Proceedings of the International Astronomical Union Colloquium 47, held in Vatican City, July 11–15, 1978. McCarthy, M. F., *et al*, eds. Vatican Observatory, 1979.

Brush, Stephen G. "Looking Up: The Rise of Astronomy in America." *American Studies* 20 (1980): 41–67.

———. *The History of Modern Science: A Guide to the Second Scientific Revolution, 1800–1950*. Ames, IA: Iowa State University Press, 1988.

Burns, D. Thorburn. "Towards a Definitive History of Optical Spectroscopy, Part I. Simple Prismatic Spectra: Newton and his Predecessors." *Journal of Analytical Atomic Spectrometry* 2 (1987): 343–347; "Towards a Definitive History of Optical Spectroscopy, Part II. Introduction of Slits and Collimator Lens: Spectroscopes Available Before and Just After Kirchoff and Bunsen's Studies." *Journal of Analytical Atomic Spectrometry* 3 (1988): 285–291.

Buys-Ballot, C. H. D. "Akustische Versuche auf der Niederländischen Eisenbahn, nebst gelegentlichen Bemerkungen zur Theorie des Hrn. Prof. Doppler. " *Annalen der Physik und Chemie* 66 (1845): 321–351.

Callandreau, Octave. "Prosper Henry." *PA* 11 (1903): 558–560.

Calver, George. "On the Working of the Speculum for Mr. Common's 37-inch Silver-on-Glass Reflector." *MNRAS* 40 (1879):17–20.

Campbell, William W. "Discovery of Asteroids by Photography." *PASP* 4 (1892): 264–265.

———. "James Edward Keeler." *ApJ* 12 (1900): 239–253.

———. "Sir William Huggins, K. C. B., O. M." *PASP* 22 (1910): 149–163.

———. "Biographical Memoir of Edward Singleton Holden, 1846–1914." *BMNAS* 8 (1916): 347–372.

———. "The Nebulae: Address of the Retiring President of the American Association for the Advancement of Science." *Science, New Series* 45 (May 25, 1917): 513–548.

———. "The Daily Influence of Astronomy." *PA* 29 (1921): 456–468.

———. "Sale of the Chile Station of the Lick Observatory." *PASP* 40 (1928): 249–252.

Cannon, Annie J. "Mrs. Henry Draper." *Science, New Series* 41 (March 12, 1915): 380–382.

Carnegie, Andrew. "The Best Fields of Philanthropy." *North American Review* 149 (1889): 682–698.

Chambers, George F. *Handbook of Descriptive Astronomy*, 3rd ed. Oxford: Clarendon Press, 1877.

Chant, Clarence A. "Work at the Lick Observatory and Improvements in its Equipment." *Journal of the Royal Astronomical Society of Canada* 1 (1907): 246–263.

Chapman, Allan. "William Lassell (1799–1880): Practitioner, Patron and 'Grand Amateur' of Victorian Astronomy." *Vistas in Astronomy* 32 (1988): 341–370.

———. *The Victorian Amateur Astronomer: Independent Astronomical Research in Britain 1820–1920*. New York: John Wiley & Sons, 1998.

Chapman, D. C. "Astronomical Photography." *British Journal of Photography* 22 (1875): 630–631.

Christianson, Gale E. *Edwin Hubble: Mariner of the Universe*. New York: Farrar, Straus and Giroux, 1995.

Christianson, Gale E. "Edwin Hubble: Reluctant Cosmologist." *Historical Development of Modern Cosmology*. Martinez, V. J., *et al*, eds. *Astronomical Society of the Pacific Conference Series* 252 (2001): 145–156.

Clerke, Agnes M. "Sidereal Photography." *Edinburgh Review* 167 (1888): 23–46.

———. *A Popular History of Astronomy During the Nineteenth Century*, 3rd ed. London: Adam & Charles Black, 1893; 4th ed., 1902.

Common, Andrew A. "Description of a Three-Feet Telescope." *Observatory* 3 (1879**a**): 167–169.

———. "Note on Large Telescopes, With Suggestion for Mounting Reflectors." *MNRAS* 39 (1879**b**): 382–386.

———. "Note on a Photograph of the Great Nebula in Orion and Some New Stars Near Theta Orionis." *MNRAS* 43 (1883): 255–257.

———. "Note on a Method of Giving Long Exposures in Astronomical Photographs." *MNRAS* 45 (1884**a**): 25–27.

———. "Note on Stellar Photography." *MNRAS* 45 (1884**b**): 22–25.

———. "Telescopes for Astronomical Photography." *Nature* 31 (1884–85): 38–40; 270–271.

———. "Photographs of Nebulae." *Observatory* 11 (1888): 390–394.

———. "Great Telescopes." *Observatory* 12 (1889**a**): 138–140.

———. "Note on an Apparatus for Correcting the Driving of the Motor Clocks of Large Equatorials for Long Photographic Exposures." *MNRAS* 49 (1889**b**): 297–300.

———. "The Photographic Chart of the Heavens." *Observatory* 13 (1890): 174–176.

———. "On the Best Form of Mounting for a Large Reflector." *MNRAS* 53 (1892**a**): 19–22.

———. "Silvering Glass Mirrors." *Observatory* 15 (1892**b**): 369–374.

———. "Two Large Telescopes." *Observatory* 15 (1892**c**): 389–392; 437–441.

Crew, Henry. "Robert Wilhelm Bunsen." *ApJ* 10 (1899): 301–305.

Crookes, William. "Early Researches on the Spectra of Artificial Light from Different Sources." *Chemical News* 3 (1861): 184–185; 303–304.

Curtius, Theodor. "Robert Wilhelm Bunsen." *Great Chemists*, Farber, Eduard, ed. New York: Interscience Publishers, 1961.

Daguerre, Louis. *Daguerréotype* (broadside). Paris: Pollet, Soupe and Guillois, 1838. [www.daguerreotypearchive.org]

———. *An Historical & Descriptive Account of the Various Processes of the Daguerréotype & the Diorama.* New York: Winter House, 1971.

Darius, Jon. "A Double Centenary in Celestial Photography." *Observatory* 103 (1983): 46–48.

"Death of Henry Fitz, the Telescope Maker." *Scientific American* 9 (November 14, 1863): 311.

De La Rue, Warren. "On the Figuring of Specula." *MNRAS* 13 (1852): 44–51.

———. "Mr. De La Rue, on Lunar Photography." *MNRAS* 18 (1857): 16–18.

———. "On the Silvering of Glass Specula." *MNRAS* 19 (1859): 171–172.

———. "Celestial Photography." *American Journal of Photography, New Series* 3 (1859–60): 180–186.

———. "The Present State of Celestial Photography in England." *Report of the 29th Meeting of the British Association for the Advancement of Science Held at Aberdeen in September 1859.* London: John Murray, 1860.

———. "Proceedings of the Chemical Society, June 20, 1861." *Chemical News* 4 (1861): 130–133.

———. "Address Delivered by the President, Warren De La Rue, Esq., on Presenting the Gold Medal of the Society to Professor G. P. Bond." *MNRAS* 25 (1865**a**): 125–137.

———. "On a Photo-Engraving of a Lunar Photograph." *MNRAS* 25 (1865**b**): 171.

———. "The Approaching Transit of Venus." *Astronomical Register* 12 (1874): 40–41.

Dennison, E. B. "The Approaching Transit of Venus." *Astronomical Register* 12 (1874): 41–43.

de Vaucouleurs, Gerard. "Discovering M31's Spiral Shape." *S&T* 74 (December 1987): 595–598.

Devons, Samuel. "Lewis Morris Rutherfurd, 1816–1892." *Applied Optics* 15 (1976): 1731–1740.

DeVorkin, David H. "Community and Spectral Classification in Astrophysics: The Acceptance of E. C. Pickering's System in 1910." *Isis*, 72 (1981): 29–49.

———. ed. *The American Astronomical Society's First Century.* Washington, DC: American Institute of Physics, 1999.

———. "In the Grip of the Big Telescope Age." *Experimental Astronomy* 25 (2009): 63–77.

DiCicco, Dennis. "Astrophotography Then and Now." *S&T* 76 (November 1988): 463–467.

———. "Astrophotography's Rise and Fall." *S&T* 78 (August 1989): 124.

"Direct Photography of the Heavens." *Astronomical Register* 24 (1886): 245–248.

Donati, G. B. "On the Striae of Stellar Spectra." *MNRAS* 23 (1863): 100–107.

Draper, Henry. "On a Reflecting Telescope for Celestial Photography, Erecting at Hastings, near New York." *Report of the Thirtieth Meeting of the British Association for the Advancement of Science: Held at Oxford in June and July 1860*. London: John Murray, 1861, 63–64.

———. *On the Construction of a Silvered Glass Telescope, Fifteen and a Half Inches in Aperture and Its Uses in Celestial Photography.* Washington, DC: Smithsonian Institution, 1864**a**.

———. "On the Photographic Use of a Silvered-Glass Reflecting Telescope." *Philosophical Magazine, 4th Series* 28 (1864**b**): 249–255.

———. "On Diffraction-Spectrum Photography." *Philosophical Magazine, 4th Series* 46 (1873): 419–25.

———. "Photographs of the Spectra of Venus and Alpha Lyrae." *American Journal of Science and Arts* 13 (1877): 95.

———. "On Photographing the Spectra of the Stars and Planets." *American Journal of Science and Arts* 18 (1879): 419–425.

———. "Photographs of the Nebula in Orion." *American Journal of Science* 20 (1880): 433.

———. "On Photographs of the Nebula in Orion, and of its Spectrum." *MNRAS* 42 (1882**a**): 367–368.

———. "On Photographs of the Spectrum of the Nebula in Orion." *American Journal of Science* 23 (1882**b**): 339–341.

———. "Researches Upon the Photography of Planetary and Stellar Spectra." *Proceedings of the National Academy of Arts and Sciences* 19 (1884): 231–261; *Researches on Astronomical Spectrum Photography*. Cambridge: John Wilson and Son, University Press, 1884.

Draper, John W. "Portraits in Daguerreotype." *Philosophical Magazine, 3rd Series* 16 (1840): 535.

———. "On the Process of Daguerreotype, and its Application to Taking Portraits from the Life." *Philosophical Magazine, 3rd Series* 17 (1840): 217–225.

———. *Scientific Memoirs: Being Experimental Contributions to a Knowledge of Radiant Energy*. New York: Harper and Bros., 1878.

Dreyer, J. L. E., and Turner, H. H., eds. *History of the Royal Astronomical Society, 1820–1920*. London: Royal Astronomical Society, 1923.

"Dr. Henry Draper's Photographs of the Moon." *Harper's Weekly* (March 19, 1864): 186–187.

Dyson, F. W. "Andrew Ainslie Common." *MNRAS* 64 (1904): 274–278.

"Earl Rosse." *Harper's Weekly* (December 12, 1867): 796.

Eddington, Arthur S. "David Gill." *MNRAS* 75 (1915): 236–247.

Evans, David. "Astronomical Institutions in the Southern Hemisphere, 1850–1950." *The General History of Astronomy*, Vol. 4: *Astrophysics and Twentieth-Century Astronomy to 1950*, Part A. Gingerich, Owen, ed. Cambridge: Cambridge University Press, 1984.

"Exhibition of Bond's Photograph of Zeta and g Ursae Majoris, and of the Transit of Alpha Lyrae." *MNRAS* 18 (1858): 18–21.

Eyre, John W. H. "Richard Leach Maddox." *Transactions of the American Microscopical Society* 25 (1904): 155–159.

Fernie, J. Donald. "The Period-Luminosity Relation: A Historical Review." *PASP* 81 (1969): 707–731.

———."The Historical Quest for the Nature of the Spiral Nebulae." *PASP* 82 (1970): 1189–1230.

———. "Hubble." *Journal of the Royal Astronomical Society of Canada, National Newsletter* 72 (1978): L3–L5.

"Finds Spiral Nebulae Are Stellar Systems: Dr. Hubbell [sic] Confirms View That They Are 'Island Universes' Similar to Our Own." *New York Times* (November 23, 1924): 6.

Fleming, Donald. *John William Draper and the Religion of Science*. Philadelphia: University of Pennsylvania Press, 1950.

Forbes, George, ed. *David Gill: Man and Astronomer*. London: John Murray, 1916. [openlibrary.org/books/OL7032214M/David_Gill_man_and_astronomer]

Foucault, Léon. "On the Simultaneous Emission and Absorption of Rays of the Same Definite Refrangibility." (Translation of excerpt from original 1849 paper in *L'Institut.*) *Philosophical Magazine, 4th Series* 19 (1860): 193–194.

Fowler, Alfred. "Spectroscopic Astronomy." *Nature* 104 (November 6, 1919): 234–235.

Fraunhofer, Joseph. "Determination of the Refractive and the Dispersive Power of Different Kinds of Glass with Reference to the Perfecting of Achromatic Telescopes." *Denkschriften der königlichen Akademie der Wissenschaften zu München* 5 (1817): 193–226.

———. "On the Refractive and Dispersive Power of Different Species of Glass, in Reference to the Improvement of Achromatic Telescopes, with an Account of the Lines or Streaks which Cross the Spectrum." *Edinburgh Philosophical Journal* 9 (1823): 288–299; 10 (1824): 26–40.

———. "On the Construction of a Large Refractor Telescope Just Completed." *Philosophical Magazine and Journal* 66 (1825): 41–47.

Freeman, Ken. "Slipher and the Nature of the Nebulae." *Origins of the Expanding Universe: 1912–1932*. Way, M. J., and Hunter, D., eds. *Astronomical Society of the Pacific Conference Series* 471 (2013): 63–70.

Frost, Edwin B. "Hermann Carl Vogel." *ApJ* 27 (1908): 1–11.

———. "Biographical Memoir of Charles Augustus Young, 1834–1908." *BMNAS* 7 (1910): 91–114.

———. "Edward Emerson Barnard, 1857–1923." *BMNAS* 21 (1927): 1–23.

———. *An Astronomer's Life*. Boston: Houghton Mifflin Co., 1933.

Gascoigne, S. C. B. "The Great Melbourne Telescope and other 19th-century Reflectors." *Quarterly Journal of the Royal Astronomical Society* 37 (1996): 101–128.

Gernsheim, Helmut and Alison. *L. J. M. Daguerre: The World's First Photographer*. New York: World Publishing Co., 1956.

Gill, David. "Notes on the Great Comet (b) 1882." *MNRAS* 43 (1882**a**): 19–21.

———. "On Photographs of the Great Comet (b) 1882." *MNRAS* 43 (1882**b**): 53–54.

———. "The Applications of Photography in Astronomy." *Observatory* 10 (1887): 267–272, 283–294.

———. "Correspondence to the Editors: The Photographic Chart of the Heavens." *Observatory* 11 (1888): 320–326.

———. "An Astronomer's Work in a Modern Observatory." *Nature* 44 (1891): 603–607.

Gill, David, and Kapteyn, J. C. *The Cape Photographic Durchmusterung. Annals of the Cape Observatory*, Vols. 3–5. London: Darling & Son, 1896.

Gingerich, Owen. "The First Photograph of a Nebula." *S&T* 60 (1980): 364–366.

———."Henry Draper's Scientific Legacy." *Annals of the New York Academy of Sciences* 395 (1982): 308–320.

———. "The Great Comet and the 'Carte.'" *S&T* 64 (1983): 237–239.

———. "The Mysterious Nebulae, 1610–1924." *Journal of the Royal Astronomical Society of Canada* 81 (1987): 113–127.

———. "Shapley, Hubble, and Cosmology." *Evolution of the Universe of Galaxies: Proceedings of the Edwin Hubble Centennial Symposium*. Kron, Richard G., ed. *Astronomical Society of the Pacific Conference Series* 10 (1990): 19–21.

———. *The Great Copernicus Chase and Other Adventures in Astronomical History.* Cambridge, MA: Sky Publishing Corp., 1992.

———. "The Scale of the Universe: A Curtain Raiser in Four Acts and Four Morals." *PASP* 108 (1996): 1068–1072.

Glass, I. S. "An Early Photographic Refractor at the Cape." *Monthly Notices of the Astronomical Society of South Africa* 48 (1989**a**): 29–34.

———. "The Beginnings of Astronomical Photography at the Cape." *Monthly Notices of the Astronomical Society of South Africa* 48 (1989**b**): 117–122.

Gorman, Jessica. "Photography at a Crossroads." *Science News* 162 (November 23, 2002): 331–333.

Gould, Benjamin A. "Celestial Photography." *Observatory* 2 (1878): 13–19.

———. "Lewis Morris Rutherfurd." *Astronomical Journal* 12 (1892): 32.

———. "Memoir of Lewis Morris Rutherfurd, 1816–1892." *BMNAS* 3 (1895): 415–441.

Grant, M. "John A. Whipple and the Daguerrean Art." *Photographic Art-Journal* 2 (1851): 94–95.

"Great Telescope and Photographs of the Moon [Henry Draper]" *Scientific American* 9 (November 21, 1863): 330.

Greenslade, Thomas B., Jr. "The First Stereoscopic Pictures of the Moon." *American Journal of Physics* 40 (1972): 536–540.

Grubb, Howard. "On Great Telescopes of the Future." *Scientific Proceedings of the Royal Dublin Society, New Series* 1 (1878): 1–3.

Hale, George E. "Note on Solar Prominence Photography." *Astronomische Nachrichten* 126 (1890**a**): 81.

———. "Photography of the Solar Prominences." *Technology Quarterly* 3 (1890**b**): 310–316.

———. "The Kenwood Physical Observatory." *Sidereal Messenger* 10 (1891**a**): 321–323.

———. "Photography and the Invisible Solar Prominences." *Sidereal Messenger* 10 (1891**b**): 257–264.

———. "Solar Photography at the Kenwood Astro-Physical Observatory." *Astronomy and Astro-Physics* 11 (1892): 407–417.

———. "The Spectroheliograph." *Astronomy and Astro-Physics* 12 (1893): 241–257.

———. "On the Comparative Value of Refracting and Reflecting Telescopes for Astrophysical Investigations." *ApJ* 5 (1897): 119–131.

———. "The Development of a New Observatory." *PASP* 17 (1905**a**): 41–52.

———. "A Study of the Conditions for Solar Research at Mount Wilson, California." *ApJ* 21 (1905**b**): 124–150.

———. "A 100-Inch Mirror for the Solar Observatory." *ApJ* 24 (1906): 214–218.

———. "Mount Wilson Solar Observatory." *Carnegie Institution of Washington Yearbook* 11 (1912): 172–213.

———. *Ten Years of Work of a Mountain Observatory.* Washington, DC: Carnegie Institution, 1915.

———. *The New Heavens.* New York, Scribner, 1922.

———. "The Possibilities of Instrumental Development." *PA* 31 (1923): 568–574.

———. "The Possibilities of Large Telescopes." *Harper's Magazine* 156 (1928**a**): 639–646.

———. "Work for the Amateur Astronomer." *PASP* 28 (1916): 53–61; 40 (1928**b**): 285–302.

———. "Building the 200-Inch Telescope." *Harper's Magazine* 159 (1929): 720–732.

Harrison, W. Jerome. *The Chemistry of Photography.* New York: Scovill and Adams, 1892.

Hartmann, Johannes. "On the Scale of Kirchhoff's Solar Spectrum." *ApJ* 9 (1899): 69–85.

Hastings, Charles S. "Biographical Memoir of James Edward Keeler, 1857–1900." *BMNAS* 5 (1903): 231–246.

Hearnshaw, John. *The Analysis of Starlight: One Hundred and Fifty Years of Astronomical Spectroscopy*. Cambridge: Cambridge University Press, 1986.

———. "The Analysis of Starlight: Some Comments on the Development of Stellar Spectroscopy, 1815–1965." *Vistas in Astronomy* 30 (1987): 319–375.

———. "Doppler and Vogel—Two Notable Anniversaries in Stellar Astronomy." *Vistas in Astronomy* 35 (1992): 157–177.

———. *The Measurement of Starlight: Two Hundred Years of Astronomical Photometry.* Cambridge: Cambridge University Press, 1996.

———. "Astrophysics in the 1890s – The Dawn of a New Age in Astronomy." *The Impact of Large-Scale Surveys on Pulsating Star Research.* Szabados, L., and Kurtz, D. W., eds. *Astronomical Society of the Pacific Conference Series* 203 (2000): 1–6.

———. *Astronomical Spectroscopes and their History*. Cambridge: Cambridge University Press, 2009.

———. "Auguste Comte's Blunder: An Account of the First Century of Stellar Spectroscopy and How it Took One Hundred Years to Prove That Comte was Wrong!" *Journal of Astronomical History and Heritage* 13 (2010): 90–104.

Helmholtz, Robert von. "A Memoir of Gustav Robert Kirchhoff." *Deutsche Rundschau* **14** (1888): 232–245. Trans. by De Perott, Joseph. *Annual Report, Smithsonian Institution* (1890): 527–540.

"The Henry Draper Memorial." *Science* 13 (April 26, 1889): 320–323.

Hentschel, Klaus. "The Culture of Visual Representations in Spectroscopic Education and Laboratory Instruction." *Physics in Perspective* 1 (1999**a**): 282–327.

———. "Photographic Mapping of the Solar Spectrum, 1864–1900." *JHA* 30 (1999**b**): 93–119; 201–224.

Herczeg, T. J., and Kinney, Anne. "*Annals of the New York Academy of Sciences* 395 (1982): 331–336.

Herrmann, Dieter B., and Krisciunas, Kevin. *The History of Astronomy from Herschel to Hertzsprung*. New York: Cambridge University Press, 1984.

Herschel, William. "On the Construction of the Heavens." *Philosophical Transactions of the Royal Society of London* 75 (1785): 213–266.

———. "Astronomical Observations and Experiments Tending to Investigate the Local Arrangement of the Celestial Bodies in Space, and to Determine the Extent and Condition of the Milky Way." *Philosophical Transactions of the Royal Society of London* 107 (1817): 302–331.

Hetherington, Norriss S. "Edwin Hubble and a Relativistic, Expanding Model of the Universe." *Astronomical Society of the Pacific Leaflets* 10 (1971): 473–480.

———. "Mid-Nineteenth-Century American Astronomy: Science in a Developing Nation." *Annals of Science* 40 (1983): 61–80.

———. "Hubble's Cosmology." *American Scientist* 78 (1990): 142–151.

Hills, Edmond H. "Address on Presenting the Gold Medal of the Society to Dr. Max Wolf."

MNRAS 74 (1914): 377–389.

Hingley, Peter D. "The First Photographic Eclipse?" *Astronomy and Geophysics* 42 (2001): 18–22.

Hockey, Thomas, ed. *Biographical Encyclopedia of Astronomers.* New York: Springer, 2007.

Hoffleit, Dorrit. *Some Firsts in Astronomical Photography.* Cambridge, MA: Harvard College Observatory, 1950.

———. *The 28-inch Draper Mirror*. Dorrit Hoffleit Papers, Radcliffe Institute for Advanced Study, Harvard University, 1953.

———. "Pioneering Women in the Spectral Classification of Stars." *Physics in Perspective* 4 (2002): 370–398.

Holden, Edward S. "Monograph of the Central Parts of the Nebula of Orion." *Astronomical and Meteorological Observations made at the U.S. Naval Observatory* 18 (1882): Appendix I, 1–230. (Includes Henry Draper's "Memorandum to Accompany the

Photograph of the Nebula in Orion Sent to Professor Holden for his Memoir," pp. 226–230.)

———. "Photography the Servant of Astronomy." *Overland Monthly* 8 (November 1886): 459–470.

———. "[H. C. Vogel's] Orbit and Mass of the Variable Star Algol." *PASP* 2 (1890): 27.

———. "Comparison of Some Photographs and Drawings of the Orion Nebula." *PASP* 3 (1891): 57–61.

———. *Memorials of William Cranch Bond, Director of the Harvard College Observatory, 1840–1859, and of his Son George Phillips Bond, Director of the Harvard College Observatory, 1859–1865*. San Francisco: C. A. Murdock & Co., 1897.

Hoskin, Michael. "The 'Great Debate': What Really Happened." *JHA* 7(1976): 169–182.

Hoskin, Michael, ed. *The Cambridge Illustrated History of Astronomy.* Cambridge: Cambridge University Press, 2000.

Howell, Julia Fitz. "Henry Fitz, 1808–1863." *Contributions from the (U. S.) Museum of History and Technology*, Bulletin No. 228 (1962): 164–170.

"How Mr. Warren De La Rue Photographed the Moon." *British Journal of Photography* 15 (May 29, 1868): 256–257; (June 5, 1868): 270–271; (June 12, 1868): 279–281.

Hubble, Edwin P. "The Variable Nebula N.G.C. 2261." *ApJ* 44 (1916): 190–197.

———. "Recent Changes in the Variable Nebula N.G.C. 2261." *ApJ* 45 (1917): 351–353.

———. "Photographic Investigations of Faint Nebulae (Ph.D. Dissertation)." *Publications of the Yerkes Observatory* 4 (1920): 69–85.

———. "A General Study of Diffuse Galactic Nebulae." *ApJ* 56 (1922**a**): 162–199.

———. "The Source of Luminosity of in Galactic Nebulae." *ApJ* 56 (1922**b**): 400–438.

———. "Cepheids in Spiral Nebulae." *PA* 33 (1925**a**): 252–255.

———. "N.G.C. 6822, A Remote Stellar System." *ApJ* 62 (1925**b**): 409–433.

———. "Extra-Galactic Nebulae." *ApJ* 64 (1926**a**): 321–369.

———. "A Spiral Nebula as a Stellar System: Messier 33." *ApJ* 63 (1926**b**): 236–274.

———. "A Relation Between Distance and Radial Velocity Among Extra-Galactic Nebulae." *Proceedings of the National Academy of Sciences* 15 (1929): 168–173.

———. "Angular Rotations of Spiral Nebulae." *ApJ* 81 (1935): 334–335.

———. *Realm of the Nebulae.* New York: Dover, 1958.

Hubble, Edwin P., and Humason, Milton. "The Velocity-Distance Relation Among Extra-Galactic Nebulae." *ApJ* 74 (1931): 43–80.

Hufbauer, Karl. "Amateurs and the Rise of Astrophysics." *Berichte zur Wissenschaftsgeschichte* 9 (1986): 183–190.

Huggins, Margaret Lindsay. "Warren De La Rue." *Observatory* 12 (1889): 244–250.

Huggins, William. "Description of an Observatory Erected at Upper Tulse Hill." *MNRAS* 16 (1856): 175–176.

———. "On the Spectra of Some of the Chemical Elements." *PRS of London* 13 (1863): 43–44; *PTRS* 154 (1864**a**): 139–160.

———. “On the Spectra of Some of the Nebulae.” *PRS* 13 (1864**b**): 492–493; *PTRS* 154 (1864**c**): 437–444.

———. “On the Physical and Chemical Constitution of the Fixed Stars and Nebulae.” *Proceedings of the Royal Institution* 4 (1865**a**): 441–449.

———. “On the Spectrum of the Great Nebula in the Sword-Handle of Orion.” *PRS* 14 (1865**b**): 39–42.

———. “Further Observations on the Spectra of Some of the Stars and Nebulae, with an Attempt to Determine Therefrom Whether These Bodies are Moving towards or from the Earth.” *PRS* 16 (1868**a**): 382–386; *PTRS* 158 (1868**b**): 529–564.

———. “On the Spectrum of the Great Nebula in Orion, and on the Motions of Some Stars towards or from the Earth.” *PRS* 20 (1872): 379–394.

———. “Note on the Photographic Spectra of Stars.” *PRS* 25 (1876): 445–446.

———. “On the Inferences to Be Drawn from the Appearance of Bright Lines in the Spectra of Irresolvable Nebulae.” *PRS* 26 (1877**a**): 179–181.

———. “The Photographic Spectra of Stars.” *Observatory* 1 (1877**b**): 4–7.

———. “On the Photographic Spectra of Stars.” *PTRS* 171 (1880): 669–690.

———. “The New Astronomy.” *The Nineteenth Century* 41 (1897): 907–929.

Huggins, William, and Huggins, Lady. “Spectroscopic Notes: On the Spectra of the Stars in the Trapezium of the Great Nebula of Orion.” *ApJ* 6 (1897): 322–327.

———. *An Atlas of Representative Stellar Spectra*. London: W. Wesley and Son, 1899.

———. *The Scientific Papers of Sir William Huggins*. London: W. Wesley and Son, 1909.

Huggins, William, and Miller, W. A. “Notes on the Lines of Some of the Fixed Stars.” *PRS* 12 (1863): 444–445.

———. “On the Spectra of Some of the Fixed Stars.” *PRS* 13 (1864): 242–244; *PTRS* 154 (1864): 413–435.

Hughes, Stefan. *Catchers of the Light: A History of Astrophotography*. 2010. [www.catchersofthelight.com/]

Humason, Milton. “Edwin Hubble.” *MNRAS* 114 (1954): 291–295.

Humphrey, Samuel D. “Lunar Daguerreotypes.” *Daguerreian Journal* 1 (1850): 14.

Hussey William J. “Report by W. J. Hussey on Certain Possible Sites for Astronomical Work in California and Arizona.” *Carnegie Institution of Washington Yearkbook*, No. 2, 1903.

Hutchins, Roger. *British University Observatories 1772–1932*. Farnham, Surrey, UK: Ashgate, 2008.

Hyde, W. Lewis. “John William Draper, 1811–1882, Photographic Scientist.” *Applied Optics* 15 (1976): 1726–1730.

———. “The Calamity of the Great Melbourne Telescope.” *Proceedings of the Astronomical Society of Australia* 7 (1987): 227–230.

“In the Observatory of Mr. Huggins.” *MNRAS* 25 (1865): 107–109.

Jackson, Myles W. "Buying the Dark Lines of the Solar Spectrum: Joseph von Fraunhofer and His Standard for Optical Glass Production." *Archimedes: New Studies in the History and Philosophy of Science and Technology* 1 (1996): 1–22.

———. "Illuminating the Opacity of Achromatic Lens Production: Joseph von Fraunhofer's Use of Monastic Architecture and Space as a Laboratory." *The Architecture of Science*, Galison, Peter, and Thompson, Emily Ann, eds. Cambridge, MA: MIT Press, 1999.

———. *Spectrum of Belief: Joseph von Fraunhofer and the Craft of Precision Optics.* Cambridge, MA: MIT Press, 2000.

Jahn, Wolfgang, *et al. Fraunhofer in Benediktbeuern: Glassworks and Workshop*. Munich: Fraunhofer-Gesellschaft, 2008.

James, Frank A. J. L. *The Early Development of Spectroscopy and Astrophysics*. (PhD Thesis) London: University of London, 1981.

———. "The Establishment of Spectro–Chemical Analysis as a Practical Method of Qualitative Analysis, 1854–1861." *Ambix* 30 (1983): 30–53.

———. "The Creation of a Victorian Myth: The Historiography of Spectroscopy." *History of Science* 23 (1985): 1–24.

———. "The Discovery of Line Spectra." *Ambix* 32 (1985): 53–70.

———. "Science as a Cultural Ornament: Bunsen, Kirchhoff and Helmholtz in Mid-Nineteenth-Century Baden." *Ambix* 42 (1995): 1–9.

James, Stephen H. G. "Dr. Isaac Roberts (1829–1904) and His Observatories." *Journal of the British Astronomical Association* 103 (1993): 120–122.

Johnson, K. L. "Andrew Ainslie Common." *Oxford Dictionary of National Biography.* Oxford, UK: Oxford University Press, 2004–11.

Jones, Bessie Zaban. "Diary of the Two Bonds: 1846–1849." *Harvard Library Bulletin* 15 (1967): 368–386; 16 (1968): 49–71, 178–207.

Jones, Bessie Zaban, and Boyd, Lyle Gifford. *Harvard College Observatory*. Cambridge, MA: Harvard University Press, 1971.

Jones, Bryn. "Isaac Roberts (1829–1904)." *A History of Astronomy in Wales*. 2009. [www.jonesbryn.plus.com/wastronhist/people/isaacroberts/p_iroberts.html]

Jungnickel, Christa, and McCormmach, Russell. *Intellectual Mastery of Nature: Theoretical Physics from Ohm to Einstein.* Chicago: University of Chicago Press, 1990.

Keeler, James E. "Spectroscopic Observations of Spica at Potsdam." *PASP* 3 (1891): 46–48.

———. "Note on a Cause of Differences Between Drawings and Photographs of Nebulae." *PASP* 7 (1895): 279–282.

———. "The Importance of Astrophysical Research and the Relation of Astrophysics to Other Physical Sciences." *ApJ* 6 (1897): 271–287.

———. "New Nebulae Discovered Photographically with the Crossley Reflector of Lick Observatory." *MNRAS* 60 (1899**a**): 128.

———. "Photograph of the Great Nebula in Orion, Taken with the Crossley Reflector of the Lick Observatory ." *PASP* 11 (1899**b**): 39–40.

———. "Photographic Efficiency of the Crossley Reflector." *PASP* 11 (1899**c**): 199–202.

———. "On the Predominance of Spiral Forms among the Nebulae." *Astronomische Nachrichten* 151 (1899**d**): 1–4.

———. "The Crossley Reflector of Lick Observatory." *ApJ* 11 (1900): 325–349.

———. E. "Photographs of Nebulae and Clusters Made with the Crossley Reflector." *Publications of the Lick Observatory* 8 (1908): 1–114.

Kendall, Phebe Mitchell, ed. *Maria Mitchell: Life, Letters, and Journals.* Freeport, NY: Books for Libraries Press, 1971.

Kevles, Daniel J. "George Ellery Hale, the First World War, and the Advancement of Science in America." *Isis* 59 (1968): 427–437.

King, Henry C. *The History of the Telescope.* London: Charles Griffin and Co., 1955.

Kirchhoff, Gustav. "On the Fraunhofer Lines." *Philosophical Magazine, 4th Series* 19 (1860): 195–196.

Kirchhoff, Gustav. "On the Chemical Analysis of the Solar Atmosphere." *Philosophical Magazine, 4th Series* 21 (1861**a**): 185–188.

———. "On a New Proposition on the Theory of Heat." *Philosophical Magazine, 4th Series* 21 (1861**b**): 241–247.

———. *Researches on the Solar Spectrum and the Spectra of the Chemical Elements.* London: Macmillan and Co., 1862.

———. "Contributions Toward the History of Spectral Analysis and the Analysis of the Solar Atmosphere." *Philosophical Magazine, 4th Series* 25 (1863): 250–262.

Kirchhoff, Gustav, and Bunsen, Robert. "Chemical Analysis by Spectrum-observations." *Philosophical Magazine, 4th Series* 20 (1860): 89–109; 22 (1861): 329–349.

Kragh, Helge. "The Solar Element: A Reconsideration of Helium's Early History." *Annals of Science* 66 (2009): 157–182.

Kragh, Helge, and Smith, Robert W. "Who Discovered the Expanding Universe?" *History of Science* 41 (2003): 41–63.

Lankford, John. "Amateurs and Astrophysics: A Neglected Aspect in the Development of a Scientific Specialty." *Social Studies of Science* 11 (1981**a**): 275–303.

———. "Amateurs versus Professionals: The Controversy over Telescope Size in Late Victorian Science." *Isis* 72 (1981**b**): 11–28.

———. "The Impact of Photography on Astronomy." *The General History of Astronomy*, Vol. 4: *Astrophysics and Twentieth-Century Astronomy to 1950*, Part A. Gingerich, Owen, ed. Cambridge: Cambridge University Press, 1984.

———. "In Search of Henry Fitz." *S&T* 68 (September 1984): 214–218.

———. "Photography and the 19th-Century Transits of Venus." *Technology and Culture* 28 (1987): 648–657.

———. *American Astronomy: Community, Careers, and Power, 1859–1940.* Chicago: University of Chicago Press, 1997**a**.

Lankford, John, ed. *History of Astronomy: An Encyclopedia.* New York: Garland Publishing, 1997**b**.

"The Largest Telescope in the Country [Henry Fitz]." *Scientific American* 4 (April 6, 1861): 216.

Lassell, William. "Extract of a Letter from Mr. Lassell to Mr. W. De la Rue." *MNRAS* 13 (1852): 14.

———. "Observations of Planets and Nebulae at Malta." *Memoirs of the Royal Astronomical Society* 36 (1867): 1–32.

———. "Mr. Lassell's Great Reflector." *Observatory* 1 (1877): 178–179.

Law, Donald. "Lewis Morris Rutherfurd." *Aiken (SC) Standard* (July 8, 1990): 6–7.

Learner, Richard. *Astronomy Through the Telescope.* New York: Van Nostrand Reinhold, 1981.

Le Conte, David. "Warren De La Rue – Pioneer Astronomical Photographer." *The Antiquarian Astronomer* 5 (February 2011): 14–35.

Lee, John. "Address Delivered by the President, Dr. Lee, on Presenting the Gold Medal of the Society to Mr. Warren De La Rue." *MNRAS* 22 (1862): 131–139.

Leggat, Robert. *A History of Photography from its Beginnings till the 1920s.* 1995–2010. [www.rleggat.com/photohistory]

Lequeux, James. "The Great Nineteenth Century Refractors." *Experimental Astronomy* 25 (2009): 43–61.

"Lewis Morris Rutherfurd." *Scientific American* (December 14, 1889): 375–376.

Lightman, Bernard. "The Visual Theology of Victorian Popularizers of Science: From Reverent Eye to Chemical Retina." *Isis* 91 (2000): 651–680.

Livio, Mario. "Mystery of the Missing Text Solved." *Nature* 479 (2011): 171–172.

Livio, Mario, and Riess, Adam G. "Measuring the Hubble Constant." *Physics Today* 66 (2013): 41–47.

Lockyer, J. Norman. "On 'Lines in the Spectra of Some of the Fixed Stars,' by Huggins and Miller." *MNRAS* 23 (1863**a**): 179–180.

———. "Stellar Spectra." *Astronomical Register* 1 (1863**b**): 54.

———. *The Spectroscope and its Applications.* London: Macmillan and Co., 1873.

———. "Celestial Chemistry." *Nature* 9 (March 26, 1874): 411–414.

———. "On Spectrum Photography." *Nature* 10 (June 11, 1874): 109–112; (July 30, 1874): 254–256.

———. *Studies in Spectrum Analysis.* London: C. Kegan Paul & Co., 1878.

———. "Solar Physics – The Chemistry of the Sun." *Nature* 24 (1881): 267–274; 296–301; 315–324; 365–370; 391–399.

———. *The Chemistry of the Sun.* London: Macmillan and Co., 1887.

Longair, Malcolm. "History of Astronomical Discoveries." *Experimental Astronomy* 25 (2009): 241–259.

Loomis, Elias. *Recent Progress in Astronomy; Especially in the United States.* New York: Harper and Brothers, 1850.

———. "Astronomical Observatories in the United States." *Harper's New Monthly Magazine* 13 (1856): 25–52.

Luyten, Willem. "On the Completion of the *Carte du Ciel*." *Astronomical Journal* 65 (1960): 232.

MacDonald, Lee T. "Isaac Roberts, E. E. Barnard and the Nebulae." *JHA* 41 (2010): 239–259.

Maddox, R. L. "An Experiment with Gelatino-Bromide." *British Journal of Photography* (Sept. 8, 1871): 422.

Marshall, Roy K. "Astronomy in the Service of Culture." *PA* 51 (1943): 67–75.

Martin, Marion. "John William Draper and the Hastings Observatory." *Hastings Historian* 21 (1992): 1–9.

Mayall, Nicholas U. "Edwin Powell Hubble, 1889–1953." *BMNAS* 41 (1970): 175–214.

———. "Milton L. Humason—Some Personal Recollections." *Mercury* (1973): 3–8.

McCarthy Martin F. "Fr. Secchi and Stellar Spectra." *PA* 58 (1950): 153–168.

McGucken, William. *Nineteenth-Century Spectroscopy: Development of the Understanding of Spectra, 1802–1897*. Baltimore: Johns Hopkins Press, 1969.

Meadows, A. J. *Early Solar Physics*. New York: Pergamon Press, 1970.

———. "The New Astronomy." *The General History of Astronomy*, Vol. 4: *Astrophysics and Twentieth-Century Astronomy to 1950*, Part A. Gingerich, Owen, ed. Cambridge: Cambridge University Press, 1984**a**.

———. "The Origins of Astrophysics." *The General History of Astronomy*, Vol. 4: *Astrophysics and Twentieth-Century Astronomy to 1950*, Part A. Gingerich, Owen, ed. Cambridge: Cambridge University Press, 1984**b**.

Melvill, Thomas. "Observations on Light and Colours." *Essays and Observations, Physical and Literary* 2 (1756): 12–90.

Milham, Willis I. "Early American Observatories." *PA* 45 (1937): 464–474; 523–539.

Miller, Howard S. *Dollars for Research: Science and Its Patrons in Nineteenth-Century America*. Seattle: University of Washington Press, 1970.

Miller, William Allen. "On Spectrum Analysis." *Pharmaceutical Journal* 3 (1862): 399–412.

Mills, Charles, and Brooke, C. F. *A Sketch of the Life of Sir William Huggins*. Richmond, Surrey: Times Printing Works, 1936.

Mills, Deborah J. "George Willis Ritchey and the Development of Celestial Photography." *American Scientist* 54 (1966): 64–93.

Mitchell, William. "The Astronomical Observatory of Harvard University." *Christian Examiner* 50 (1851): 264–279.

Morris-Rutherfurd family papers, 1717–1889 (microfilm). University of South Carolina, Columbia, SC.

Morus, Iwan Rhys. *When Physics Became King*. Chicago: University of Chicago Press, 2005.

"Mountain Homes Shut to Holden: The Scandals of the Lick Observatory." *San Francisco Chronicle* (May 23, 1897): 23.

Moyer, Albert E. *A Scientist's Voice in American Culture: Simon Newcomb and the Rhetoric of Scientific Method.* Berkeley, CA: University of California Press, 1992.

"Mr. Rutherfurd's Photography and Diffraction Gratings." *Sidereal Messenger* 1 (1882): 44–45.

Murray, C. A. "David Gill and Celestial Photography." *Mapping the Sky: Past Heritage and Future Directions*, Proceedings of the 133rd Symposium of the International Astronomical Union, held in Paris, France, June 1–5, 1987. Débarbat, Suzanne, *et al.*, eds. Dordrecht: Kluwer, 1988.

Mussell, James. *Science, Time and Space in the Late Nineteenth-Century Periodical Press.* Farnham, Surrey, UK: Ashgate, 2007.

Nasmyth, James. *James Nasmyth, Engineer*. Smiles, Samuel, ed. London: John Murray, 1883.

Newall, H. F. "Dame Margaret Lindsay Huggins." *MNRAS* 76 (1916): 278–282.

Newcomb, Simon. "Aspects of American Astronomy." *ApJ* 6 (1897): 289–309.

Newell, Andrew. *Darkness at Noon: Or, the Great Solar Eclipse of the 16th of June, 1806.* Boston: D. Carlisle and A. Newall, 1806.

Newhall, Beaumont. *The Daguerreotype in America.* New York: New York Graphic Society, 1961.

———. *The History of Photography*. New York: Museum of Modern Art, 1982.

Newton, Isaac. "A Letter of Mr. Isaac Newton's . . . Containing his New Theory about Light and Colors." *PTRS* 6 (1671): 3075–3087.

Nicholson, Don, and Eklund, Bob. "First-Light Doubts on Mount Wilson." *S&T* 90 (July 1995): 86–88.

Norman, Daniel. "The Development of Astronomical Photography." *Osiris* 5 (1938): 560–594.

———. "John William Draper's Contributions to Astronomy." *The Telescope* 5 (1938): 11–16.

Noyes, Alfred. *Watchers of the Sky*. New York: Frederick A. Stokes Co., 1922.

O'Connor, John J., and Robertson, Edmund F. *MacTutor History of Mathematics Archive.* 2013. [www-history.mcs.st-and.ac.uk]

Oesper, Ralph E. "Robert Wilhelm Bunsen." *Journal of Chemical Education* 4 (1927): 431–439.

Official Catalogue of the Great Exhibition of the Works of Industry of All Nations, 1851. London: Spicer Brothers, W. Clowes and Sons, 1851.

Oldham, Kalil T. Swain. *The Doctrine of Description: Gustav Kirchhoff, Classical Physics, and the "Purpose of All Science" in 19th-Century Germany*. Ann Arbor, MI: ProQuest, 2008.

Olmsted, Denison. *Letters on Astronomy*. New York: Harper and Brothers, 1858.

Olson, Richard G. "The Gould Controversy at Dudley Observatory: Public and Professional Values in Conflict." *Annals of Science* 27 (1971): 265–276.

O'Raifeartaigh, Cormac. "The Contributions of V. M. Slipher to the Discovery of the Expanding Universe." *Origins of the Expanding Universe: 1912–1932.* Way, M. J., and Hunter, D. eds. *Astronomical Society of the Pacific Conference Series* 471 (2013): 49–62.

Osterbrock, Donald E. "The California-Wisconsin Axis in American Astronomy." *S&T* 51 (January 1976): 9–14; (February 1976): 91–97.

———. *James E. Keeler, Pioneer American Astrophysicist, and the Early Development of American Astrophysics*. New York: Cambridge University Press, 1984**a**.

———. "The Rise and Fall of Edward S. Holden." *JHA* 15 (1984**b**): 81–127, 151–176.

———. "The Quest for More Photons: How Reflectors Supplanted Refractors as the Monster Telescopes of the Future at the End of the Last Century." *Astronomy Quarterly* 5 (1985): 87–95.

———. *Pauper & Prince: Ritchey, Hale, and Big American Telescopes*. Tucson, AZ: University of Arizona Press, 1993.

———. "Founded in 1895 by George E. Hale and James E. Keeler: The Astrophysical Journal." *ApJ* 438 (1995): 1–7.

Osterbrock, Donald E., Brashear, Ronald S., and Gwinn, Joel A. "Self-Made Cosmologist: The Education of Edwin Hubble." *Evolution of the Universe of Galaxies: Proceedings of the Edwin Hubble Centennial Symposium*. Kron, Richard G., ed. *Astronomical Society of the Pacific Conference Series* 10 (1990): 1–18.

Osterbrock, Donald E., Gustafson, John R., Unruh, W. J. Shiloh. *Eye on the Sky: Lick Observatory's First Century*. Berkeley: University of California Press, 1988.

Pang, Alex Soojung-Kim. "Victorian Observing Practices, Printing Technology, and Representations of the Solar Corona (1): The 1860s and 1870s." *JHA* 25 (1994): 249–274.

———. "Victorian Observing Practices, Printing Technology, and Representations of the Solar Corona (2): The Age of Photomechanical Reproduction." *JHA* 26 (1995): 63–75.

———. "'Stars should henceforth register themselves': Astrophotography at the Early Lick Observatory." *British Journal for the History of Science* 30 (1997): 177– 202.

Parker, Richard. *Stellafane 2011: Testing a Henry Draper Mirror & Some Testing History*. 2011. [www.youtube.com/watch?v=-kb6R5tHVyQ]

Pasachoff, Jay, Olson, Roberta J. M., and Hazen, Martha L. "The Earliest Comet Photographs: Usherwood, Bond, and Donati 1858." *JHA* 27 (1996): 127–145.

Paterson, John A. "Edward Emerson Barnard; His Life and Work." *Journal of the Royal Astronomical Society of Canada* 18 (1924): 309–318.

Peacock, John A. "Slipher, Galaxies, and Cosmological Velocity Fields." *Origins of the Expanding Universe: 1912–1932*. Way, M. J., and Hunter, D., eds. *Astronomical Society of the Pacific Conference Series* 471 (2013): 3–24.

Phillips, John. "On Photographs of the Moon." *Report of the Twenty-Third Meeting of the British Association for the Advancement of Science*. London: John Murray, 1854.

"Photography in the United States." *Photographic Art-Journal* 5 (1853): 334–341.

Pickering, Edward C. *A Plan for Securing Observations of the Variable Stars*. Cambridge, MA: John Wilson and Sons, 1882.

———. *An Investigation in Stellar Photography at the Harvard College Observatory*. Cambridge, MA: John Wilson and Sons, 1886.

———. "Correspondence to the Editors: Photographic Chart of the Heavens." *Observatory* 12 (1889): 375.

———. "On the Spectrum of Zeta Ursae Majoris." *American Journal of Science, 3rd Series* 39 (1890): 46–47.

Pierce, Sally. *Whipple and Black: Commercial Photographers in Boston*. Boston: Boston Athenaeum, 1987.

Plotkin, Howard. *Henry Draper: A Scientific Biography*. (PhD. Dissertation.) Baltimore: Johns Hopkins University, 1972.

———. "Henry Draper, the Discovery of Oxygen in the Sun, and the Dilemma of Interpreting the Solar Spectrum." *JHA* 8 (1977): 44–51.

———. "Edward C. Pickering, the Henry Draper Memorial, and the Beginnings of Astrophysics in America." *Annals of Science* 35 (1978): 365–377.

———. "Edward C. Pickering's 'Diary of a Visit to the Harvard College Observatory, 14 November 1861.'" *Harvard Library Bulletin* 28 (1980): 282–290.

———. "Henry Draper, Edward C. Pickering, and the Birth of American Astrophysics." *Annals of the New York Academy of Sciences* 395 (1982): 321–330.

———. "Harvard College Observatory." *The General History of Astronomy*, Vol. 4: *Astrophysics and Twentieth-Century Astronomy to 1950*, Part A. Gingerich, Owen, ed. Cambridge: Cambridge University Press, 1984.

———. "Edward Charles Pickering." *JHA* 21 (1990): 47–58.

Pockels, Friedrich. "Gustav Robert Kirchhoff." *Heidelberg Professors from the 19th Century: Festschrift for the University Centenary of its Renewal by Karl Friedrich*. 2 (1903): 243–263. [www.ub.uni-heidelberg.de/helios/fachinfo/www/math/htmg/Pockels.htm]

Proctor, Richard. *The Moon: Her Motions, Aspect, Scenery and Physical Condition*. London: Longmans, Green and Co., 1873.

———. "The Approaching Transit of Venus." *Astronomical Register* 12 (1874): 39–40.

"Professor R. W. Bunsen." *Journal of the American Chemical Society (Proceedings)* 22 (1900): 88–107.

Rae, Ian D. "Spectrum Analysis: The Priority Claims of Stokes and Kirchhoff." *Ambix* 44 (1997): 131–144.

Ranyard, A. Cowper. "The Great Nebula in Andromeda." *Knowledge* 12 (1889): 75–77.

Rees, John K. "Lewis Morris Rutherfurd." *Astronomy and Astro-Physics* 11 (1892): 689–697.

———. "The Rutherfurd Photographic Measures." *Contributions from the Observatory of Columbia University*, No. 1 and No. 2 (1906).

Reingold, Nathan, ed. *Science in Nineteenth Century America: A Documentary History*. New York: Hill and Wang, 1964.

Rigge, William F. "Father Angelo Secchi." *PA* 26 (1918): 589–598.

Rinhart, Floyd, and Rinhart, Marion. *The American Daguerreotype*. Athens, GA: University of Georgia Press, 1981.

Ritchey, George W. "Celestial Photography with the 40-Inch Visual Telescope of the Yerkes Observatory." *ApJ* 12 (1900): 352–360.

———. "The Two-Foot Reflecting Telescope of the Yerkes Observatory." *ApJ* 14 (1901): 217–233.

———. "The 60-Inch Reflector of the Mount Wilson Solar Observatory." *ApJ* 29 (1909): 198–210.

———. "The Modern Reflecting Telescope and the New Astronomical Photography." *Transactions of the Optical Society* 29 (1927–28): 197–224.

Roberts, Isaac. "Photographic Maps of the Stars." *MNRAS* 46 (1886): 99–103.

———. "Photographs of Nebulae in Orion and in the Pleiades." *MNRAS* 47 (1887): 89–91.

———. "Photograph of the Nebula M51 Canum Venaticorum." *MNRAS* 49 (1889**a**): 399–390.

———. "Photographic Analyses of the Great Nebula M42 and 43 and h 1180 in Orion." *MNRAS* 49 (1889**b**): 295–297.

———. "Isaac Roberts' New Observatory on Crowborough Hill, Sussex." *MNRAS* 51 (1891): 118–119.

———. "Photograph of the Nebula Near 15 Monocerotis." *MNRAS* 55 (1895): 398–399.

———. ***Photographs of Stars, Star-clusters and Nebulae** Together with Information Concerning the Instruments and the Methods Employed in the Pursuit of Celestial Photography*. London: Knowledge Office, 1899. (Original edition: ***A Selection of Photographs of Stars, Star-clusters and Nebulae,*** 1893) [openlibrary.org/books/OL7222579M/A_selection_of_photographs_of_stars_star-clusters_and_nebulae]

———. "Herschel's Nebulous Regions." *MNRAS* 63 (1902): 26–34; *Astronomische Nachrichten* 160 (1903**a**): 337–344; *ApJ* 17 (1903**b**): 72–76.

———. "Meetings of the Royal Astronomical Society, Friday, 1903 March 13." *Observatory* 26 (1903**d**): 153–167.

Robinson, William F. *A Certain Slant of Light: The First Hundred Years of New England Photography.* Boston: New York Graphic Society/ Little, Brown, 1980.

Rohr, Moritz von. "Fraunhofer's Work and its Present-Day Significance." *Transactions of the Optical Society* 27 (1926): 277–294.

Root, M. A. *The Camera and the Pencil, or, The Heliographic Art: its Theory and Practice.* Philadelphia: M. A. Root, 1864.

Roscoe, Henry. "On Bunsen and Kirchhoff's Spectrum Observations, March 1, 1861." *Notices of the Proceedings of the Meetings of the Members of the Royal Institution of Great Britain* 3 (1862): 323–328.

———. "Bunsen Memorial Lecture." *Journal of the Chemical Society, London* 77 (1900): 513–554.

———. *The Life and Experiences of Sir Henry Enfield Roscoe*. London: Macmillan and Co., 1906.

Roth, Gunter D. *Joseph von Fraunhofer*. Stuttgart: Wissenschaftliche Verlagsgesellschaft mbH, 1976.

Rothenberg, Marc. "Organization and Control: Professionals and Amateurs in American Astronomy, 1899–1918." *Social Studies of Science* 11 (1981): 305–325.

———. "History of Astronomy." *Osiris, 2nd Series* 1 (1985): 117–131.

———. "Patronage of the Harvard College Observatory, 1839–1851." *JHA* 21 (1990): 337–46.

———. "National History: Understanding Education and Patronage in the 19th Century." Biennial History of Astronomy Workshop, Notre Dame, IN. 2001. [www.nd.edu/~histast4/exhibits/papers/rothenberg.html]

Rothermel, Holly. "Images of the Sun: Warren De La Rue, George Biddell Airy and Celestial Photography." *British Journal for the History of Science* 26 (1993): 137–169.

Rudisill, Richard. *Mirror Image: The Influence of the Daguerreotype on American Society*. Albuquerque, NM: University of New Mexico Press, 1971.

Rufus, W. C. "Astronomical Observatories in the United States Prior to 1848." *Scientific Monthly* 19 (1924): 120–139.

Russell, J. Scott. "On Certain Effects Produced on Sound by the Rapid Motion of the Observer." *Report of the 18th Meeting of the British Association for the Advancement of Science, August 1848*. London: John Murray, 1849.

Rutherfurd, Lewis M. "Observations during the Lunar Eclipse, September 12, 1848." *American Journal of Science* 6 (1848): 435–437.

———. "Companion to Sirius." *American Journal of Science and Arts* 34 (1862): 294–295.

———. "Astronomical Observations with the Spectroscope." *American Journal of Science and Arts* 35 (1863**a**): 71–77.

———. "Letter on Companion to Sirius, Stellar Spectra and the Spectroscope." *American Journal of Science and Arts* 35 (1863**b**): 407–408.

———. "Memoir of Henry Fitz." *National Academy of Sciences, Deceased Members File* (1863**c**).

———. "Observations on Stellar Spectra." *American Journal of Science and Arts* 36 (1863**d**): 154–157.

———. "On the Construction of the Spectroscope." *American Journal of Science and Arts* 39 (1865**a**): 129–132.

———. "Astronomical Photography." *American Journal of Science and Arts* 39 (1865**b**): 304–309.

———. "On the Stability of the Collodion Film." *American Journal of Science and Arts* 4 (1872): 430–433.

———. "Meeting of the Royal Astronomical Society, May 10, 1878." *Observatory* 2 (1878): 42–43.

"Rutherfurd's Photograph of the Moon." *Philadelphia Photographer* 3 (1866): 36–39.

Sandage, Allan. *The Hubble Atlas of Galaxies*. Washington, DC: Carnegie Institution of Washington, 1961.

———. "Edwin Hubble 1889–1953." *Journal of the Royal Astronomical Society of Canada* 83 (1989): 351–362.

Scheiner, J. and Frost, E. B. *A Treatise on Astronomical Spectroscopy.* Boston: Ginn & Co., 1894.

Schucking, E. L. "Henry Draper: The Unity of the Universe." *Annals of the New York Academy of Sciences* 395 (1982): 299–307.

Schuster, Arthur. "The Teachings of Modern Spectroscopy." *Popular Science Monthly* 19 (1881): 468–482.

Seares, Frederick H. "George Ellery Hale: The Scientist Afield." *Isis* 30 (1939): 241–267.

Serviss, Garrett P. "Celebrated American Astronomers." *Harper's Weekly* (December 1, 1894): 1143–1146.

Seton, William. "The Century's Progress in Science." *Catholic World* 69 (1899): 146–167.

Shane, C. Donald. "Lick Observatory: The First 75 Years." *PASP* 76 (1964): 77–87.

Shapley, Harlow. "On the Nature and Cause of Cepheid Variation." *ApJ* 40 (1914): 448–465.

———. "Studies Based on the Colors and Magnitudes in Stellar Clusters. VI. On the Determination of the Distances of Globular Clusters." *ApJ* 48 (1918**a**): 89–123.

———. "Studies Based on the Colors and Magnitudes in Stellar Clusters. VII. The Distances, Distribution in Space, and Dimensions of 69 Globular Clusters." *ApJ* 48 (1918**b**): 154–181.

———. "Studies Based on the Colors and Magnitudes in Stellar Clusters. XII. Remarks on the Arrangement of the Sidereal Universe." *ApJ* 49 (1919): 311–336.

———. *Star Clusters.* New York: McGraw-Hill, 1930.

———. *Through Rugged Ways to the Stars.* New York: Scribner, 1969.

Shapley, Harlow, and Curtis, Heber D. "The Scale of the Universe." *Bulletin of the National Research Council* 2 (1921): 171–217.

Shapley, Harlow, and Howarth, H. E. *A Source Book in Astronomy.* New York: McGraw-Hill, 1929.

Sheehan, William. *The Immortal Fire Within: The Life and Work of Edward Emerson Barnard.* New York: Cambridge University Press, 2007.

Sheehan, William, and Osterbrock, Donald E. "Hale's 'Little Elf': The Mental Breakdowns of George Ellery Hale." *JHA* 31 (2000): 93–144.

"Sketch of William Cranch Bond." *Popular Science* 47 (July 1895): 400–408.

Slipher, Vesto M. "Spectrographic Observations of Nebulae." *PA* 23 (1915): 21–24.

———. "Nebulae." *Proceedings of the American Philosophical Society* 56 (1917): 403–409.

Slocum, Frederick. "George Ellery Hale: America's Foremost Solar Physicist." *Scientific American* 105 (July 8, 1911): 23.

Smith, Robert W. "The Origins of the Velocity–Distance Relation," *JHA* 10(1979): 133–165.

———. "Edwin P. Hubble and the Transformation of Cosmology." *Physics Today* 43 (1990): 52–58.

———. "Beyond The Galaxy: The Development Of Extragalactic Astronomy 1885–1965." *JHA* 39 (2008): 91–119; 40 (2009): 71–107.

Smith, Robert W., and Baum, Richard. William Lassell and the Ring of Neptune: A Case Study in Instrumental Failure." *JHA* 15 (1984): 1–16.

Smyth, Charles Piazzi. "On Astronomical Drawing." *Memoirs of the Royal Astronomical Society* 15 (1846): 71–82.

Snelling, H. H. "Looking Back; Or, the Olden Days in Photography." *Anthony's Photographic Bulletin* 19 (1888): 559–563.

"Some Scientific Centres. IV. The Heidelberg Physical Laboratory." *Nature* 65 (April 24, 1902): 587–590.

Stanley, Matthew. "Spectroscopy – So What?" *Journal of Astronomical History and Heritage* 13 (2010): 105–111.

Stebbins, Joel. "The American Astronomical Society, 1897–1947." *PA* 55 (1947): 404–413.

Stephens, Carlene E. "Partners in Time: William Bond & Son of Boston and the Harvard College Observatory." *Harvard Library Bulletin* 35 (1987): 351–384.

———. "Astronomy as Public Utility: The Bond Years at the Harvard College Observatory." *JHA* 21 (1990): 21–36.

Stokes, G. G. "On the Simultaneous Emission and Absorption of Rays of the same definite Refrangibility; being a translation of a portion of a paper by M. Leon Foucault." *Philosophical Magazine, 4th Series* 19 (1860): 193–197.

Stone, Edmund. "Address, Delivered by the president on Presenting the Gold Medal of the Society to Mr. Common." *MNRAS* 44 (1884): 221–223.

Struve, Otto. "Fifty Years of Progress in Astronomy." *PA* 51 (1943): 469–481.

———. "The Yerkes Observatory, 1897–1947." *PA* 55 (1947**a**): 413–417.

———. "The Yerkes Observatory: Past, Present, and Future." *Science, New Series* 106 (1947**b**): 217–220.

Struve, Otto, *et al.* "The Spectroscopic Binary Alpha Virginis (Spica)." *ApJ* 128 (1958): 310–327.

Struve, Otto, and Zebergs, Velta. *Astronomy of the 20th Century*. New York: Macmillan, 1962.

Sutton, M. A. "Spectroscopy, Historiography and Myth: The Victorians Vindicated." *History of Science* 24 (1986): 425–431.

Taft, Robert. *Photography and the American Scene: A Social History 1839–1889*. New York: Dover Publications, 1964.

Talbot, William Henry Fox. "Some Experiments on Coloured Flames." *Edinburgh Journal of Science* 5 (1826): 77–81.

Teichmann, Jürgen, and Stinner, Arthur. "From William Hyde Wollaston to Alexander von Humboldt—Star Spectra and Celestial Landscape." *Annals of Science* 70 (2013): 2–34.

Tenn, Joseph S. "The Hugginses, the Drapers, and the Rise of Astrophysics." *Griffith Observer* 50 (1986): 2–15.

———. "David Gill: The Third Bruce Medalist." *Mercury* 19 (1990**a**): 84–85.

———. "Herman Carl Vogel: The Sixth Bruce Medalist." *Mercury* 19 (1990**b**): 172–173, 191.

———. "William Huggins: The Fifth Bruce Medalist." *Mercury* 19 (1990**c**): 148–149, 153.

———. "Edward Pickering: The Seventh Bruce Medalist." *Mercury* 20 (1991): 26–27, 30.

———. "Edward Emerson Barnard: The Fourteenth Bruce Medalist." *Mercury* 21 (1992**a**): 164–166.

———. "George Ellery Hale: The Thirteenth Bruce Medalist." *Mercury* 21 (1992**b**): 94–96, 110.

———. "Wallace Campbell: The Twelfth Bruce Medalist." *Mercury* 21 (1992**c**): 62–63, 75.

———. "Max Wolf: The Twenty-Fifth Bruce Medalist." *Mercury* 23 (1994): 27–28.

Tobin, William. "Foucault's Invention of the Silvered-Glass Reflecting Telescope and the History of his 80-cm Reflector at the Observatoire de Marseille." *Vistas in Astronomy* 30 (1987): 153–184.

Tobin, William, and Holberg, J. B. "A Newly-Discovered Accurate Early Drawing of M51." *Journal of Astronomical History and Heritage* 11(2008): 107–115.

Todd, Mabel Loomis. *Total Eclipses of the Sun.* Boston: Roberts Brothers, 1894.

Trimble, Virginia. "The 1920 Shapley–Curtis Discussion: Background, Issues, and Aftermath." *PASP* 107(1995): 1133–1144.

Turner, H. H. "Andrew Ainslie Common." *Observatory* 26 (1903): 304–308.

———. "Andrew Ainslie Common." *PRS* 75 (1904): 313–318.

———. "Some Reflections Suggested by the Application of Photography to Astronomical Research." *PA* 13 (1905): 72–82.

———. *The Great Star Map.* New York: E. P. Dutton, 1912.

———. "Sir William Henry Mahoney Christie." *Observatory* 45 (1922): 77–81.

Turner, H. H., and Common, A. A. "Photographic Chart of the Heavens." *Observatory* 11 (1888): 224–226; 333–334.

———. "Another Photographic Chart of the Heavens." *Observatory* 12 (1889): 308–311.

Turner, H. H., and Hollis, H. P. "Notes: The Lick Observatory." *Observatory* 20 (1897): 296, 299–300.

Urban, Sean E., and Corbin, Thomas E. "The Astrographic Catalogue: A Century of Work Pays Off." *S&T* 95 (1998): 40–44.

Utzschneider, Joseph von. *Kurzer Umriss der Lebens-Geschichte des Herrn Dr. Joseph von Fraunhofer*. Munich, 1826.

van Helden, Albert. "Telescope Building, 1850–1900." *The General History of Astronomy*, Vol. 4: *Astrophysics and Twentieth-Century Astronomy to 1950*, Part A. Gingerich, Owen, ed. Cambridge: Cambridge University Press, 1984.

van Maanen, Adriaan. "Internal Motions in Spiral Nebulae." ApJ *81* (1935): 336–337.

Vinter Hansen, Julie M. "Life and Work at the Lick Observatory." *PA* 55 (1947): 186–197.

Vogel, Hermann C. "Determination of the Motions in the Line of Sight by Means of Photography." *MNRAS* 50 (1890): 239–242.

———. "On the Spectroscopic Method of Determining the Velocity of Stars in the Line of Sight." *MNRAS* 52 (1891): 87–96.

———. "List of the Proper Motions in the Line of Sight of Fifty-one Stars." *MNRAS* 52 (1892): 541–543.

———. "On the Progress Made in the Last Decade in the Determination of Stellar Motions in the Line of Sight." *ApJ* 11 (1900): 373–392.

———. "The Spectroscopic Binary *Mizar*." *ApJ* 13 (1901): 324–328.

Vogel, Hermann W. "Astronomical Photography in America." *The Photographic News* 15 (January 20, 1871): 31–32.

———. Vogel, Hermann W. *The Chemistry of Light and Photography*. New York: D. Appleton and Co., 1875.

"War Once More on Mt. Hamilton: Holden and Hussey are at Loggerheads." *San Francisco Chronicle* (May 22, 1897): 14.

Warburg, Emil. "Zur Erinnerung an Gustav Kirchhoff." *Die Naturwissenschaften* 13 (1925): 205–212.

Warner, Brian. "Sir David Gill (1843–1914)." *Transactions of the Royal Society of South Africa* 49 (1994): 147–153.

Warner, Deborah Jean. "The American Photographical Society and the Early History of Astronomical Photography in America." *Photographic Science and Engineering* 11 (1967): 342–347.

———. *Alvan Clark & Sons: Artists in Optics*. Washington, DC: Smithsonian Institution Press, 1968.

———. "Lewis M. Rutherfurd: Pioneer Astronomical Photographer and Spectroscopist." *Technology and Culture* 12 (1971): 190–216.

———. "Astronomy in Antebellum America." *The Sciences in the American Context: New Perspectives*. Reingold, Nathan, ed. Washington, DC: Smithsonian Institution Press, 1979.

Way, Michael. "Dismantling Hubble's Legacy." *Origins of the Expanding Universe: 1912–1932, Astronomical Society of the Pacific Conference Series* 471 (2013): 97–134.

Way, Michael, and Nussbaumer, Harry. "Lemaître's Hubble Relationship." *Physics Today* 64 (2011): 8.

Weaver, Harold. "The Development of Astronomical Photometry (Parts I–III)." *PA* 54 (1946): 211–230; 287–298; 339–350.

Webb, T. W. "Dr. Draper's Telescope." *Intellectual Observer* 7 (1865): 368–373.

Wells, David Ames, and Bliss, Jr., George. "Lunar Daguerreotypes." Annual of Scientific Discovery (1850): 141–142.

Werge, John. *The Evolution of Photography*. London: Piper and Carter, 1890.

Whipple, John A. "Letter to the Editor." *Photographic Art-Journal* (July 1853): 66.

Whiting, Sarah F. "Lady Huggins." *Science, New Series* 41 (June 11, 1915): 853–855.

Whitney, Charles. *The Discovery of our Galaxy*. New York: Alfred A. Knopf, 1971.

———. "Henry Draper." *Dictionary of Scientific Biography*. Gillespie, C. C., ed. New York: Scribner, 1972, 178–180.

Whittingdale, W. "Address on the Occasion of the Sir David Gill Centenary Celebration." *Monthly Notices of the Astronomical Society of South Africa* 2 (1943): 73–78.

Williams, Mari E. W. "Astronomy in London: 1860–1900." *Quarterly Journal of the Royal Astronomical Society* 28 (1987): 10–26.

Winlock, William C. "Recent Advances in Astronomy." *Harper's Weekly* (December 1, 1894): 1146–1147.

Wolf, Max. "On Three of Sir William Herschel's Observed Nebulous Regions." *MNRAS* 63 (1903): 303–304.

Wollaston, William H. "A Method of Examining Refractive and Dispersive Powers, by Prismatic Reflection." *PTRS* 92 (1802): 365–380.

Wood, R. Derek. *The Arrival of the Daguerreotype in New York*. New York: American Photographic Historical Society, 1995.

Woolf, Harry. "The Beginnings of Astronomical Spectroscopy." *Melanges Alexandre Koyré: publiés à l'occasion de son soixante-dixième anniversaire*, Vol. 1. Paris: Hermann, 1964, 619–634.

Wright, Helen. *Explorer of the Universe: A Biography of George Ellery Hale*. New York, Dutton, 1966.

———. *James Lick's Monument*. Cambridge: Cambridge University Press, 1987.

———. *Sweeper in the Sky: The Life of Maria Mitchell*. Clinton Corners, NY: College Avenue Press, 1997.

Youmans, E. L. "Professor Henry Draper." *Harper's Weekly* (December 2, 1882): 756–757.

Young, Charles A. "History of Astronomy in the United States During the Past Century." *Proceedings of the American Association for the Advancement of Science, 25th Meeting, August 1876* (1877): 35–48.

———. "The Late Prof. Henry Draper." *The Critic* 2 (December 2, 1882): 333.

———. "The Late Dr. Henry Draper." *Science* 1 (February 16, 1883): 29–34.

———. "Astronomical Photography." *New Princeton Review* 3 (1887): 354–369.

———. "Address at the Dedication of the Kenwood Observatory." *Sidereal Messenger* 10 (1891): 312–321.

———. *The Sun*. New York: D. Appleton & Co., 1890.

Yowell, Everett. "The Debt Which Astronomy Owes to Ormsby Macknight Mitchel." *PA* 21 (1913): 70–74.

Acknowledgments

THE COMPLETION OF A BOOK, especially one that has occupied more than a decade of an author's life, entails the assistance, generosity, and forbearance of many people. I am grateful to Erika Goldman at Bellevue Literary Press, who nurtured this project from its inception and granted me the necessary number of pages to tell the story; Leslie Hodgkins at Bellevue Literary Press, who so ably guided the book from manuscript to final form; copy editor Kate McKay, who reined in my occasional tendency toward breathless prose; and Joe Gannon, for his masterful touch with layout and production. For their help in acquiring the many period illustrations that enliven the text, I acknowledge Alison Doane, Owen Gingerich, and Maria McEachern at Harvard's Center for Astrophysics, Mark Hurn at Cambridge University's Institute of Astronomy, John Grula at the Carnegie Observatories, Earl Taylor at the Dorchester Historical Society, Catherine Wehrey at the Huntington Library, Barbara Gilbert at the University of Chicago Library, and David Allen at the Royal Society of Chemistry. I also thank Harvard University for providing me my long-standing appointment as associate of the Harvard College Observatory; the University of Massachusetts Dartmouth for allowing me a sabbatical to complete the book; my colleagues in the Physics Department for making my "day job" so enjoyable; and my fellow night-sky devotees at the Astronomical Society of Southern New England, who are the living embodiment of the nineteenth-century amateur enthusiasts that populate this book. Finally, my gratitude to Sasha, Josh, and Gabe for their support and assistance along the way.

Illustration Sources and Permissions

Pages 14, 26, 144, 145, 242, 269 (top), 279 — Schweiger-Lerchenfeld, Amand. *Atlas der Himmelskunde.* Vienna, Austria: A. Hartleben's Verlag, 1898.

Page 15 — Courtesy of National Library of Ireland.

Page 16 — Nichol, J. P. *Thoughts on Some Important Points Relating to the System of the World.* Edinburgh: William Tait, 1846. Courtesy of Owen Gingerich.

Page 17 — NASA, ESA, S. Beckwith (STScI), and The Hubble Heritage Team (STScI/AURA).

Pages 24, 239, 259, 262, 269 (bottom), 288, 292, 294, 299, 305 (bottom), 308, 316 — By permission, University of Chicago Library.

Pages 29, 34, 38, 63, 139, 228 — Courtesy of Harvard College Observatory.

Page 33 — Courtesy of Dorchester Historical Society, Massachusetts.

Pages 47, 67, 124, 127 (top), 130, 209, 272, 276, 277 — By permission, University of Cambridge, Institute of Astronomy Library.

Page 54 — *Photographic Art Journal*, Vol. 2, August 1851.

Pages 74, 81 — *British Journal of Photography*, Vol. 15, May 29 and June 12, 1868.

Page 83 — Nasmyth, James. *The Moon: Considered as a Planet, a World, and a Satellite*, 2nd ed. London: John Murray, 1874.

Page 88 — Archives Center, National Museum of American History, Smithsonian Institution, image #44594.

Pages 97, 104, 108, 115, 116, 117 — By permission, Hastings Historical Society, New York.

Page 102 — Draper, Henry. *On the Construction of a Silvered Glass Telescope, Fifteen and a Half Inches in Aperture and Its Uses in Celestial Photography.* Washington, DC: Smithsonian Institution, 1864.

Pages 126 (bottom), 127 (bottom), 129, 305 (top) — Roberts, Isaac. *A Selection of Photographs of Stars, Star-clusters and Nebulae.* London: The Universal Press, 1893–1899. Courtesy of Wolbach Library, Harvard College Observatory.

Pages 135, 266 — Clerke, Agnes M. *History of Astronomy During the Nineteenth Century*, 3rd ed. London: Adam & Charles Black, 1893.

Page 140 — Observatorio Astrofisico di Torino.

Page 161 — Reproduced courtesy of the Library of The Royal Society of Chemistry.

Pages 164, 190 — Guillemin, Amédée. *The Forces of Nature.* London: Macmillan and Co., 1877.

Page 165 — Wollaston, William H. "A Method of Examining Refractive and Dispersive Powers, by Prismatic Reflection." *Philosophical Transactions of the Royal Society* 92 (1802): 365–380.

Page 175 — Roscoe, Henry. *Spectrum Analysis.* London: Macmillan and Co., 1873.

Pages 195, 197, 217 — Huggins, William. *The Scientific Papers of Sir William Huggins*. London: W. Wesley and Son, 1909. Courtesy of Wolbach Library, Harvard College Observatory.

Pages 202, 220, 222 — Huggins, William, and Huggins, Lady. *An Atlas of Representative Stellar Spectra.* London: W. Wesley and Son, 1899. Courtesy of Wolbach Library, Harvard College Observatory.

Page 253 — "The Great 36-Inch Equatorial of the Lick Observatory." *Knowledge* (December 1, 1888).

Pages 265 (top & bottom) — Hale, George Ellery. "The Kenwood Physical Observatory." *Sidereal Messenger* 10 (1891): 321–323.

Page 280 — Wellcome Library, London.

Pages 284, 289, 291, 297, 325 — By permission, Huntington Digital Library.

Page 314 — Courtesy of the Carnegie Observatories.

Page 326 — Hubble, Edwin. "A Relation Between Distance and Radial Velocity Among Extra-Galactic Nebulae." *Proceedings of the National Academy of Sciences* 15 (1929): 168–173.

Page 328 — Humason, Milton. "The Apparent Radial Velocities of 100 Extra-Galactic Nebulae." *Astrophysical Journal* 83 (1936): 10–22.

Page 330 — Courtesy of the Archives, California Institute of Technology.

Index

Bellevue Literary Press is devoted to publishing literary fiction and nonfiction at the intersection of the arts and sciences because we believe that science and the humanities are natural companions for understanding the human experience. With each book we publish, our goal is to foster a rich, interdisciplinary dialogue that will forge new tools for thinking and engaging with the world.

To support our press and its mission, and for our full catalogue of published titles, please visit us at blpress.org.

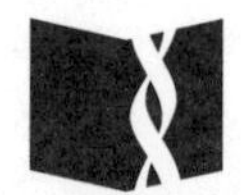

Bellevue Literary Press
New York